Rigid Cohomology

Dating back to work of Berthelot, rigid cohomology appeared as a common generalization of Monsky–Washnitzer cohomology and crystalline cohomology. It is a p-adic Weil cohomology suitable for computing Zeta and L-functions for algebraic varieties on finite fields. Moreover, it is effective, in the sense that it gives algorithms to compute the number of rational points of such varieties.

This is the first book to give a complete treatment of the theory, from full discussion of all the basics to descriptions of the very latest developments. Results and proofs are included that are not available elsewhere, local computations are explained, and many worked examples are given. This accessible tract will be of interest to researchers working in arithmetic geometry, p-adic cohomology theory, and related cryptographic areas.

Rigid Cohomology

BERNARD LE STUM
Université de Rennes I, France

CAMBRIDGE
UNIVERSITY PRESS

CAMBRIDGE
UNIVERSITY PRESS

University Printing House, Cambridge CB2 8BS, United Kingdom

Cambridge University Press is part of the University of Cambridge.

It furthers the University's mission by disseminating knowledge in the pursuit of education, learning and research at the highest international levels of excellence.

www.cambridge.org
Information on this title: www.cambridge.org/9780521875240

First published 2007

A catalogue record for this publication is available from the British Library

ISBN 978-0-521-87524-0 Hardback

À Pierre Berthelot.

Une mathématique bleue,
Sur cette mer jamais étale
D'où me remonte peu à peu
Cette mémoire des étoiles

(LÉO FERRÉ)

Contents

Preface

In 2004, I was asked by Professor King Fai Lai to come to Peking University in order to give a course on rigid cohomology. We agreed on the last two weeks of January 2005. I want to thank here Professor Zhao Chunlai for the organization of my visit as well as Professor Zhou Jian and his wife for showing us the city. My family and I will always remember it.

While preparing this course, I realized that there was no introductory book on rigid cohomology. Actually, there was no available material in English and only an old document in French, *Cohomologie rigide et cohomologie rigide à support propre*, by Pierre Berthelot. A revised version of the first part of this document appeared as an official preprint in 1996 but the second part is not fully written yet and, therefore, not really available to the mathematical community. Fortunately, Berthelot was kind enough to answer my questions on this second part and point out some articles where I could find some more information.

Rigid cohomology was introduced by Berthelot as a p-adic analogue of l-adic cohomology for lisse sheaves, generalizing Monsky–Washnitzer theory as well as crystalline cohomology (up to torsion). Recently, it appeared that this theory may be used in order to derive new algorithms for cryptography. The first result in this direction is due to Kiran Kedlaya who has also done incredible work on the theoretical aspect of the theory.

I knew that it was impossible to cover the full story in twenty one-hour lectures. I decided to first introduce the theory from the cryptography point of view (Introduction), then describe the basics of the theory with complete proofs (heart of the the course), and conclude with an overview of the development of the theory in the last 20 years (Conclusion). In particular, the main part of this book is quite close to Berthelot's original document. I hope that this will be useful to the students who want to learn rigid cohomology and, eventually, improve on our results.

I insist on the fact that there is no original matter here and that almost everything is due to Pierre Berthelot, apart from the mistakes arising from my misunderstanding.

One can split the so-far short life of rigid cohomology into three periods: (1) foundations, (2) cohomology of varieties and (3) cohomology of F-isocrystals.

It is a wonderful idea of Berthelot's to generalize crystalline and Monsky–Washnitzer cohomology into one theory. The principle is to compactify the variety X, embed it into some smooth formal scheme P and compute the limit de Rham cohomology of "strict" neighborhoods V of X in the generic fiber P_K of P. The astonishing fact is that the result does not depend on the choices. Better: there exists a category of coefficients for this theory. This is simply the limit category of differential modules on strict neighborhoods V that have the good idea to be "overconvergent". What makes this possible is a deep geometrical result, the strong fibration theorem. It tells us that even the geometry of V is essentially independent of the choices.

Some time after Berthelot had laid the foundations, the theory got a kick thanks to Johan de Jong's alterations theorem. He made it possible to use rigid cohomology as a bridge between crystalline cohomology and Monsky–Washnitzer cohomology in order to show that the latter is finite dimensional. More generally, de Jong's theorem, which states that one can solve singularity in characteristic p if one is ready to work with étale topology, can be used to show that rigid cohomology of varieties satisfies the formalism of Bloch–Ogus (finiteness, Poincaré duality, Künneth formula, cycle class, etc.).

The third period started with three almost simultaneous proofs of the so called conjecture of Crew. Even if he never stated this as an explicit conjecture, Crew raised the following question: is a differential module with a strong Frobenius structure on a Robba ring automatically quasi-unipotent? The first proof, due to Yves André is an application of representation theory using results of Christol and Mebkhout on "slopes". The second one, due to Zogman Mebkhout, is derived from his previous work. More important for us, the third one, due to Kiran Kedlaya, is a direct construction and can be generalized to higher dimension. From this, he can derive all the standard properties of rigid cohomology of overconvergent F-isocrystals, and in particular finite dimensionality.

In this course, we will focus on foundations. The main results of the second period will be mentioned in the Introduction where we will try to give a historical introduction based on Weil's conjectures and recent results in cryptography. In the Conclusion, we will try to evaluate the state of the art and, at the same time, review the main results of the third period.

Acknowledgments

During the writing of this course, I had useful conversations with Gweltaz Chatel, Michel Gros, Ke-Zheng Li, David Lubicz, Laurent Moret-Bailly, Richard Crew and, of course, Pierre Berthelot. I want to thank them here. Thank you also to Marion Angibaud for helping with the drawings.

Also, this work was partially supported by the European Network Arithmetic Algebraic Geometry.

Of course, I thank PKU University for its invitation and hospitality.

Finally, I want to thank Roger Astley from Cambridge University Press who made it possible to turn this course into a real book.

Outline

Chapter 1 is an introduction to the theory from the cryptography viewpoint. More precisely, several decades after André Weil had stated his conjectures, it appeared that an effective proof of these results would be useful to cryptography. Actually, the p-adic approach gives better results in some cases. We recall the discrete logarithm problem and explain why it is useful to explicitly compute the number of points of algebraic varieties. This is the purpose of Weil conjectures which predict the existence of arithmetic cohomology theories that will compute Zeta functions. Rigid cohomology is such a theory and we give its properties. Finally, in Weil's proof of the diagonal hypersurface case, the necessity of introducing coefficients for the theory already appears. This leads to the notion of L-functions.

Chapter 2 is devoted to the study of non-archimedean tubes. After fixing the setting (we assume that the reader is familiar with rigid analytic geometry), we introduce successively the notion of open tube of radius one and then the notion of tube of smaller radius. If we are given a subvariety of the special fiber of a formal scheme, the tube is simply the set of points in the generic fiber that specialize into the given subvariety. The idea is to see this tube as a lift of the algebraic variety (of positive characteristic) to an analytic variety (of characteristic zero). Such a tube is not quasi-compact in general, and it is therefore necessary to introduce smaller tubes which are quasi-compact and whose increasing union is the original tube. The main result of this chapter is the Weak Fibration Theorem (Corollary 2.3.16) which says that a smooth morphism of formal scheme around an algebraic variety induces locally a fibration by open balls (or better said, polydiscs). This will imply that the de

Rham cohomology of the tube does not depend on the embedding into the formal scheme.

Chapter 3 is rather technical but fundamental. The de Rham cohomology of a closed ball is infinite dimensional and it is therefore necessary to work in its neighborhood if one is looking for something interesting. Unfortunately, rigid topology is not a usual topology and it is necessary to refine the notion of neighborhood into that of strict neighborhood. We introduce the notion of frame which is a sequence of an open immersion of algebraic varieties and a closed immersion into a formal scheme. We then study the strict neighborhoods of the tube of the first variety into the tube or the second one. For future computations, it is essential to have a deep understanding of these strict neighborhoods. The first idea is to remove a small tube of the complement (the locus at infinity). But this does not give sufficiently general strict neighborhoods. We really need to play around a little more with tubes in order to define the so-called standard strict neighborhoods. It is then possible to extend the Weak Fibration Theorem to strict neighborhoods and obtain the Strong Fibration Theorem (Corollary 3.4.13). This is it for the geometrical part.

Chapter 4 is supposed to be a break. After recalling the basics about modules with integrable connections and their cohomology, we study them in the context of strict neighborhoods and show that this is closely related to the notion of radius of convergence. Actually, modules with integrable connections, $\mathcal{D}$-modules, stratified modules and crystals are simply different ways of seeing the same objects. We try to make this clear in the rigid geometric setting. We work out some examples, Dwork, Kummer, superelliptic curves, Legendre family, hypergeometric equations, etc. Then, we introduce the overconvergence condition for an integrable connection. It means that the Taylor series is actually defined on some strict neighborhood of the diagonal. Locally, there is an explicit description of this condition and this is the main result of the chapter (Theorem 4.3.9). We then introduce the notion of radius of convergence of an integrable connection with respect to a given set of étale coordinates and use it to rewrite the overconvergence condition when the geometry is not too bad. Finally, we also do the case of weakly complete algebras and Robba ring. They will appear to be very important in the future when we try to compute rigid cohomology.

Chapter 5 introduces the notion of overconvergent sheaf. The idea is to work systematically with sections defined on a strict neighborhood. This notion of overconvergence is actually very general and works in any topos but we quickly specialize to the case of frames. In the case of abelian sheaves, we introduce also the notion of sections with overconvergent support as well as the more classical notion of sections with support. Next, we consider the sheaf of overconvergent sections of the structural sheaf on the tube of a frame, and modules on this ring, which we call dagger modules. The main result of this chapter (Theorem 5.4.4) shows that the category of coherent dagger modules is equivalent to the limit category of coherent modules on strict neighborhoods. In this chapter, we also give a geometric meaning to weakly complete algebras and Robba rings and prove Serre Duality in this context.

Chapter 6 studies dagger modules with integrable connections and their cohomology. We simply apply usual calculus as explained earlier to dagger modules. Rigid cohomology is just usual de Rham cohomology and it can be extended to rigid cohomology with support in a closed subset by using sections with overconvergent support. There is also the alternative theory of cohomology with compact support which is made out of usual sections with support. We give comparison theorems with Monsky-Washnitzer cohomology and de Rham cohomology of Robba rings. The main result of the chapter (Theorem 6.5.2) shows that rigid cohomology of coherent dagger modules with integrable connection is invariant under a morphism of frames which is the identity at the first level, proper at the second and smooth at the third. We also prove the analogous results for cohomology with support. These theorems will prove fundamental later.

Chapter 7 gives a crystalline interpretation of the theory. We define a (finitely presented) overconvergent isocrystal on a frame as a family of coherent dagger modules on all frames above it with some compatibility conditions. We prove in Proposition 7.1.8 that the category of overconvergent isocrystals is invariant under a morphism of frames which is the identity at the first level, proper at the second and smooth at the third. We also show in Proposition 7.2.13 that one recovers exactly the notion of overconvergent integrable connection introduced earlier. In particular, we can define the rigid cohomology of an overconvergent isocrystal as the rigid cohomology of the corresponding module with connection. Next, we consider what we call a virtual frame. This is simply an open immersion of algebraic varieties but we want to see it as an incomplete frame. One defines overconvergent isocrystals on a virtual frame exactly as above and shows that we do get the same category when the virtual frame extends to a smooth frame. Moreover, rigid cohomology is then independent of the chosen extension thanks to our comparison theorems. Thus, it makes sense to talk about the rigid cohomology of an overconvergent isocrystal on a virtual frame.

Chapter 8 rewards us because we may now define overconvergent isocrystals on an algebraic variety and their cohomology in a functorial way. First of all, exactly as above, an overconvergent isocrystal on an algebraic variety is just a family of dagger modules on each frame above the variety with some compatibility conditions. If we embed successively our variety as an open subset of a proper variety and then as a closed subset of a smooth formal scheme, we get an equivalence with the category of overconvergent isocrystals on the frame as shown in Corollary 8.1.9. Moreover, the cohomology is independent of the choice of the embeddings as showed in Proposition 8.2.3. We finish with the study of Frobenius action. We show that rigid cohomology is fully compatible with Monsky–Washnitzer theory and, in particular, prove that overconvergent F-isocrystals correspond exactly to coherent modules with an integrable connection and a strong Frobenius.

Chapter 9 gives some informal complements. We recall what crystalline cohomology is and how it may be used to compute rigid cohomology. This comparison theorem could have been included with a complete proof in the main part of the course but it did not seem reasonable to assume that the reader was familiar with crystalline cohomology. Then, we explain how alterations can be used to derive finiteness of rigid cohomology without coefficient from this comparison theorem. Again, finiteness of rigid cohomology with compact support could have been included with full proof. Unfortunately, the proof for cohomology without support relies on a Gysin isomorphism that requires the theory of arithmetic $\mathcal{D}$-modules. We also explain the Crew conjecture and Kedlaya's methods to solve it. We end with Shiho's theory of convergent log site and his monodromy conjecture which may be seen as a generalization of the conjecture of Crew.

Conventions and notations

When there is no risk of confusion, we will use standard multi-index notations, namely

$$\underline{i} := i_1, \ldots, i_n, \quad |\underline{i}| := i_1 + \cdots + i_n, \quad \underline{i} \leq \underline{j} \Leftrightarrow \forall k, i_k \leq j_k$$

$$\underline{i}! = i_1! \cdots i_n!, \quad \binom{\underline{i}!}{\underline{j}!} := \frac{\underline{i}!}{\underline{j}!(\underline{i} - \underline{j})!},$$

$$\underline{t}^{\underline{i}} := t_1^{i_1} \cdots t_n^{i_n}, \quad \underline{t}^{[\underline{i}]} := \frac{\underline{t}^{\underline{i}}}{\underline{i}!},$$

and so on. Also, if $X = \cup_{i \in I} X_i$ and $J \subset I$, then $X_J := \cap_{i \in J} X_i$ and if $\lambda_i : X_i \hookrightarrow X$ denotes the inclusion map, we will write $\lambda_J : X_J \hookrightarrow X$ for the inclusion of the intersection.

Throughout this book, we will work on a complete ultrametric field K with a non trivial absolute value. We will denote by $\mathcal{V}$ its valuation ring, $\mathfrak{m}$ its maximal ideal and by k its residue field. Also, π will be a non zero element of $\mathfrak{m}$. There is no harm in assuming that K has characteristic zero even if this is almost never used in the theory. The reader will get a better intuition if he also assumes that k has positive characteristic p.

Positive real numbers are always assumed to live in $|K^*| \otimes \mathbf{Q} \subset \mathbf{R}_{>0}$.

As usual, if S is any scheme (or ring), then $\mathbf{A}_S^N$ (resp. $\mathbf{P}_S^N$) will denote the affine (resp. projective) space of dimension N over S. Also, when K is a complete ultrametric field and $\rho > 0$, then $\mathbf{B}^N(0, \rho^+)$ (resp. $\mathbf{B}^N(0, \rho^-)$) will denote the closed (resp. open) ball (or polydisc) of radius ρ. It is the rigid analytic open subset of $\mathbf{A}_K^{N,\mathrm{rig}}$ defined by $|t_i| \leq \rho$ (resp. $|t_i| < \rho$). We may allow $\rho = 0$ in the $+$ case and $\rho = \infty$ in the $-$ case. When $N = 1$, we drop it from the notations. Finally, if $0 < \epsilon < \rho$, then

$$\mathbf{A}_K(0, \epsilon^\pm, \rho^\pm) := \mathbf{B}(0, \rho^\pm) \setminus \mathbf{B}(0, \epsilon^\mp)$$

will denote the annulus off radii ϵ and ρ. Again, we may allow $\epsilon = \rho$ in the $++$ case as well as $\epsilon = 0$ in the $+\pm$ case and $\rho = \infty$ in the $\pm-$ case.

Since it is sometimes needed in applications, we choose not to assume that varieties or formal schemes are quasi-compact. Many results and definitions are however invalid without this assumption. It is therefore necessary to add this hypothesis form time to time. The reader who so wishes may assume that all (formal) schemes are quasi-compact in order to turn many assertions into a simpler form.

1
Introduction

1.1 Alice and Bob

Suppose Alice wants to send a secret message s to Bob. If Eve intercepts the message, then she can read it and it will not be secret anymore. Thus, Alice and Bob should agree on a two ways protocol that will turn the secret message s into a public message p. This is called encryption. Reversing the operations will allow Bob to recover s from p. For example, Alice would shift the letters of the message in alphabetical order and Bob will simply do the same thing in the reverse order (Caesar cipher). The Advanced Encryption Standard (AES) protocol does the same thing in a more complicated way, but this is not the subject of this course.

If Eve knows the two ways protocol, then she can derive s from p as easily as Bob does and the message will not stay secret anymore. The solution is to use a protocol with a parameter, the *key*. Then, Alice and Bob can make their protocol public as long as they keep secret their key k. For example, the protocol could be "replacing each letter in the message with the letter that is k places further down the alphabet". Again, AES does the same thing in a more complicated way.

Still, Alice and Bob should agree on their common key k. If Alice chooses the key, it can be intercepted by Eve when Alice sends it to Bob. This problem can be fixed as was shown by Diffie, Hellman and Merkle: Alice and Bob can make public the choice of a finite order element g in a group G. Alice chooses a private key $a \in \mathbf{N}$ from which she derives her public key $A := g^a$. Bob does the same thing and obtains also a public key $B := g^b$. Then Alice chooses as common key $k := B^a$. She does not have to send it to Bob because he can derive the same key in the same way. More precisely, Bob knows his private key b, he knows the public key A of Alice and we have $k = A^b$.

At this point, I should mention that if Eve chooses a private key a' and publishes a fake public key $A' := g^{a'}$ for Alice, then Bob might use it to code his message. If Eve intercepts the message, she can then use her private key a' to read it. Thus, there is still a weakness in this system but we do not want to discuss this here. So we will assume that Alice and Bob can trust each other's public key.

Thus, Eve knows the group G and the generator g and she also knows the public keys A and B. But, in order to discover k, she needs to solve the *Diffie–Hellman problem*: recovering g^{ab} from g^a and g^b. Of course, it is sufficient to be able to derive x from $X := g^x$. This last question is called the *discrete logarithm* problem. Even if they cannot prove it, specialists think that, in fact, the Diffie–Hellman problem is as hard as the discrete logarithm problem. And, in practice, it takes way more time, given g, to recover x from X than to derive X from x. We will try to explain this below.

1.2 Complexity

How long does it take to make a computation and how much room do we need to store the data? This is called a *complexity* question. For example, what is the complexity of (discrete) exponentiation? If $|G| = n$, it takes at most $2\log_2 n$ elementary operations to get $X = g^x$ from x: this is derived from the 2-adic expansion of x. One says that the complexity of exponentiation is linear (in $\log_2 n$). On the other hand, since exhaustive search of x from X might need n elementary operations, and there is no clear alternative, it seems that the complexity of the discrete logarithm is exponential (in $\log_2 n$). Actually, there are many other methods that compute discrete logarithms which are more efficient than exhaustive search but their complexity is still exponential in general.

However, note that in the case $G := (\mathbf{Z}/n\mathbf{Z})^*$, there exists a sub-exponential algorithm, the so called *Index-Calculus method*. Roughly speaking, one first solves the discrete logarithm problem simultaneously for the small primes. In order to do that, one looks at primary decompositions of random powers g^r until we get enough linear relations. Then, it is sufficient to consider the primary decomposition of $g^s X$ for random s until only small prime factors appear. Although not polynomial, this is better than exponential. If G is no longer equal to $(\mathbf{Z}/n\mathbf{Z})^*$ but is the set of rational points of a more general algebraic group, there is no real equivalent to the Index-Calculus method, and there is no known sub-exponential method to compute discrete logarithms in G.

Actually, the best-known technique is the Pohlig–Hellman algorithm which is exponential but still quite fast when the order of the group has only small prime factors. In order to discard such a group, it is necessary to compute its order. Thus, the crucial point becomes the determination of the number of rational points of an algebraic variety over a finite field (see [22] for a recent survey of the problem). As far as we stick to elliptic curves, the l-adic methods work perfectly. But already in the hyperelliptic case, the p-adic methods are way more effective for small primes p.

1.3 Weil conjectures

Concerning Weil conjectures, the case of diagonal hypersurfaces is completely worked out in André Weil's original article [85]. Also, if you want to go a little further into the p-adic point of view, you are encouraged to look at Paul Monsky's course [70]. For the l-adic approach, Milne's book [68] is the reference.

We want to compute the number of points of an algebraic variety X over a finite field $\mathbf{F}_q$ with $q := p^f$ elements, p a prime. For example, we are given

$$F_1, \ldots, F_d \in \mathbf{Z}[T_1, \ldots, T_n]$$

and we want to compute the number of solutions of

$$\begin{cases} F_1(a_1, \ldots, a_n) = 0 \mod p \\ \vdots \\ F_d(a_1, \ldots, a_n) = 0 \mod p \end{cases}$$

More generally, given an algebraic variety X over $\mathbf{F}_q$, we want to compute

$$N_r(X) :- |X(\mathbf{F}_{q^r})|$$

for all r. The main result is the conjecture made by André Weil in 1949 ([85]) and proved in several steps, starting with *rationality* by Bernard Dwork in 1960 ([39]), using p-adic methods:

If X is an algebraic variety of dimension d over $\mathbf{F}_q$ there exists finitely many algebraic integers α_i and β_i such that for all r, we have

$$N_r(X) = \sum \beta_i^r - \alpha_i^r,$$

and ending with *purity* (also called *Riemann hypothesis*) by Pierre Deligne in 1974 ([36]), using l-adic methods (recall that a *Weil number of weight m* is an algebraic integer whose archimedean absolute values are of the form $q^{\frac{m}{2}}$):

The algebraic integers α_i and β_i are Weil numbers with weight in $[0, 2d]$, with many other results in between and, in particular, the *functional equation* by Grothendieck in 1965 ([49]):

If X is proper and smooth, the application $\gamma \mapsto q^d/\gamma$ induces a permutation of the α_i's and a permutation of the β_i's.

As an example, we can consider for $p \neq 2$, an affine hyperelliptic curve of equation $y^2 = F(x)$ with F separable of degree $2g + 1$. Then,

$$N_r(X) = q^r - \sum_{i=1}^{2g} \alpha_i^r$$

where each α_i is a Weil number of weight 1 and $\alpha_i \alpha_{i+g} = q$.

1.4 Zeta functions

Weil conjectures are easier to deal with if we form the generating function

$$\zeta(X, t) := \exp\left(\sum_{r=1}^{\infty} N_r(X) \frac{t^r}{r}\right) = \prod_{x \in |X|} \frac{1}{1 - t^{\deg x}},$$

which is called the *zeta function* of X. The above results are then better reformulated in the following way:

Rationality: the function $\zeta(X, t)$ is rational with coefficients in $\mathbf{Q}$.

Purity: its zeros and poles are Weil numbers with weights $\in [-2d, 0]$.

Functional equation: if X is proper and smooth, then

$$\zeta\left(X, \frac{1}{q^d t}\right) = -q^{dE/2} t^E \zeta(X, t)$$

with $E \in \mathbf{Z}$ (Euler characteristic).

Actually, this can be rewritten in the more precise form (recall that a *Weil Polynomial* is a monic polynomial with integer coefficients whose roots are all Weil numbers):

We have

$$\zeta(X, t) = \prod_{i=0}^{2d} P_i(t)^{(-1)^{i+1}}$$

where P_i is a Weil Polynomial with non-negative weights between $2(i - d)$ and i. Moreover, if X is proper and smooth, P_i is pure of weight i and

$$P_{2d-i}(t) = C_i t^{B_i} P_i\left(\frac{1}{q^d t}\right)$$

with $C_i \in \mathbf{Z}$ and $B_i \in \mathbf{N}$ (Betti numbers).

As an example, one can show that if X is an abelian variety, then $|X(\mathbf{F}_q)| = P_1(1)$. Actually, if X is a projective non singular curve and J is the jacobian of

X, we also have $|J(\mathbf{F}_q)| = P_1(1)$. In other words, if we can compute P_1 for a curve, then we can tell the number of rational points of its jacobian which is an algebraic group.

1.5 Arithmetic cohomology

If an algebraic variety X over $\mathbf{F}_q$ lifts to a compact manifold V over $\mathbf{C}$, then the numbers B_i that appear in the functional equation are the *Betti number* of V (and E is its *Euler characteristic*). In other words, B_i is the rank of $H^i(V, \mathbf{Z})$. Actually, as André Weil already knew, the whole story can be told using a suitable cohomology theory with a Fixed Point Lefschetz trace formula for Frobenius.

Actually, if l is any prime (even $l = p$), there exists a finite extension K of $\mathbf{Q}_l$ such that the following holds (an operator ϕ on a finite-dimensional vector space will be called a *Weil operator* if its characteristic polynomial is a Weil polynomial):

We have (rationality)

$$\zeta(X, t) = \prod_{i=0}^{2d} \det(1 - t\varphi_i)^{(-1)^{i+1}}$$

where φ_i is a Weil operator (purity) with non-negative weights inside $[2(i - d), i]$ *on a finite-dimensional K-vector space $H^i_c(X)$. Moreover, If X is proper and smooth (in which case we write $H^i := H^i_c$), then φ_i is pure of weight i and there is a perfect pairing (functional equation or* Poincaré duality*)*

$$H^i(X) \times H^{2d-i}(X) \to K(-d)$$

compatible with the operators (with multiplication by q^d on the right).

In the case $l \neq p$, l-adic cohomology with compact support has all these properties. As already mentioned, a good introduction to l-adic cohomology is Milne's book [68].

We will now stick to the case $l = p$. There is a good theory for proper and smooth algebraic varieties that was developed by Pierre Berthelot in the late 1960s. It is called *crystalline cohomology* (see for example [20]). For smooth affine varieties, *Monsky–Washnitzer cohomology* was also available at that time (see for example [83], but also [72], [69] and [71]) although finite dimensionality was not known. *Rigid cohomology* is a theory that generalizes both crystalline and Monsky–Washnitzer theories and was developed by Pierre Berthelot in the 1980s (see [11]).

1.6 Bloch–Ogus cohomology

Rigid cohomology is a Bloch–Ogus cohomology ([74]). We explain below what it means. We fix a complete ultrametric field K of characteristic zero with valuation ring $\mathcal{V}$ and residue field k. Although it does not seem necessary, according to a yet unpublished result of Vladimir Berkovich based on an idea of Offer Gabber, we prefer assuming that the valuation is discrete.

There exists a contravariant functor

$$(Y \hookrightarrow X) \mapsto H^i_{Y,\mathrm{rig}}(X)$$

from the category of closed embeddings of algebraic varieties over k (with cartesian diagrams as morphisms) to the category of finite dimensional vector spaces over K. This cohomology only depends on a neighborhood of Y in X. If we are given a sequence of closed immersions

$$Z \hookrightarrow Y \hookrightarrow X,$$

there is a functorial long exact sequence

$$\cdots \to H^i_{Z,\mathrm{rig}}(X) \to H^i_{Y,\mathrm{rig}}(X) \to H^i_{Y\setminus Z,\mathrm{rig}}(X \setminus Z) \to \cdots$$

There is another "functor"

$$X \mapsto H^i_{\mathrm{rig},c}(X)$$

from the category of algebraic varieties over k to the category of finite-dimensional vector spaces over K. Actually, this is only covariant with respect to open immersions and contravariant with respect to proper maps (and these two functorialities are compatible). If we are given a closed immersion $Y \hookrightarrow X$, there is a functorial long exact sequence

$$\cdots \to H^i_{\mathrm{rig},c}(X \setminus Y) \to H^i_{\mathrm{rig},c}(X) \to H^i_{\mathrm{rig},c}(Y) \to \cdots$$

For $Y \hookrightarrow X$ a closed immersion, there is a cup-product

$$H^i_{\mathrm{rig},c}(Y) \times H^j_{Y,\mathrm{rig}}(X) \to H^{i+j}_{\mathrm{rig},c}(X)$$

which is functorial with respect to proper morphisms. For X irreducible of dimension d, there exists a trace map

$$\mathrm{tr} : H^{2d}_{\mathrm{rig},c}(X) \to K$$

which is functorial with respect to open immersions. In the case X is smooth, the *Poincaré pairing*

$$H^i_{\mathrm{rig},c}(Y) \times H^{2d-i}_{Y,\mathrm{rig}}(X) \to H^{2d}_{\mathrm{rig},c}(X) \to K$$

is perfect and compatible with the long exact sequences (*Poincaré duality*).

We can say a little more about rigid cohomology. First of all, by definition, we have

$$H^i_{\text{rig}}(X) := H^i_{X,\text{rig}}(X)$$

and there is a canonical morphism

$$H^i_{\text{rig},c}(X) \to H^i_{\text{rig}}(X)$$

which is an isomorphism for X proper. Note that $H^i_{\text{rig},c}(X)$ and $H^i_{\text{rig}}(X)$ are both 0 unless $0 \leq i \leq 2d$. Also, there are *Künneth formulas*

$$H^i_{\text{rig},c}(X) \otimes H^i_{\text{rig},c}(X') \simeq H^i_{\text{rig},c}(X \times X')$$

in general and

$$H^i_{\text{rig}}(X) \otimes H^i_{\text{rig}}(X') \simeq H^i_{\text{rig}}(X \times X')$$

when X and X' are both smooth. And finally, rigid cohomology and rigid cohomology with compact support both commute to isometric extensions of K.

1.7 Frobenius on rigid cohomology

Before introducing the Frobenius map, I want to recall that the *Chow group* of an algebraic variety X is defined as the quotient $A(X)$ of the cycle group $Z(X)$ modulo rational equivalence. More precisely, a *cycle* T on X is a closed integral subvariety and the *cycle group* is the free abelian group on cycles. A cycle is *rationally equivalent to* 0 if it is of the form $f_*(D)$ where D is a principal Cartier divisor and

$$f : X' \to X$$

a proper map. In [74], Denis Pétrequin shows that in the situation of the previous paragraph, there is a canonical pairing

$$\begin{array}{ccc} A_i(X) \times H^{2i}_{\text{rig},c}(X) & \longrightarrow & K \\ (T, \omega) & \longmapsto & \int_T \omega := tr(\omega_{|T}). \end{array}$$

From this, one can derive a *Lefschetz trace formula*. More precisely, if $\varphi : X \to X$ is an endomorphism with a finite number of fixed points N (counting multiplicities), then

$$N = \sum_{i=0}^{2d} (-1)^{i+1} \text{tr}\varphi^*_{|H^i_{\text{rig},c}(X)}.$$

When k is a field of characteristic p, then the Frobenius map $x \mapsto x^q$ with $q = p^f$ acts by functoriality on $H^i_{Y,\text{rig}}(X)$ and $H^i_{\text{rig},c}(X)$. Actually, we need to choose a continuous lifting σ of Frobenius to K and we get semi-linear maps. Anyway, it can be shown that the above Bloch–Ogus formalism is compatible with the Frobenius actions up to a twist by q^d in the trace map.

When $k = \mathbf{F}_q$, we can derive from the Lefschetz trace formula the following equalities:

$$\zeta(X,t) = \prod_{i=0}^{2d} \det(1 - tF^*_{|H^i_{\text{rig},c}(X)})^{(-1)^{i+1}}$$

and, for $Y \subset X$ smooth,

$$\zeta(Y,t) = \prod_{i=0}^{2d} \det(1 - tq^d(F^*)^{-1}_{|H^i_{Y,\text{rig}}(X)})^{(-1)^{i+1}}$$

One can show (see for example [23]) that the Frobenius is a Weil operator on $H^i_{\text{rig},c}(X)$ with positive weights between $2(i-d)$ and i. By duality again, we see that when X is smooth, Frobenius is also a Weil operator on $H^i_{Y,\text{rig}}(X)$ with weights between i and $2i$ but less than $2d$.

1.8 Slopes of Frobenius

Up to this point, l-adic cohomology does as well as, and in general better than, rigid cohomology. However, the latter becomes essential when it comes to the computation of slopes. So, let us assume that the valuation is discrete and that σ fixes a uniformizer π. Then, Dieudonné–Manin theorem tells us that, up to a finite extension of k, any σ-linear operator has a basis $\{e_{ij}\}$ with

$$e_{i1} \mapsto e_{i2} \mapsto \cdots \mapsto e_{is} \mapsto \pi^r e_{i1}$$

in which case

$$\lambda := \frac{1}{[K:\mathbf{Q}_p]} \frac{r}{s}$$

is called a *slope*. It is shown in [23] that any slope λ of $H^i_{\text{rig},c}(X)$ satisfies $0 \le \lambda \le d$ and $0 \le i - \lambda \le d$. By duality again, the same is true for $H^i_{Y,\text{rig}}(X)$ when X is smooth.

The above p-adic cohomological formula for the Zeta function shows that the slopes of rigid cohomology of X determine the p-adic absolute values of the algebraic integers α_i and β_i such that

$$|X(\mathbf{F}_q^r)| = \sum_i \alpha_i^r - \beta_i^r.$$

1.9 The coefficients question

In his marvellous article [85], Weil proves his conjecture for diagonal hypersurfaces $\sum x_i^r = 0$. The method consists in projecting to the diagonal hyperplane $\sum x_i = 0$. In other words, if $f : Y \to X$ is the projection $x \mapsto x^r$, we have

$$N_r(Y) = \sum_{x \in X(\mathbf{F}_{q^r})} N_r(Y_x)$$

with $Y_x := f^{-1}(x)$ and therefore, we are led to compute more complicated sums (than just counting points) on simpler algebraic varieties.

In order to generalize this, we must define in a functorial way, for each algebraic variety X, a category of coefficients E on X. Moreover, we need to associate to E, at each closed point $x \in X$, some $S(X, E, x)$. Then, we will define

$$S_r(X, E) = \sum_{x \in X(\mathbf{F}_{q^r})} S(X, E, x)$$

and

$$L(X, E, t) := \exp(\sum_{r-1}^{\infty} S_r(X, E)\frac{t^r}{r}).$$

Of course, the cohomology theory should take into account these coefficients and provide a cohomological formula for computing L-functions.

Ideally, one looks for a "constructible" theory of coefficients satisfying Grothendieck six operations formalism. The p-adic candidate is the theory of arithmetic $\mathcal{D}$-modules of Berthelot ([14], [18] and [19]) and its study is beyond the scope of this course. We will consider here the "lisse" p-adic theory and introduce the category or *overconvergent F-isocrystals*.

1.10 F-isocrystals

For an algebraic variety X over k, we will define the category

$$F\text{−isoc}^\dagger(X/K)$$

of overconvergent F-isocrystals over X/K. This is an abelian category with $\otimes$, internal *Hom* and rank. If $f : Y \to X$ is any morphism, there is an inverse image functor

$$f^* : F\text{-isoc}^\dagger(X/K) \to F\text{-isoc}^\dagger(Y/K).$$

If f is proper and smooth, there should be a direct image functor

$$f_* : F\text{-isoc}^\dagger(Y/K) \to F\text{-isoc}^\dagger(X/K).$$

This is the case when f is finite or if it has a "nice" lifting.

Let us concentrate for a while on the case of a closed point. First of all, $F\text{-isoc}^\dagger(\mathrm{Spec}k/K)$ is identical to the category $F\text{-isoc}(K)$ of strong finite dimensional F*-isocrystals* over K: an F-isocrystal over K is a vector space E with a σ-linear endomorphism φ. It is said to be *strong* when the Frobenius endomorphism is actually bijective. More generally, let k' be a finite extension of k. If K' is an unramified extension of K with residue field k' and σ' a frobenius on K' compatible with σ, then $F\text{-isoc}^\dagger(\mathrm{Spec}k'/K)$ is equivalent to $F\text{-isoc}(K')$.

Assume that $k = \mathbf{F}_q$ and let $d := [k' : k]$. If (E, ϕ) is an F-isocrystal on K', then ϕ^d is linear and one sets

$$S(k', E) := Tr\phi^d.$$

Now, if X is any algebraic variety over $\mathbf{F}_q$ and E is an overconvergent F-isocrystal over X/K, then for each closed point $x \in X$, the inverse image E_x of E on x is an overconvergent F-isocrystal on $k(x)$ and one sets

$$S(X, E, x) = S(k(x), E_x).$$

Using the formulas of Section 1.9, we can define $S_r(X, E)$ and $L(X, E, t)$. Note that, as before, if $K(x)$ denotes an unramified extension of K having $k(x)$ as residue field, we have

$$L(X, E, t) = \prod_{x \in |X|} \frac{1}{\det_{K(x)}(1 - t^{\deg x}\phi_x^{\deg x})}.$$

One can define the rigid cohomology of overconvergent F-isocrystals. These vector spaces come with a Frobenius automorphism and one can prove cohomological formulas

$$L(X, E, t) = \prod_{i=0}^{2d} \det(1 - tF^*_{|H^i_{\mathrm{rig},c}(X,E)})^{(-1)^{i+1}}$$

and, for $Y \subset X$ smooth,

$$L(Y, E, t) = \prod_{i=0}^{2d} \det(1 - tq^d (F^*)^{-1}_{|H^i_{Y,\mathrm{rig}}(X, \check{E})})^{(-1)^{i+1}}$$

as before (see [43]).

2

Tubes

Up to the next to last chapter, I will mainly follow Berthelot's preprint on rigid cohomology ([13] and [8]).

2.1 Some rigid geometry

2.1.1 Ultrametric fields

We assume that the reader is familiar with rigid analytic geometry such as in [21] or [45]. We fix a complete ultrametric field K (complete for a non trivial non archimedean absolute value) with valuation ring $\mathcal{V}$, maximal ideal $\mathfrak{m}$ and residue field k. We will denote by π a non zero, non invertible element of $\mathcal{V}$. Finally, when we pick up a positive real number, it is always assumed to live in $|K^*| \otimes \mathbf{Q} \subset \mathbf{R}_{>0}$.

We will be mainly interested in the discrete valuation case. Then, unless otherwise specified, we will implicitly assume that π is a uniformizer. In the equicharacteristic situation, we have $K \simeq k((t))$. Actually, we will be mainly concerned with the mixed characteristic case, namely $\operatorname{Char} K = 0$ and $\operatorname{Char} k = p > 0$.

Note that, starting with a field k of characteristic p, there always exists a *Cohen ring* for k. This is a minimal discrete valuation ring $\mathcal{V}$ of mixed characteristic with uniformizer p and residue field k. Actually, any complete discrete valuation ring with residue field k contains a Cohen ring of k. When k is perfect, there is essentially one Cohen ring for k and this is the ring of Witt vectors $W := W(k)$ of k. One usually denotes by $\mathbf{Z}_q := W(\mathbf{F}_q)$ the Witt vectors ring of $k = \mathbf{F}_q$ and let $\mathbf{Q}_q$ be its fraction field.

Note that even when k is perfect, $k(t)$ is not and it is therefore necessary to consider non perfect fields. Also, since having a discrete valuation and being algebraically closed are incompatible, it is also necessary to consider non discrete valuations. The main example is the completion $\mathbf{C}_p$ of the algebraic closure of $\mathbf{Q}_p$. Finally, even if we stick to discrete valuations, it is not reasonable to consider only Cohen rings. For example, the naive exponential function $\exp t$ has radius of convergence $|p|^{\frac{1}{p-1}}$ which is too small and has to be replaced by $\exp \pi t$ where π is a $(p-1)$-th root of $-p$. This function plays an essential role in the Fourier transform and is necessary if one wants to interpolate additive characters.

2.1.2 Fibers of formal schemes

We fix some complete ultrametric field K with $\mathcal{V}$, k and π as explained just before.

We call *algebraic variety* any scheme locally of finite type over a field. We call *formal scheme* over $\mathcal{V}$ any π-adic formal scheme locally topologically of finite presentation over $\mathcal{V}$. When the valuation is discrete, an algebraic k-variety is a very special kind of formal $\mathcal{V}$-scheme, more precisely algebraic k-varieties form a full subcategory of formal $\mathcal{V}$-schemes. Unfortunately, with our conventions, this is not true in general because k itself need not be finitely presented over $\mathcal{V}$. However, this would still be valid if we considered a broader category of formal schemes.

Any formal $\mathcal{V}$-scheme P has a *special fiber* P_k which is an algebraic k-variety with an "embedding" $P_k \hookrightarrow P$ which is a homeomorphism. We will often identify the underlying spaces of P and P_k. For example, if $X \to P_k$ is a morphism of k-variety, we might also call the composite $X \to P_k \hookrightarrow P$ a morphism. As a particular case, if $X \hookrightarrow P_k$ is a (locally closed) immersion, we will simply say that $X \hookrightarrow P$ is an immersion. Of course, this is a straight generalization from the discrete valuation case and compatible with the usual terminology if we consider more general formal schemes.

The formal $\mathcal{V}$-scheme P also has a *generic fiber* P_K which is a quasi-separated rigid analytic variety over K with a *specialization map*

$$sp : P_K \to P$$

which is continuous. Actually, if $P' \subset P$ is an open subset, then $P'_K = sp^{-1}(P')$. We will denote by

$$\begin{array}{ccc} P_K\text{Ê} & \longrightarrow & P_k \\ x & \longmapsto & \overline{x} \end{array}$$

the map induced by sp via the identification of the underlying spaces of P and P_k. Sometimes, we will just write $sp(x) = \overline{x}$. And all these constructions are functorial in P.

When $P = \mathrm{Spf} A$, with A a complete π-adic $\mathcal{V}$-algebra topologically of finite presentation, then $P_k = \mathrm{Spec} A_k$ with $A_k := k \otimes_{\mathcal{V}} A$ and the embedding $P_k \hookrightarrow P$ is given by

$$x \mapsto \ker(A \to A_k \to k(x))$$

where $k(x)$ denotes the residue field of x. We also have $P_K = \mathrm{Spm}(A_K)$ with $A_K := K \otimes_{\mathcal{V}} A$. If $x \in P_K$ and $K(x)$ denotes its residue field, the reduction map $A_K \to K(x)$ induces a morphism $A \to \mathcal{V}(x)$ where $\mathcal{V}(x)$ is the valuation ring of $K(x)$. Then, the specialization map is given by

$$x \mapsto \ker(A \to \mathcal{V}(x) \to k(x)).$$

Note that, by functoriality, if $g \in A$ and $x \in P_K$, we have $\overline{g}(\overline{x}) = \overline{g(x)}$ where $\overline{g}$ denotes the image of g in A_k.

Recall that, to any algebraic K-variety V is associated a rigid analytic variety V^{rig} whose underlying space is the set of closed points of V. Recall also that any $\mathcal{V}$-scheme locally of finite presentation X has a formal completion $\widehat{X}$ which is a formal $\mathcal{V}$-scheme and a generic fiber X_K which is an algebraic K-variety. There is a canonical morphism $\widehat{X}_K \to X_K^{\mathrm{rig}}$ which is an open immersion when X is separated and an isomorphism when X is proper.

As a first example, we may consider the affine space $\mathbf{A}^N$. If we identify the underlying space of $\widehat{\mathbf{A}}_{\mathcal{V}}^N$ with $\mathbf{A}_k^N$ and if we denote by $x \mapsto \overline{x}$ the reduction mod $\mathfrak{m}$, then

$$\widehat{\mathbf{A}}_K^N := (\widehat{\mathbf{A}}_{\mathcal{V}}^N)_K = \mathbf{B}^N(0, 1^+)$$

is the closed ball of radius one. And the specialization map is given on rational points by

$$sp : (x_1, \ldots, x_N) \mapsto (\overline{x}_1, \ldots, \overline{x}_N).$$

The other important example is the projective space $\mathbf{P}_{\mathcal{V}}^N$ which gives

$$\widehat{\mathbf{P}}_K^N := (\widehat{\mathbf{P}}_{\mathcal{V}}^N)_K = \mathbf{P}_K^{N,\mathrm{rig}}.$$

If, as usual, we identify the underlying space of $\widehat{\mathbf{P}}_{\mathcal{V}}^N$ with $\mathbf{P}_k^N$, the specialization map is given on rational points by

$$sp : (x_0, \ldots, x_N) \mapsto (\overline{x}_0, \ldots, \overline{x}_N) \quad \text{with} \quad \max((|x_0|, \ldots, |x_N|) = 1.$$

2.1.3 Base change

We fix some complete ultrametric field K with $\mathcal{V}$, k and π as before.

If $\sigma : K \hookrightarrow K'$ is an isometric embedding of complete ultrametric fields and V is a rigid analytic variety, we can consider its pull back $V_{K'}$ (also written V^σ) along σ. When $V = \mathrm{Spm} A$, with A a Tate algebra, then $V_{K'} := \mathrm{Spm} A_{K'}$ with $A_{K'} := K'\widehat{\otimes}_K A$. This functor $V \mapsto V_{K'}$ is continuous in the sense that it preserves admissible open subsets and admissible covering. If we denote by $\widetilde{V}$ the *rigid topos of* V, which is just the category of sheaves of sets on V, we can consider the functor

$$\varpi_* : \widetilde{V_{K'}} \to \widetilde{V}$$

defined by

$$(\varpi_* \mathcal{F})(W) := \mathcal{F}(W_{K'}).$$

The functor ϖ_* has a left adjoint

$$\varpi^{-1} : \widetilde{V} \to \widetilde{V_{K'}}$$

which is exact. In other words, we have a morphism of topos

$$\varpi : \widetilde{V_{K'}} \to \widetilde{V}.$$

Moreover, there is a canonical morphism $\varpi^{-1}\mathcal{O}_V \to \mathcal{O}_{V_{K'}}$ and we define for any $\mathcal{O}_V$-module $\mathcal{E}$,

$$\varpi^*\mathcal{E} = \mathcal{O}_{V_{K'}} \otimes_{\varpi^{-1}\mathcal{O}_V} \varpi^{-1}\mathcal{E}.$$

We will also write $\mathcal{E}_{K'}$ or $\mathcal{E}^\sigma$ for $\varpi^*\mathcal{E}$.

When $V = \mathrm{Spm} A$, we can consider the canonical morphism of rings $A \to A_{K'}$. If $\mathcal{E}$ is a coherent $\mathcal{O}_V$-module with $\Gamma(V, \mathcal{E}) = M$, then $\mathcal{E}_{K'}$ is the coherent $\mathcal{O}_{V_{K'}}$-module with

$$\Gamma(V', \mathcal{E}_{K'}) = K'\widehat{\otimes}_K M$$

(also written $M_{K'}$ or M^σ).

Note that σ induces morphism of rings $\sigma : \mathcal{V} \hookrightarrow \mathcal{V}'$ where $\mathcal{V}'$ is the ring of integers of K'. Then, the construction of generic fiber and specialization are compatible with base extension $P \mapsto P^\sigma$ in the category of formal schemes. More precisely, if P is a $\mathcal{V}$-formal scheme, there is a canonical isomorphism $(P_K)^\sigma \simeq P^\sigma_K$. Note also that, if V is an algebraic variety over K, then $(V^\sigma)^{\mathrm{rig}} = (V^{\mathrm{rig}})^\sigma$.

2.2 Tubes of radius one

We fix some (non trivial) complete ultrametric field K with $\mathcal{V}$, k and π as before.

Definition 2.2.1 *A* formal embedding $\iota : X \hookrightarrow P$ *is a locally closed immersion of an algebraic k-variety into a formal $\mathcal{V}$-scheme. Then, the* tube *of X in P is*

$$]X[_P := sp^{-1}(\iota(X)).$$

Even if this notion is purely set-theoretic, we want to reserve this terminology to varieties. Actually, the tube will inherit a natural structure of rigid analytic variety. Note also that, later on, when we introduce the notion of tube of radius η, the scheme structure of X will play an important role.

Proposition 2.2.2 *Any formal embedding $\iota : X \hookrightarrow P$ factors through a closed formal embedding $\iota' : X \hookrightarrow P'$ where P' is an open formal subscheme of P and we have $]X[_P =]X[_{P'}$. When $\iota(X)$ is open in P_k, we may choose P' with $P'_k = X$, and then $]X[_P = P'_K$.*

Proof Since P and P_k have the same underlying space, if X' is an open subset of P_k, we can write $X' \simeq P'_k$ with P' an open subset of P. And we know that $P'_K = sp^{-1}(P')$. Both assertions follow. More precisely, we may factor ι as

$$X \hookrightarrow X' \simeq P'_k \hookrightarrow P' \hookrightarrow P$$

with X closed in X' and P' open in P. □

Proposition 2.2.3 *Let P be a formal $\mathcal{V}$-scheme. If a subvariety X of P_k is the union $X = \cup_i X_i$ (resp. intersection $X = \cap_i X_i$) of subvarieties of P_k, then*

$$]X[_P = \cup_i]X_i[_P \quad (\text{resp.}]X[_P = \cap_i]X_i[_P).$$

Proof This is an immediate consequence of the fact that inverse image commutes with arbitrary union and intersection. □

In the future, when we consider finite union and finite intersection of subvarieties, we will always endow them with the induced scheme structure. We shall come back to this point later on.

Corollary 2.2.4 *Let S be a formal $\mathcal{V}$-scheme. If $\iota : X \hookrightarrow P$ and $\iota' : X' \hookrightarrow P'$ are two formal embeddings over S, we have*

$$]X \times_{S_k} X'[_{P\times_S P'} = (]X[_P \times_{S_K} P'_K) \cap (P_K \times_{S_K}]X'[_{P'}).$$

Proof Follows from the fact that

$$X \times_{S_k} X' = (X \times_{S_k} P'_k) \cap (P_k \times_{S_K} X')$$

and Proposition 2.2.3. □

As already mentioned, the tube is independent of the scheme structure of X. However, if $u : P' \to P$ is a morphism of formal S-schemes, we will always endow $u^{-1}(\iota(X))$ with the inverse image scheme structure. In other words, we set

$$u^{-1}(X) := u^{-1}(\iota(X)) := P' \times_P X.$$

Definition 2.2.5 *A morphism between two formal embeddings $\iota : X \hookrightarrow P$ and $\iota' : X' \hookrightarrow P'$ is a commutative diagram of morphisms*

$$\begin{array}{ccc} X' & \overset{\iota'}{\hookrightarrow} & P' \\ \downarrow f & & \downarrow u \\ X & \overset{\iota}{\hookrightarrow} & P. \end{array}$$

Proposition 2.2.6 *If*

$$\begin{array}{ccc} X' & \hookrightarrow & P' \\ \downarrow f & & \downarrow u \\ X & \hookrightarrow & P \end{array}$$

is a morphism of formal embeddings, we have

$$]X'[_{P'} \subset u_K^{-1}(]X[_P).$$

Moreover, we get an equality when the diagram is cartesian ($X' \simeq P' \times_P X$).

Proof This follows from the functoriality of specialization. More precisely, if we denote by ι and ι' the immersions, we have for all $x' \in]X'[_{P'}, sp(x') \in \iota'(X')$ and it follows that

$$(u \circ sp)(x') = (sp \circ u_K)(x') \in u(\iota'(X')) = \iota(f(X')) \subset \iota(X)$$

which is what we want.

Assume now that the diagram is cartesian so that

$$u^{-1}(\iota(X)) = \iota'(X').$$

If $x' \in P_K$ satisfies $u_K(x') \in]X[_P$, then, we have

$$(u \circ sp)(x') = (sp \circ u_K)(x') \in \iota(X)$$

and it follows that $sp(x') \in u^{-1}(\iota(X)) = \iota'(X')$ and we are done. □

We do not want to say much about it but *formal modifications* are morphisms of formal schemes that induce an isomorphism on generic fibers and *formal blowing up* are particular explicit formal modifications.

Corollary 2.2.7 *If $\iota : X \hookrightarrow P$ is a formal embedding, $u : P' \to P$ a formal blowing up (or more generally, a formal modification) and $X' = u^{-1}(X)$, we have*

$$]X'[_{P'} \simeq]X[_P.$$

Proof As already mentioned, formal blowing up induce an isomorphism on the generic fibers and the construction of generic fibers is functorial. □

Proposition 2.2.8 *If $\sigma : K \hookrightarrow K'$ is an isometric embedding and $\iota : X \hookrightarrow P$ is a formal embedding, we have*

$$]X^\sigma[_{P^\sigma} =]X[^\sigma_P.$$

Proof If X is open in P, then $X = P'_k$ with $P' \subset P$ open and

$$]X^\sigma[_{P^\sigma} = P'^\sigma_{K'} =]X[^\sigma_P.$$

If X is closed, its complement U is open and $]X[_P$ is the complement of $]U[_P$ in P_K. Thus, the closed case follows from the open case. In general, X is the intersection of a closed subvariety and an open subvariety. □

Definition 2.2.9 *A formal embedding $\iota : X \hookrightarrow P$ is* affine *if P is affine.*

Note that the immersion need not be closed and we do not assume that X itself is affine.

If $P = \mathrm{Spf} A$ is affine and $E \subset A$ is any subset, we will denote by $V(E)$ the closed formal subscheme defined by E and by $D(E)$ its open complement. If $\iota : X \hookrightarrow P$ is a formal embedding, we call an equality of the form

$$\iota(X) = V(f_1, \ldots, f_r) \cap D(g_1, \ldots, g_s) \cap P_k$$

a *finite presentation of X inside P.*

Lemma 2.2.10 *Let $\iota : X \hookrightarrow P$ be an affine formal embedding and let*

$$\iota(X) = V(f_1, \ldots, f_r) \cap D(g_1, \ldots, g_s) \cap P_k$$

be a finite presentation of X inside P. Then

$$\begin{aligned}]X[_P = \{x \in P_K, |f_1(x)|, \ldots, |f_r(x)| < 1 \\ \text{and} \quad \exists j \in \{1, \ldots, s\}, |g_j(x)| = 1\}. \end{aligned}$$

Proof We write $P = \mathrm{Spf} A$ and first assume that $j = 0$ and $i = 1$ so that $\iota(X) = V(f) \cap P_k$. Then, we have

$$x \in]X[_P \Leftrightarrow \overline{x} \in \iota(X) \Leftrightarrow \overline{f}(\overline{x}) = 0 \Leftrightarrow \overline{f(x)} = 0 \Leftrightarrow |f(x)| < 1.$$

Thus, we see that

$$]X[_P = \{x \in P_K, |f(x)| < 1\}.$$

In the case $i = 0$ and $j = 1$ so that $X = D(g) \cap P_k$, we get since $|g(x)| \leq 1$ for $g \in A$,

$$]X[_P = \{x \in P_K, |g(x)| > 1\} = \{x \in P_K, |g(x)| = 1\}$$

which is also open in P_K. In general, we write X as union and intersection of such subvarieties and the result follows from Proposition 2.2.3. □

As a first example, we see that

$$]0[_{\widehat{\mathbf{A}}^N_{\mathcal{V}}} = \mathbf{B}^N(0, 1^-)$$

is not quasi-compact (unless $N = 0$).

Another very important example is given by the Monsky–Washnitzer setting. We start developing now the geometric ground for this theory.

Proposition 2.2.11 *Let P be a (locally closed) formal subscheme of $\widehat{\mathbf{P}}^N_{\mathcal{V}}$ and $X := \mathbf{A}^N_k \cap P_k$. Then, we have*

$$]X[_P = \mathbf{B}^N(0, 1^+) \cap P_K.$$

Proof By definition, there is a cartesian diagram

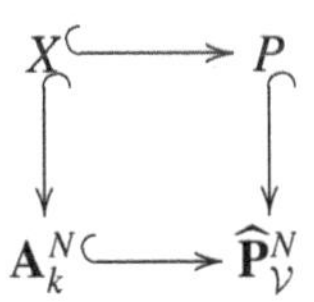

and it follows from Proposition 2.2.6 that

$$]X[_P = \mathbf{B}^N(0, 1^+) \cap P_K.$$

□

Corollary 2.2.12 *Let A be a $\mathcal{V}$-algebra of finite presentation, $X := \mathrm{Spec} A$ and $V := X_K^{\mathrm{rig}}$. Let*

$$\mathcal{V}[T_1, \ldots, T_N] \to A$$

be a presentation of A, $X \subset \mathbf{A}^N_{\mathcal{V}}$ *the corresponding inclusion,* Y *the algebraic closure of* X *in* $\mathbf{P}^N_{\mathcal{V}}$ *and* $P := \widehat{Y}$. *Then, we have*

$$]Y_k[_{\widehat{Y}} = \widehat{Y}_K = Y_K^{\mathrm{rig}} \supset X_K^{\mathrm{rig}}$$

and

$$]X_k[_{\widehat{Y}} = \widehat{X}_K = \mathbf{B}^N(0, 1^+) \cap X_K^{\mathrm{rig}} \subset Y_K^{\mathrm{rig}}.$$

Proof By definition, we have $]Y_k[_{\widehat{Y}} = \widehat{Y}_K$ and since Y is proper,

$$\widehat{Y}_K = Y_K^{\mathrm{rig}}.$$

Now, by functoriality, $Y_K^{\mathrm{rig}} \supset X_K^{\mathrm{rig}}$. Proposition 2.2.11 tells us that

$$]X_k[_{\widehat{Y}} = \mathbf{B}^N(0, 1^+) \cap Y_K^{\mathrm{rig}}$$

and our assertion follows from the fact that

$$\mathbf{B}^N(0, 1^+) \cap Y_K^{\mathrm{rig}} \subset \mathbf{A}_K^{N,\mathrm{rig}} \cap Y_K^{\mathrm{rig}} = X_K^{\mathrm{rig}}.$$

□

In the next proposition, we need the notion of *quasi-Stein* morphism. Recall that a *Weierstrass domain* in an affinoid variety is an admissible open subset defined by

$$|f_1(x)|, \ldots, |f_r(x)| \leq 1.$$

A rigid analytic variety is *quasi-Stein* if it is an increasing union of Weierstrass domains. A morphism of rigid analytic varieties

$$v : W \to V$$

is *quasi-Stein* if there exists an admissible affinoid covering $V = \cup V_i$ such that $v^{-1}(V_i)$ is quasi-Stein for each i. If this is the case and $\mathcal{E}$ is a coherent $\mathcal{O}_W$-module, it follows from the main result of [56] that

$$R^q v_* \mathcal{E} = 0 \quad \text{for} \quad q > 1.$$

Proposition 2.2.13 *If* $\iota : X \hookrightarrow P$ *is a formal embedding, then* $]X[_P$ *is an admissible open subset of* P_K. *If moreover* ι *is a closed immersion, then* $]X[_P \hookrightarrow P_K$ *is quasi-Stein.*

Proof Both questions are local on P_K and therefore also on P which we may assume affine. We may also assume that X is actually a subset of P. We can choose a finite presentation

$$X = V(f_1, \ldots, f_r) \cap D(g_1, \ldots g_s) \cap P_k$$

of X as a formal subscheme of P so that

$$]X[_P = \{x \in P_K, |f_1(x)|, \ldots, |f_r(x)| < 1 \\ \text{and} \quad \exists j \in \{1, \ldots, s\}, |g_j(x)| = 1\}.$$

This is easily seen to be an admissible open subset. More precisely, admissible open subsets are stable under finite union and finite intersections and, we know that for each $g = 1, \ldots, s$,

$$\{x \in P_K, |g_j(x)| = 1\}$$

is a Weierstrass domain and, in particular, is an admissible open subset. We are therefore reduced to the case $s = 0$, which means that X is closed. In this case,

$$]X[_P := \{x \in P_K, |f_1(x)|, \ldots, |f_r(x)| < 1\}$$

is not only an admissible open subset but also a quasi-Stein rigid analytic variety. More precisely, $]X[_P$ is the increasing union of the Weierstrass domains

$$\{x \in P_K, |\frac{1}{\pi} f_1^N(x)|, \ldots, |\frac{1}{\pi} f_r^N(x)| \leq 1\}.$$

□

Definition 2.2.14 *A morphism of formal embeddings*

$$\begin{array}{ccc} X' & \hookrightarrow & P' \\ \downarrow & & \downarrow \\ X & \hookrightarrow & P \end{array}$$

is an immersion *(resp. an* open immersion, *resp. a* closed immersion *if both vertical arrows are immersions (resp. open immersions, resp. closed immersions). It is called* mixed *if the first vertical arrow is a closed immersion and the second one is open.*

A family of immersions (resp. open immersions, resp. mixed immersions) of formal embeddings

$$\begin{array}{ccc} X_i & \hookrightarrow & P_i \\ \downarrow & & \downarrow \\ X & \hookrightarrow & P \end{array}$$

is a covering *(resp. an* open covering, *resp. a* mixed covering*) if it is so at all steps.*

Proposition 2.2.15 *If*

$$\begin{array}{ccc} X_i & \hookrightarrow & P_i \\ \downarrow & & \downarrow \\ X & \hookrightarrow & P \end{array}$$

is an open covering or a finite mixed covering of formal embeddings, then $]X[_P = \cup_i]X_i[_{P_i}$ *is an admissible covering.*

Proof First of all, since the tube only depends on an open neighborhood, we may assume that $P_i = P$ for each i. Now, for open coverings, our assertion is a direct consequence of the continuity of the specialization map. We focus now on finite mixed coverings. First of all, the assertion is local on P_K and we may therefore assume that P is affine. By induction on the number of components and using formal properties of admissible coverings, we may assume that $X = X_1 \cup X_2$ with $X_1 = V(f) \cap P_K$ and $X_2 = V(g) \cap P_k$. We have to show that the covering

$$\{x \in P_K, |(fg)(x) < 1\} = \{x \in P_K, |f(x)| < 1\} \cup \{x \in P_K, |g(x)| < 1\}$$

is admissible, which is an easy exercise in rigid analytic geometry and is left to the reader. □

Definition 2.2.16 *A morphism of formal embeddings*

$$\begin{array}{ccc} X' & \hookrightarrow & P' \\ \downarrow{\scriptstyle f} & & \downarrow{\scriptstyle u} \\ X & \hookrightarrow & P \end{array}$$

is flat *(resp.* smooth, *resp.* étale*) if u is flat (resp. smooth, étale) in the neighborhood of X'.*

Proposition 2.2.17 *If*

$$\begin{array}{ccc} X' & \hookrightarrow & P' \\ \downarrow{\scriptstyle f} & & \downarrow{\scriptstyle u} \\ X & \hookrightarrow & P \end{array}$$

is a flat (resp. smooth, étale) morphism of formal embeddings, then the induced map $]X'[_{P'} \to]X[_P$ *is also flat (resp. smooth, étale).*

Proof We may replace P by an open neighborhood of X and assume that u is flat (resp. smooth, étale). It follows that u_K is flat (resp. smooth, étale) and so is the induced map because a tube is an admissible open subset of the generic fiber. □

2.3 Tubes of smaller radius

We fix a complete ultrametric field K with $\mathcal{V}$, k and π as before.

A tube is not quasi-compact in general – not even locally on P – and it is therefore necessary to introduce the more subtle notion tube with given radius. Recall that when we pick up a positive real number, it is always assumed to live in $|K^*| \otimes \mathbf{Q} \subset \mathbf{R}_{>0}$. In other words, it is of the form $|\pi|^{n/m}$.

Lemma 2.3.1 *Let $P = \mathrm{Spf} A$ be a formal affine $\mathcal{V}$-scheme,*

$$X := V(f_1, \ldots, f_r) \cap D(g_1, \ldots g_s) \cap P_k \subset P$$

and X' a subvariety of X defined in P as

$$X' := V(f'_1, \ldots, f'_{r'}) \cap D(g'_1, \ldots g'_{s'}) \cap P_k.$$

Then, there exists $\eta_0 < 1$ such that for all $\eta_0 \leq \eta < 1$, the subset

$$[X']_{P\eta} := \{x \in P_K, |f'_1(x)|, \ldots, |f'_{r'}(x)| \leq \eta \text{ and } \exists j \in \{1, \ldots, s'\}, |g'_j(x)| = 1\}$$

is contained in

$$[X]_{P\eta} := \{x \in P_K, |f_1(x)|, \ldots, |f_r(x)| \leq \eta \text{ and } \exists j \in \{1, \ldots, s\}, |g_j(x)| = 1\}.$$

When the valuation is discrete, we may choose $\eta_0 = |\pi|$.

Proof The question is local on P_K and therefore also on P. Thus, we may assume that $s = s' = 0$. More precisely, we can replace P by some $D(g_j g'_{j'})$. Also, we may clearly assume that $r = 1$ so that

$$X = V(f) \cap P_k$$

and

$$X' := V(f'_1, \cdots, f'_{r'}) \cap P_k.$$

We can write

$$f = \sum h_i f'_i + \alpha k$$

with $h_i, k \in A$ and $\alpha \in \mathfrak{m}$. For $x \in P_K$, we have

$$|f(x)| \leq \sup\{|h_i(x)||f_i'(x)|, |\alpha||k(x)|\} \leq \sup\{|f_i'(x)|, |\alpha|\}.$$

Therefore, if for each i, we have $|f_i'(x)| \leq \eta$ with $\eta \geq |\alpha|$, then, necessarily, $|f(x)| \leq \eta$. Of course, this last condition is always satisfied when the valuation is discrete and $\eta \geq |\pi|$. □

Proposition 2.3.2 *Let $P = \mathrm{Spf} A$ be a formal affine $\mathcal{V}$-scheme and*

$$X := V(f_1, \cdots, f_r) \cap D(g_1, \cdots g_s) \cap P_k \subset P$$

a finite presentation of X as a formal subscheme of P. Define as before

$$[X]_{P\eta} := \{x \in P_K, |f_1(x)|, \ldots, |f_r(x)| \leq \eta \quad \text{and} \quad \exists j \in \{1, \ldots, s\}, |g_j(x)| = 1\}$$

and also

$$]X[_{P\eta} := \{x \in P_K, |f_1(x)|, \ldots, |f_r(x)| < \eta \quad \text{and} \quad \exists j \in \{1, \ldots, s\}, |g_j(x)| = 1\}.$$

Then we have:

(i) *If $P' \subset P$ is an open subset, then for all $\eta < 1$,*

$$[X \cap P_k']_{P'\eta} = [X]_{P\eta} \cap P_K' \quad \text{and} \quad]X \cap P_k'[_{P'\eta} =]X[_{P\eta} \cap P_K'.$$

(ii) *When the valuation is discrete, $[X]_{P\eta}$ and $]X[_{P\eta}$ only depend on X and not on the choice of the presentation as soon as $\eta \geq |\pi|$ in the first case and $\eta > |\pi|$ in the second case.*

(iii) *In general, if*

$$X = V(f_1', \cdots, f_{r'}') \cap D(g_1', \cdots g_{s'}') \cap P_k$$

is another presentation of X as a formal subscheme of P, and $[X]'_{P\eta}$ and $]X['_{P\eta}$ denote the corresponding rigid analytic varieties, there exists $\eta_0 < 1$ such that for all $\eta_0 \leq \eta < 1$, we have

$$[X]'_{P\eta} = [X]_{P\eta} \quad \text{and} \quad]X['_{P\eta} =]X['_{P\eta}.$$

Proof We consider only the case of $[X]_{P\eta}$. Since

$$]X[_{P\eta} = \cup_{\delta < \eta} [X]_{P\delta},$$

the other case will follow.

For the first assertion, we can write $P' := D(h_1, \dots, h_t)$ and our statement follows from the fact that

$$P'_K =]P'_k[_P = \{x \in P_K, \exists l \in \{1, \dots, t\}, |h_l(x)| = 1\}.$$

In order to prove the last assertion, it is sufficient, by symmetry, to prove that $[X]'_{P\eta} \subset [X]_{P\eta}$. This results from Lemma 2.3.1.

The second assertion also follows because we can choose $\eta_0 := |\pi|$. □

If $X \hookrightarrow P$ is a formal embedding (with X necessarily quasi-compact), a *finite presentation of X as a formal subscheme of P* is an affine covering $P = \cup P_i$ such that $X_i := X \cap P_i$ is almost always empty, and for each i, a finite presentation of X_i as a formal subscheme of P_i.

Definition 2.3.3 *Let $X \hookrightarrow P$ be a formal embedding.*

When the valuation is discrete, the closed and open tubes of radius η of X in P *are the rigid analytic varieties $[X]_{P\eta}$ and $]X[_{P\eta}$ locally defined in the proposition for $\eta \geq |\pi|$ and $\eta > |\pi|$ respectively.*

In general, if X is quasi-compact and we are given a finite presentation of X as a formal subscheme of P, the rigid analytic varieties $[X]_{P\eta}$ and $]X[_{P\eta}$ locally defined in the proposition for $\eta_0 \leq \eta < 1$ for some $\eta_0 < 1$, are also called the closed and open tubes of radius η of X in P.

It is important to note that the tubes are only defined when $\eta_0 \leq \eta < 1$ for some $\eta_0 < 1$ and that this η_0 depends on the presentation of X as a formal subscheme of P. However, given another presentation, we may always choose η_0 in such a way that the tubes coincide for $\eta_0 \leq \eta < 1$. We will always implicitly assume that X comes with a fixed presentation. We will also implicitly assume that η is chosen in such a way that the tubes are well defined.

When the valuation is discrete, it is sufficient to assume that $\eta > |\pi|$. Note also that, in this case, most results below are still valid even in the non-compact case. However, later on, we will need the quasi-compact hypothesis for other purposes (for example in Theorem 5.4.4) even in the case of a discrete valuation.

Before going any further, we may consider our usual example. We get

$$[0]_{\widehat{\mathbf{A}}^N_{\mathcal{V}}\eta} = \mathbf{B}^N(0, \eta^+) \quad \text{and} \quad]0[_{\widehat{\mathbf{A}}^N_{\mathcal{V}}\eta} = \mathbf{B}^N(0, \eta^-).$$

Proposition 2.3.4 *Let $X \hookrightarrow P$ be a formal embedding of a quasi-compact algebraic variety. Then,*

(i) *If X is open in P_k, we have for $\eta < 1$,*

$$[X]_{P\eta} =]X[_{P\eta} =]X[_P.$$

(ii) *More generally, if the image of X is contained in P' which is open in P, we have for $\eta < 1$,*

$$[X]_{P'\eta} = [X]_{P\eta}, \quad]X[_{P'\eta} =]X[_{P\eta} \quad \text{and} \quad]X[_{P'} =]X[_P.$$

(iii) *Finally, if X' is a subvariety of X, there exists $\eta_0 < 1$ such that for all $\eta_0 \leq \eta < 1$, we have $[X']_{P\eta} \subset [X]_{P\eta}$ and $]X'[_{P\eta} \subset]X[_{P\eta}$.*

Proof All the assertions are local on P which we may assume affine. Assertion (iii) follows from Lemma 2.3.1 and assertion (i) is a particular case of assertion (ii) which follows from the first part of Proposition 2.3.2. □

Recall that if $\{X_i\}_i$ is a finite family of subvarieties of an algebraic variety X, there is a homeomorphism

$$\prod_X X_i \simeq \cap X_i$$

and we implicitly endow $\cap X_i$ with the corresponding subvariety structure.

Proposition 2.3.5 *Let P be a formal $\mathcal{V}$-scheme.*

If $X = \cap_i X_i$ is a finite intersection of quasi-compact algebraic subvarieties of P_k, endowed with the scheme structure of the intersection, there exists $\eta_0 < 1$ such that for all $\eta_0 \leq \eta < 1$,

$$[X]_{P\eta} = \cap_i [X_i]_{P\eta} \quad \text{and} \quad]X[_{P\eta} = \cap_i]X_i[_{P\eta}.$$

Proof This is a local question on P that we may assume affine. Moreover, since any subvariety is the intersection of a closed and an open subvariety, we may consider only open or closed subvarieties. Also, by induction, it is sufficient to consider the intersection of two algebraic varieties X and X'. If one of them is open, then the assertion results from the second part of Lemma 2.3.4. Otherwise, they are both closed and, by induction again, we may assume that X and X' are hypersurfaces. If we write

$$X = V(f) \cap P_k \quad \text{and} \quad X' = V(g) \cap P_k,$$

we have

$$X \cap X' = V(f, g) \cap P_k$$

and the assertion follows from the definition. □

Recall that if $\{X_i\}_i$ is a finite family of closed subvarieties of an algebraic variety X, the underlying space of the schematic image of the canonical map

$$\coprod_i X_i \to X$$

is $\cup X_i$. This union is implicitly endowed with this structure.

Proposition 2.3.6 *Let P be a formal $\mathcal{V}$-scheme and $X = \cup_i X_i$ a finite union of quasi-compact closed subvarieties of P_k, endowed with the scheme structure of the union. If each $\eta_i < 1$ and $\eta := \prod \eta_i$, then*

$$[X]_{P\eta} \subset \cup_i [X_i]_{P\eta_i} \quad \text{and} \quad]X[_{P\eta} \subset \cup_i]X_i[_{P\eta_i}$$

and these unions are admissible coverings.

Proof As usual, it is sufficient to consider the case of the closed tubes. Moreover, this is a local question on P that we may assume affine. Also, by induction, it is sufficient to consider the case of two algebraic varieties X_1 and X_2. If $X_1 := V(I_1)$ and $X_2 = V(I_2)$, then $X = V(I_1 \cap I_2)$ is a subvariety of $X' := V(I_1 I_2)$. Using assertion (iii) of Proposition 2.3.4, we are reduced to showing that

$$[X']_{P\eta} \subset [X_1]_{P\eta_1} \cup [X_2]_{P\eta_2}.$$

If we write

$$X_1 = V(f_1, \ldots, f_r) \cap P_k \quad \text{and} \quad X_2 = V(g_1, \ldots, g_s) \cap P_k,$$

we have

$$X' - V(\{f_i g_j\}_{i,j}) \cap P_k.$$

And the assertion follows from the definition; more precisely, if for all i, j, we have $|f_i g_j|(x) \leq \eta$ but some $|f_{i_0}|(x) > \eta_1$, then necessarily, for all j, we have $|g_j|(x) \leq \eta_1$. Note also that the covering is admissible as a finite affinoid covering. □

Proposition 2.3.7 *If $\iota : X \hookrightarrow P$ is a formal embedding of a quasi-compact algebraic variety, the open (resp. closed) tubes of radius $\eta < 1$ form an admissible covering of $]X[_P$.*

Proof This question is local on P_K. We may therefore assume that P is affine and $X = V(f_1, \cdots, f_r) \cap P_k$ and we are reduced to the well-known results that the subsets

$$\{x \in P_K, |f_1(x)|, \ldots, |f_r(x)| \leq \eta\}$$

as well as the subsets

$$\{x \in P_K, |f_1(x)|, \ldots, |f_r(x)| < \eta\}$$

form an admissible covering of

$$\{x \in P_K, |f_1(x)|, \ldots, |f_r(x)| < 1\}$$

when $\eta < 1$. And the reader can easily verify this assertion. □

As a corollary, we see that the tubes of radii η are essentially independent of the scheme structure of X.

Corollary 2.3.8 *Let $\iota : X \hookrightarrow P$ and $\iota' : X' \hookrightarrow P$ be two formal embedding of quasi-compact algebraic varieties into the same formal $\mathcal{V}$-scheme P. Assume that $\iota(X)$ and $\iota(X')$ have the same underlying subspace. Then for each $\eta' < 1$ there exists $\eta < 1$ such that $[X']_{P\eta'} \subset [X]_{P\eta}$ and $]X'[_{P\eta'} \subset]X[_{P\eta}$.*

Proof The case of open tubes immediately results from the case of closed tubes. Given η', we have

$$[X']_{P\eta'} \subset]X'[_P =]X[_P = \cup_\eta [X]_{P\eta}.$$

Since X is quasi-compact, so is $[X']_{P\eta'}$. Since the covering is admissible, then necessarily, $[X']_{P\eta'} \subset [X]_{P\eta}$ for some η. □

Proposition 2.3.9 *If*

$$\begin{array}{ccc} X' & \hookrightarrow & P' \\ \downarrow f & & \downarrow u \\ X & \hookrightarrow & P \end{array}$$

is a morphism of formal embeddings of quasi-compact algebraic varieties, there exists $\eta_0 < 1$ such that for each $\eta_0 \leq \eta < 1$, we have

$$[X']_{P'\eta} \subset u_K^{-1}([X]_{P\eta}) \quad \text{and} \quad]X'[_{P'\eta} \subset u_K^{-1}(]X[_{P\eta}).$$

When the diagram is cartesian, we even get equalities.

Proof The questions are local on P'_K and we may therefore assume X closed in P affine and X' closed in P' affine. There are only two cases to consider, the cartesian case and the case $P' = P$ (and X' subvariety of X). The first case directly follows from the definition because X' is defined in P' by the same equations as X in P. And the second one has already been dealt with in Lemma 2.3.1. □

Corollary 2.3.10 *In the situation of the proposition, if the diagram is cartesian and u is a formal blowing up (or a formal modification), we get isomorphisms*

$$[X']_{P'\eta} \simeq [X]_{P\eta} \quad \text{and} \quad]X'[_{P'\eta} \simeq]X[_{P\eta}.$$

Proof Again this is simply because a formal blowing-up induces an isomorphism on the generic fibers. □

If $\eta < 1$, the inclusion map

$$\mathbf{D}(0, \eta^+) \hookrightarrow \mathbf{D}(0, 1^+)$$

with $\eta = |\pi|^{\frac{m}{n}}$ has a formal model

$$\mathrm{Spf}\, A\{T, U\}/(T^n - \pi^m U) \to \mathrm{Spf}\, A\{T\}$$

(which is not an open immersion). This is a general situation and we have the following:

Corollary 2.3.11 *Let $X \hookrightarrow P$ be a formal embedding of a quasi-compact algebraic variety, $u : P' \to P$ a formal model for the open immersion $[X]_{P\eta} \hookrightarrow P_K$ and $X' := u^{-1}(X)$. Then, we have*

$$P'_K = [X']_{P'\eta} \simeq [X]_{P\eta} \hookrightarrow P_K.$$

Proof Since u_K is the open immersion of $[X]_{P\eta}$ in P_K, it follows from Proposition 2.3.9 that $[X']_{P'\eta} \simeq [X]_{P\eta}$. □

This applies in particular when $P = \mathrm{Spf}\, A$ is a formal affine $\mathcal{V}$-scheme and

$$X = V(f_1, \ldots, f_r) \cap P_k \subset P.$$

Then, u is the morphism associated to

$$A \to A[T_1, \ldots, T_r]/(f_i^n - \pi^m T_i)$$

with $\eta = |\pi|^{m/n}$.

Corollary 2.3.12 *Let S be a formal $\mathcal{V}$-scheme. If $\iota : X \hookrightarrow P$ and $\iota' : X' \hookrightarrow P'$ are two formal embeddings of quasi-compact algebraic varieties, we have*

$$[X \times_{S_k} X']_{P \times_S P'\eta} = \left([X]_{P\eta} \times_{S_K} P'_K\right) \cap \left(P_K \times_{S_K} [X']_{P'\eta}\right)$$

and

$$]X \times_{S_k} X'[_{P \times_S P'\eta} = \left(]X[_{P\eta} \times_{S_K} P'_K\right) \cap \left(P_K \times_{S_K}]X'[_{P'\eta}\right).$$

Proof We know that

$$X \times_{S_k} X' = \left(X \times_{S_k} P'_k\right) \cap \left(P_k \times_{S_K} X'\right)$$

and we saw in Proposition 2.3.5 that tubes of radius η commute to intersections. We also know from Proposition 2.3.9 that they commute with cartesian morphisms. We are done. □

Proposition 2.3.13 *If $\sigma : K \hookrightarrow K'$ is an isometric embedding and $\iota : X \hookrightarrow P$ is a formal embedding of a quasi-compact algebraic variety, we have*

$$[X^\sigma]_{P^\sigma\eta} = [X]^\sigma_{P\eta} \quad \text{and} \quad]X^\sigma[_{P^\sigma\eta} =]X[^\sigma_{P\eta}.$$

Proof The question is local and we may therefore assume P affine and X closed in P. Then X^σ is defined in P^σ by the same equations as X in P and the result follows easily. □

Lemma 2.3.14 *Let*

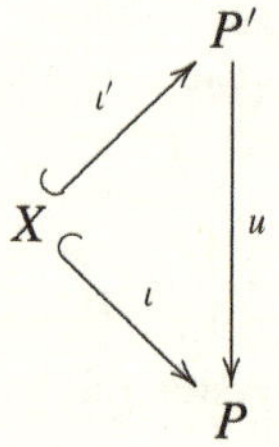

be a morphism of formal embeddings of X. Then, any coverings of P and P' have affine refinements $P = \cup P_i$ and $P' = \cup P'_i$ with $u(P'_i) \subset P_i$ and $\iota'^{-1}(P'_i) = \iota^{-1}(P_i)$.

Proof We can first refine the covering of P' into an affine covering $P' = \cup P'_i$. Now, the question is local on P and therefore also on X which has the induced topology. Thus, we may assume that P is affine, that $P_i = P$ and $X \subset P'_i$ for each i, and we are done. □

The next proposition is the first important result of the theory. This is the firs step towards the fibration theorems.

Proposition 2.3.15 *Let*

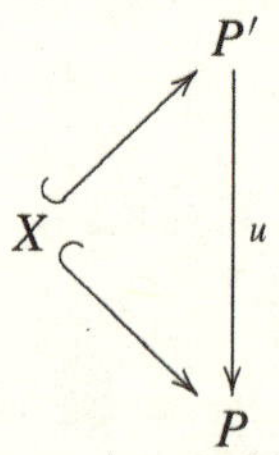

be an étale morphism of formal embeddings of X. Then, u_K induces an isomorphism

$$]X[_{P'} \simeq]X[_P.$$

Actually, when X is quasi-compact, for $\eta < 1$, u_K induces an isomorphism

$$[X]_{P'\eta} \simeq [X]_{P\eta} \quad \text{(and also} \quad]X[_{P'\eta} \simeq]X[_{P\eta}).$$

Proof We may assume that X is closed in P and in P' and that u is an étale morphism. Moreover, the map $u^{-1}(X) \to X$ induced by u is an étale retraction of the inclusion $X \hookrightarrow u^{-1}(X)$. It follows that X is open (and closed) in $u^{-1}(X)$. Thus, we may assume that $X = u^{-1}(X)$. Finally, since the question is local on X, we may assume, thanks to Lemma 2.3.14, that $P = \mathrm{Spf} A$ and $P' = \mathrm{Spf} A'$.

It is sufficient to consider the case of closed tubes of radius η and we can write $\eta = |\pi|^{m/n}$. From $X = u^{-1}(X)$, we see that if X is defined in P by the ideal I of A, then it is defined in P' by IA'. Thus, our morphism u comes from some étale map $A \to A'$ which induces an isomorphism

$$A/I \simeq A'/IA'.$$

If we write $I := (f_1, \ldots, f_r, \mathfrak{m})$, then for all $N, m, n \in \mathbf{N}$, the morphism

$$A\{T_1, \ldots, T_r\}/(f_i^n - \pi^m T_i, \pi^N) \to A'\{T_1, \ldots, T_r\}/(f_i^n - \pi^m T_i, \pi^N).$$

is also étale and induces the isomorphism

$$(A/I)[T_1, \ldots, T_r] \simeq (A'/IA')[T_1, \ldots, T_r]$$

modulo I. Since I is a nilideal in $A\{T_1, \ldots, T_r\}/(f_i^n - \pi^m T_i, \pi^N)$, this means that the original morphism was already an isomorphism. Taking the limit on all N gives an isomorphism

$$A\{T_1, \ldots, T_r\}/(f_i^n - \pi^m T_i) \to A'\{T_1, \ldots, T_r\}/(f_i^n - \pi^m T_i).$$

And looking at the generic fibers gives the expected morphism

$$[X]_{P'\eta} \simeq [X]_{P\eta}.$$

□

For example, we can consider the embedding of the point $X = \mathrm{Spec} k$ in the formal affine line $P' = P = \mathbf{A}^1_{\mathcal{V}}$ and the morphism

$$u : t \mapsto (t+1)^2 - 1.$$

Then, u_K is the map given by the same formula

$$]X[_{P'} = \mathbf{D}(0, 1^-) \to]X[_{P} = \mathbf{D}(0, 1^-).$$

If the characteristic of k is not equal to 2, this is an automorphism whose inverse is given by

$$t \mapsto \sqrt{t+1} - 1 = \sum_{k=1}^{\infty} \frac{(-1)^{k-1}(2k-3)!}{2^{2k-2}k!(k-2)!} t^k.$$

Here is the weak fibration theorem:

Corollary 2.3.16 *(Weak fibration theorem) Let*

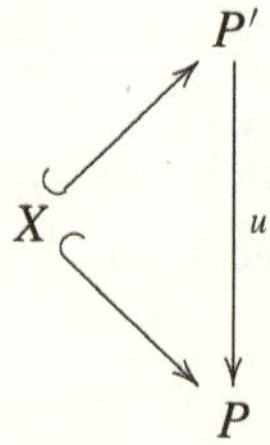

be a smooth morphism of formal embeddings of X. Then, locally on X, the morphism u_K induces an isomorphism

$$]X[_{P'} \simeq]X[_P \times \mathbf{B}^d(0, 1^-).$$

Actually, locally on X, there exists $\eta_0 < 1$ such that for each $\eta_0 \leq \eta < 1$, we have an isomorphism

$$[X]_{P'\eta} \simeq [X]_{P\eta} \times \mathbf{B}^d(0, \eta^+)$$

and an isomorphism

$$]X[_{P'\eta} \simeq]X[_{P\eta} \times \mathbf{B}^d(0, \eta^-).$$

Proof The question being local on X, it is also local on P and P' thanks to Lemma 2.3.14 and we may therefore assume that X is quasi-compact and u factors as an étale map $P' \mapsto \widehat{\mathbf{A}}^d_P$ followed by the projection $\widehat{\mathbf{A}}^d_P \to P$. We can use the zero section of $\widehat{\mathbf{A}}^d_V$ in order to embed X in $\widehat{\mathbf{A}}^d_P$. Then, it follows from the proposition that there is an isomorphism

$$[X]_{P'\eta} \simeq [X]_{\widehat{\mathbf{A}}^d_P\eta} = [X]_{P\eta} \times \mathbf{B}^d(0, \eta^+).$$

and similar results for open tubes. □

A first non trivial example, is the fact that if P is smooth at a rational point x, then

$$]x[_P \simeq \mathbf{B}^d(0, 1^-).$$

We will need a more precise version of this fibration theorem:

Corollary 2.3.17 *Let*

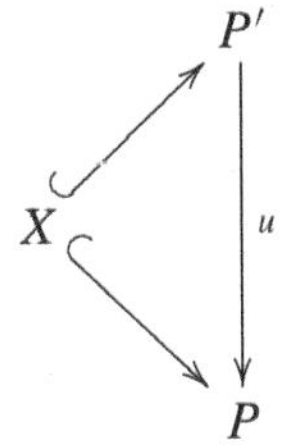

be a smooth morphism of formal embeddings into affine formal schemes.

Assume that there exist $t_1, \ldots, t_d \in A'$, *where* $P' = \mathrm{Spf}\, A'$, *inducing a basis* $(\bar{t}_1, \ldots, \bar{t}_d)$ *of the conormal sheaf* $\omega_{X'/X}$ *of* X *in* $X' := u^{-1}(X)$.

Then, the morphism

$$(t_1, \ldots, t_d) : P' \to \widehat{\mathbf{A}}^d_P$$

induces an isomorphism

$$]X[_{P'} \simeq]X[_P \times \mathbf{B}^d(0, 1^-).$$

Actually, there exists $\eta_0 < 1$ *such that for each* $\eta_0 \leq \eta < 1$, *we have an isomorphism*

$$[X]_{P'\eta} \simeq [X]_{P\eta} \times \mathbf{B}^d(0, \eta^+)$$

and an isomorphism

$$]X[_{P'\eta} \simeq]X[_{P\eta} \times \mathbf{B}^d(0, \eta^-).$$

We first need a lemma:

Lemma 2.3.18 *In the situation of the corollary, if we embed* X *in* $\widehat{\mathbf{A}}^d_P$ *using the zero section, the morphism*

$$(t_1, \ldots, t_d) : P' \to \widehat{\mathbf{A}}^d_P$$

is étale in a neighborhood of X.

Proof Note first that we may assume that u is smooth and, in particular, that $\Omega^1_{P'/P}$ is locally free. Also, we know that $dt_1, \ldots, dt_d$ form a basis of $\Omega^1_{P'/P}$ on some open subset $P'' \subset P'$. Moreover, this property can be checked pointwise. If $P'' \neq \emptyset$, then u is étale on P''. In order to finish the proof, it is therefore sufficient to check that $X \subset P''$.

Since u induces a smooth retraction $X' \to X$ of the embedding $X \hookrightarrow X'$, the map $\bar{t}_i \mapsto \overline{dt_i}$ is an isomorphism between $\omega_{X'/X}$ and $(\Omega^1_{X'/X})_{|X}$. It follows that

$\overline{dt_1}, \ldots, \overline{dt_d}$ form a basis of $\Omega^1_{X'/X} = (\Omega^1_{P'/P})_{|X}$. In other words, $dt_1, \ldots, dt_d$ form a basis of $\Omega^1_{P'/P}$ at each point of X. And we are done. □

Proof (of Corollary 2.3.17) We can use the zero section of $\widehat{\mathbf{A}}^d_V$ in order to embed X in $\widehat{\mathbf{A}}^d_P$. Then, the corollary follows from the proposition thanks to Lemma 2.3.18. □

3

Strict neighborhoods

We fix a complete ultrametric field K with $\mathcal{V}$, k and π as usual.

3.1 Frames

If we are only interested in proper algebraic varieties, the notion of tube is sufficient to develop the whole theory. More precisely, in this case, convergent cohomology (see [73]) coincides with rigid cohomology and the former theory can be developed without introducing the notion of strict neighborhood. However, in general, it is absolutely necessary to use strict neighborhoods in order to take into account the situation at infinity.

We first start with a very general notion of strict neighborhood.

Definition 3.1.1 *Let V be a rigid analytic variety and $T \subset V$ an admissible open subset. An admissible open subset V' of V is called a* strict neighborhood *of $V \setminus T$ in V if the covering $V = V' \cup T$ is admissible.*

For example, if $V := \mathbf{A}_K^{1,\mathrm{rig}}$ and $T := \mathbf{A}(0, 1^-, \infty)$, then

$$V \setminus T = \mathbf{B}(0, 1^+),$$

and $V' := \mathbf{B}(0, \rho^+)$ will be a strict neighborhood of $V \setminus T$ in V.

We now show that strict neighborhoods form a direct system. More precisely, we have the following:

Proposition 3.1.2 *Let V be a rigid analytic variety and $T \subset V$ an admissible open subset.*

(i) *If V' is a strict neighborhood of $V \setminus T$ in V, then any admissible open subset of V that contains V' is also a strict neighborhood of $V \setminus T$ in V.*

(ii) *If* $\{V_\alpha\}$ *is a finite family of strict neighborhoods of* $V \setminus T$ *in* V*, then their intersection* $\bigcap_\alpha V_\alpha$ *is also a strict neighborhood of* $V \setminus T$ *in* V*.*

Proof The first assertion results from the fact that a covering by admissible open subsets that has an admissible refinement is already admissible.

Concerning the second assertion, it is sufficient to consider the case of two strict neighborhoods V' and V''. And this is completely formal. We want to show that

$$V = (V' \cap V'') \cup T$$

is admissible. But this is a local question on V. Using the admissibility of $V = V' \cup T$, it is therefore sufficient to show that both coverings

$$V' = (V' \cap V'') \cup (V' \cap T) \quad \text{and} \quad T = (V' \cap V'' \cap T) \cup T$$

are admissible. But the first one is induced on V' by the admissible covering $V = V'' \cup T$ and the other one can be refined into the trivial covering. □

Functoriality is as follows:

Proposition 3.1.3 *Let* V *be a rigid analytic variety,* $T \subset V$ *an admissible open subset and* V' *is a strict neighborhood of* $V \setminus T$ *in* V*. Then*

(i) *If* $u : W \to V$ *a morphism of rigid analytic varieties and* S *an admissible open subset of* V *containing* $u^{-1}(T)$*, then* $u^{-1}(V')$ *is a strict neighborhood of* $W \setminus S$ *in* W*.*
(ii) *If* $\sigma : K \to K'$ *is an isometric extension, then* V'^{σ} *is a strict neighborhood of* $V^\sigma \setminus T^\sigma$ *in* V^σ*.*

Proof Considering the first assertion, since the covering $V = V' \cup T$ is admissible, so is the covering

$$W = u^{-1}(V') \cup u^{-1}(T).$$

And it follows that the covering $W = u^{-1}(V') \cup S$ is also admissible since it has an admissible refinement. Thus, we see that $u^{-1}(V')$ is a strict neighborhood of $W \setminus S$ in W.

The second assertion immediately follows from the fact that the covering $V^\sigma = V'^{\sigma} \cup T^\sigma$ is admissible. □

We also want to show that the notion of strict neighborhood is local in the following sense:

Proposition 3.1.4 *Let* V *be a rigid analytic variety,* $T \subset V$ *an admissible open subset and* $V := \bigcup_i V_i$ *an admissible open covering. An admissible open subset*

V' of V is a strict neighborhood of $V \setminus T$ in V if and only if for all i, $V' \cap V_i$ is a strict neighborhood of $V_i \setminus (V_i \cap T)$ in V_i.

Proof It just means that the covering

$$V = V' \cup T$$

is admissible if and only if for each i, the covering

$$V_i = (V' \cap V_i) \cup (T \cap V_i)$$

is admissible. □

This is it for general results on strict neighborhood and we now turn to the definition of frames. Following Kedlaya, we give the following definition (corresponding to the notion of triple of Chiarellotto and Tsuzuki in [26]):

Definition 3.1.5 *A* frame *(or a K-frame to be more precise) is a diagram*

$$X \hookrightarrow Y \hookrightarrow P$$

made of an open immersion of an algebraic k-variety X into another algebraic k-variety Y and a closed immersion of Y into a formal $\mathcal{V}$-scheme P. The trivial frame *associated to a formal scheme S is the frame*

$$S_k = S_k \hookrightarrow S.$$

The standard way to build a frame is the following: starting with a quasi-projective variety X over k, we embed it in some projective space $\mathbf{P}^n_k$. Then, we let $Y = \overline{X}$, the algebraic closure of X, and $P := \widehat{\mathbf{P}}^n_{\mathcal{V}}$.

A very important case is the Monsky–Washnitzer setting which is used to study smooth affine varieties. Any smooth affine variety over k lifts to a smooth affine scheme X over $\mathcal{V}$ (see for example [4]). We can see X as a closed subscheme of $\mathbf{A}^N_{\mathcal{V}}$, and denote by Y its algebraic closure in $\mathbf{P}^N_{\mathcal{V}}$. We consider the frame $(X_k \subset Y_k \subset \widehat{Y})$.

It is also worth mentioning the case of a smooth curve over k. The recent progress in the field show that most difficulties already appear in the one-dimensional case. However, this is a very nice geometric situation because any smooth curve has a smooth projective compactification that lifts to a smooth projective curve over $\mathcal{V}$. We will therefore be interested in frames $(X \subset Y \subset \mathcal{Y})$ where $\mathcal{Y}$ a flat formal $\mathcal{V}$-scheme whose special fiber Y is a connected curve (and X will be a non empty open subset of Y). We will generally also assume that all points $x \in Y \setminus X$ are rational. This can always be achieved after a finite extension of k.

Finally, the simplest non trivial example is given as follows: take $P = \widehat{\mathbf{A}}^2_{\mathcal{V}}$ with variables t and s, and let Y (resp. X) be the subset defined in P_k by $t = 0$ (resp. $t = 0$ and $s \neq 0$).

Definition 3.1.6 *A* morphism of frames *is a commutative diagram*

$$\begin{array}{ccccc} X' & \hookrightarrow & Y' & \hookrightarrow & P' \\ \downarrow{\scriptstyle f} & & \downarrow{\scriptstyle g} & & \downarrow{\scriptstyle u} \\ X & \hookrightarrow & Y & \hookrightarrow & P. \end{array}$$

Unless necessary, we will only mention u. A morphism of frames is said to be cartesian *if the left-hand square is cartesian and* strictly cartesian, *or* strict *for short, if they both are. If S is a formal $\mathcal{V}$-scheme, an S*-frame *is a frame over the trivial frame associated to S.*

Note that, by construction, a morphism of frames as above induces a morphism

$$]f[_u :]X'[_{P'} \to]X[_P$$

as well as

$$]g[_u :]Y'[_{P'} \to]Y[_P$$

which are just the restrictions of u_K. When there is no risk of confusion, we will write u_K instead of $]f[_u$ or $]g[_u$.

Proposition 3.1.7 *Frames form a category with finite inverse limits which are built term by term.*

Recall that a finite inverse limit is just a kernel of finite products.

Proof It is clear that we do have a category and that if we are given a finite diagram of frames $(X_i \subset Y_i \subset P_i)$, then its inverse limit is exactly $(X \subset Y \subset P)$ where P (resp. Y, resp. X) is the inverse limit of the diagram P_i (resp. Y_i, resp. X_i). Of course inverse limits of open or closed embeddings are open or closed embeddings. □

If we fix a frame, we can always consider the category of frames over the fixed one. In particular, we will consider the category of S-frames when S is a fixed formal $\mathcal{V}$-scheme.

If $v : S' \to S$ is a morphism of formal $\mathcal{V}$-schemes, there is an obvious forgetful functor from S'-frames to S-frames as well as a pull back functor in the other direction. Also, if $\sigma : K \hookrightarrow K'$ is an isometric embedding, there is an obvious extension functor

$$(X \subset Y \subset P) \mapsto (X^\sigma \subset Y^\sigma \subset P^\sigma).$$

Proposition 3.1.8

(i) *Given a frame $(X \subset Y \subset P)$, any morphism of formal $\mathcal{V}$-schemes $u : P' \to P$ extends uniquely to a strict morphism of frames.*

(ii) *Given a frame $(X \subset Y \subset P)$, any morphism*

$$\begin{array}{ccc} Y' & \hookrightarrow & P' \\ \downarrow{\scriptstyle g} & & \downarrow{\scriptstyle u} \\ Y & \hookrightarrow & P \end{array}$$

of closed embeddings extends uniquely to a cartesian diagram of frames.

(iii) *Any morphism of frames can be split into a morphism with $u = \mathrm{Id}$, followed by a strict morphism.*

(iv) *Any morphism of frames can be split into a morphism with $u = \mathrm{Id}$ and $g = \mathrm{Id}$ (and f an open immersion), followed by a cartesian morphism.*

Proof The first two assertions are completely trivial : pulling back a frame by u gives

$$\begin{array}{ccccc} u^{-1}(X) & \hookrightarrow & u^{-1}(Y) & \hookrightarrow & P' \\ \downarrow & & \downarrow & & \downarrow{\scriptstyle u} \\ X & \hookrightarrow & Y & \hookrightarrow & P \end{array}$$

and pulling back by g in the second case gives

$$\begin{array}{ccccc} g^{-1}(X) & \hookrightarrow & Y' & \hookrightarrow & P' \\ \downarrow & & \downarrow{\scriptstyle g} & & \downarrow{\scriptstyle u} \\ X & \hookrightarrow & Y & \hookrightarrow & P. \end{array}$$

For the last two, note that any morphism splits as

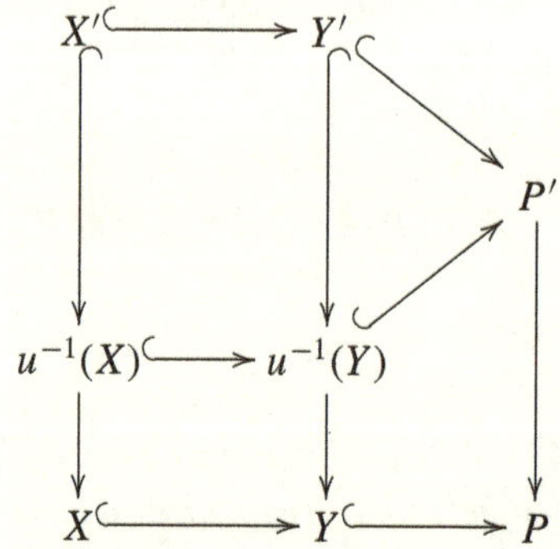

and also as

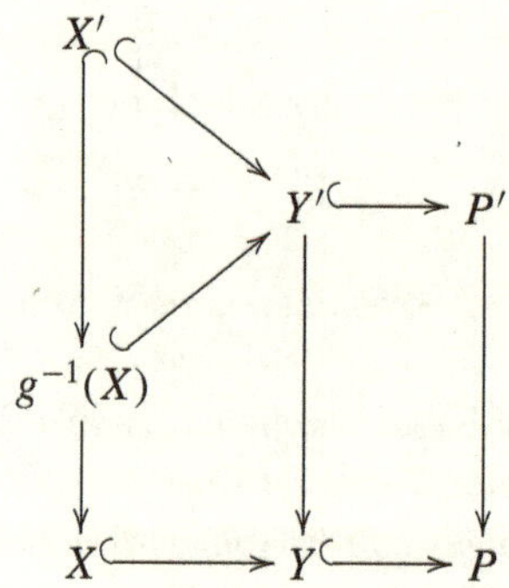

□

If $(X \subset Y \subset P)$ is a frame, then the covering of $]Y[_P$ by $]X[_P$ and $]Y \setminus X[_P$ is not admissible in general. If we consider for example the frame

$$(\mathbf{A}^1_k \setminus 0 \subset \mathbf{A}^1_k \subset \widehat{\mathbf{A}}^1_{\mathcal{V}}),$$

we get the covering

$$\mathbf{D}(0, 1^+) = \mathbf{A}(0, 1^+, 1^+) \cup \mathbf{D}(0, 1^-)$$

which is not admissible (recall that

$$\mathbf{A}(0, \eta^+, \lambda^+) := \mathbf{D}(0, \lambda^+) \setminus \mathbf{D}(0, \eta^-)$$

for $\eta \leq \lambda$). Note however that the covering

$$\mathbf{D}(0, 1^+) = \mathbf{A}(0, \eta^+, 1^+) \cup \mathbf{D}(0, 1^-)$$

with $\eta < 1$ is admissible. This is why we need strict neighborhoods.

More precisely, when we are given a frame $(X \subset Y \subset P)$, we will consider strict neighborhoods V of $]X[_P$ in $]Y[_P$. If Z denotes the closed complement of X in Y, it means that the covering

$$]Y[_P = V \cup]Z[_P$$

is admissible. More generally, if W is an admissible open subset of $]Y[_P$, we will also have to consider strict neighborhoods of $W \cap]X[_P$ in W.

Proposition 3.1.9

(i) *If*

$$\begin{array}{ccccc} X' & \hookrightarrow & Y' & \hookrightarrow & P' \\ \downarrow f & & \downarrow g & & \downarrow u \\ X & \hookrightarrow & Y & \hookrightarrow & P \end{array}$$

is a morphism of frames and V a strict neighborhood of $]X[_P$ in $]Y[_P$, then $]g[_u^{-1}(V) = u_K^{-1}(V) \cap]Y'[_{P'}$ is a strict neighborhood of $]X'[_{P'}$ in $]Y'[_{P'}$.

(ii) *If $\sigma : K \hookrightarrow K'$ is an isometric embedding, $(X \subset Y \subset P)$ a K-frame and V a strict neighborhood of $]X[_P$ in $]Y[_P$, then V^σ is a strict neighborhood of $]X^\sigma[_{P_{V'^\sigma}}$ in $]Y[_{P^\sigma}$.*

Proof Since any morphism of triples is the composite of a cartesian one and another with $u = \mathrm{Id}_P$ and $g = \mathrm{Id}_Y$, the first assertion follows from Proposition 3.1.3.

Since base extension preserves the tubes, the second assertion is also an immediate consequence of Proposition 3.1.3. □

The next point is to show that the notion of strict neighborhood only depends on the closure of X in P:

Proposition 3.1.10 *Let $(X \subset Y \subset P)$ be a frame and $Y' \subset Y$ a closed subvariety containing X. Then, a subset V of $]Y'[_P$ is a strict neighborhood of $]X[_P$ in $]Y[_P$ if and only if it is a strict neighborhood of $]X[_P$ in $]Y'[_P$.*

Proof It follows from Proposition 3.1.9 that the condition is necessary. For the converse, we denote by Z a closed complement of X in Y and let $Z' := Z \cap Y'$. We want to show that a strict neighborhood V of $]X[_P$ in $]Y[_P$ is necessarily a strict neighborhood of $]X[_P$ in $]Y[_P$. In other words, we assume that

$$]Y'[_P = V \cup]Z'[_P$$

is an admissible covering and we want to show that

$$]Y[_P = V \cup]Z[_P$$

is also admissible. But it follows from Proposition 2.2.15 that the closed covering

$$Y = Y' \cup Z$$

induces an admissible covering

$$]Y[_P =]Y'[_P \cup]Z[_P.$$

Since our question is local on $]Y[_P$, we are reduced to check it on $]Y'[_P$ where it follows from our hypothesis and on $]Z[_P$ where it is trivial. □

Finally, we show that the notion of strict neighborhood has a local nature. But, before that, we need a definition.

Definition 3.1.11 *An* immersion of frames *is a morphism of frames*

$$\begin{array}{ccccc} X' & \hookrightarrow & Y' & \hookrightarrow & P' \\ \downarrow & & \downarrow & & \downarrow \\ X & \hookrightarrow & Y & \hookrightarrow & P \end{array}$$

where all vertical maps are immersions. It is said to be open *when all maps are open immersions. It is said* mixed *when the first two vertical arrows are closed immersions and the last one is an open immersion.*

We will also say that $(X' \subset Y' \subset P')$ *is a* (open, mixed) subframe *of* $(X \subset Y \subset P)$.

A covering *is a family of immersions of frames*

$$\begin{array}{ccccc} X_i & \hookrightarrow & Y_i & \hookrightarrow & P_i \\ \downarrow & & \downarrow & & \downarrow \\ X & \hookrightarrow & Y & \hookrightarrow & P \end{array}$$

which is a covering at all steps. It is called open, mixed, cartesian *or* strict *if all immersions have this property.*

Proposition 3.1.12 *Let*

$$\begin{array}{ccccc} X_i & \hookrightarrow & Y_i & \hookrightarrow & P_i \\ \downarrow & & \downarrow & & \downarrow \\ X & \hookrightarrow & Y & \hookrightarrow & P \end{array}$$

be a cartesian covering which is either open or finite mixed. Then, a subset V *of* $]Y[_P$ *is a strict neighborhood of* $]X[_P$ *in* $]Y[_P$ *if and only if* $V_i := V \cap]Y_i[_{P_i}$ *is a strict neighborhood of* $]X_i[_{P_i}$ *in* $]Y_i[_{P_i}$ *for each* i.

Proof First of all, the condition is necessary thanks to Proposition 3.1.9. Since the covering is open or finite mixed, it follows from Proposition 2.2.15 that we have an admissible covering $V := \cup_i V_i$. Moreover, since it is cartesian, we have for each $i \in I$,

$$V_i \backslash]X_i[_{P_i} = V_i \cap (V \backslash]X[_P).$$

Then, we can apply Proposition 3.1.4. □

Proposition 3.1.13 *Let*

$$\begin{array}{ccccc} X' & \hookrightarrow & Y' & \hookrightarrow & P' \\ \downarrow f & & \downarrow g & & \downarrow u \\ X & \hookrightarrow & Y & \hookrightarrow & P \end{array}$$

be a strict morphism of frames with u a formal blowing up (or more generally, a formal modification). An admissible open subset V of P_K is a strict neighborhood of $]X[_P$ in $]Y[_P$ if and only if $u_K^{-1}(V)$ is a strict neighborhood of $]X'[_{P'}$ in $]Y'[_{P'}$.

Proof We know that u_K is an isomorphism and that it induces an isomorphism $]Y'[_{P'} \simeq]Y[_P$ as well as $]X'[_{P'} \simeq]X[_P$ and $]Z'[_{P'} \simeq]Z[_P$ if Z (resp. Z') denote as usual a closed complement of X in Y (resp. X' in Y'). □

Using the last two propositions, one sees that one can localize directly on P_K. More precisely, if P is quasi-compact, any admissible covering has a refinement induced by an open covering of P' with $P' \to P$ a formal blowing up.

3.2 Frames and tubes

This section is rather technical but necessary to achieve a better understanding of strict neighborhoods. We first introduce the following definition:

Definition 3.2.1 *A morphism of frames*

$$\begin{array}{ccccc} X' & \hookrightarrow & Y' & \hookrightarrow & P' \\ \downarrow f & & \downarrow g & & \downarrow u \\ X & \hookrightarrow & Y & \hookrightarrow & P \end{array}$$

is quasi-compact (resp. affine) *if u is quasi-compact (resp. affine).*

A frame $(X \subset Y \subset P)$ *is* quasi-compact (resp. affine) *if the final morphism of frames*

$$\begin{array}{ccccc} X & \hookrightarrow & Y & \hookrightarrow & P \\ \downarrow & & \downarrow & & \downarrow \\ \operatorname{Spec} k & \hookrightarrow & \operatorname{Spec} k & \hookrightarrow & \operatorname{Spf} \mathcal{V} \end{array}$$

is quasi-compact (resp. affine).

Of course, the latter simply means that P is quasi-compact (resp. affine).

Let $(X \subset Y \subset P)$ be a quasi-compact frame and Z a closed complement of X in Y. If $\lambda < 1$, we will write

$$V^{\lambda} :=]Y[_P \setminus]Z[_{P\lambda}.$$

Moreover, if $\eta < 1$, we will write

$$V_{\eta}^{\lambda} := [Y]_{P\eta} \cap V^{\lambda}.$$

Note that, V_{η}^{λ} increases with η but decreases with λ.

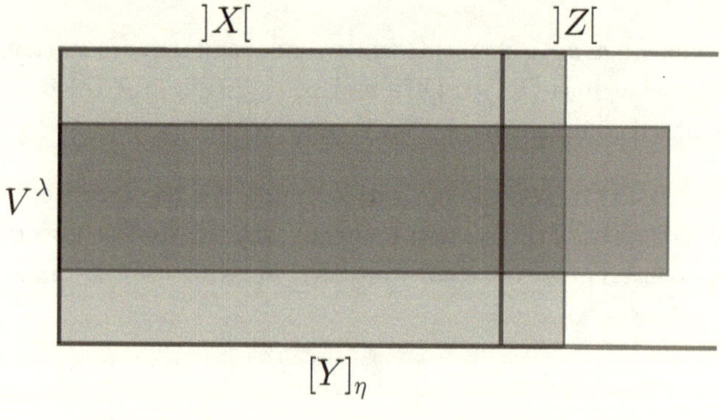

Fig. 3.1 V^{λ} and V_{η}^{λ}

The next lemma gives a better understanding of the nature of V^{λ} in the affine case.

Lemma 3.2.2 *Let* $(X \subset Y \subset P)$ *be an affine frame and*

$$Z = V(g_1, \ldots, g_r) \cap P_k \subset P$$

a closed complement for X *in* Y.

(i) *If* $\lambda < 1$, *then* V^{λ} *is an admissible open subset of* $]Y[_P$ *and we have*

$$V^{\lambda} = \{x \in]Y[_P, \exists i \in \{1, \ldots, r\}, |g_i(x)| \geq \lambda\}.$$

Moreover we have an admissible covering

$$]Y[_P = V^\lambda \cup [Z]_{P\lambda}.$$

Also if, for $i = 1, \ldots, r$, *we let*

$$V_i^\lambda := \{x \in]Y[_P, |g_i(x)| \geq \lambda, |g_1(x)|, \ldots, |g_r(x)|\},$$

then we have an admissible covering

$$V^\lambda = \cup_{i=1}^r V_i^\lambda.$$

(ii) *If* $\lambda, \eta < 1$, *then* V_η^λ *is the quasi-compact admissible open subset*

$$V_\eta^\lambda = \{x \in [Y]_{P\eta}, \exists i \in \{1, \ldots, r\}, |g_i(x)| \geq \lambda\},$$

and we have an admissible covering

$$[Y]_{P\eta} := V_\eta^\lambda \cup ([Y]_{P\eta} \cap [Z]_{P\lambda}).$$

Also, if we set for each $i = 1, \ldots, r$,

$$V_{\eta,i}^\lambda := \{x \in [Y]_{P\eta}, |g_i(x)| \geq \lambda, |g_1(x)|, \ldots, |g_r(x)|\},$$

we have an affinoid covering

$$V_\eta^\lambda = \cup_{i=1}^r V_{\eta,i}^\lambda.$$

Proof By definition, we have

$$]Z[_{P\lambda} = \{x \in P_K, \forall i \in \{1, \ldots, r\}, |g_i(x)| < \lambda\}.$$

It follows that

$$V^\lambda = \{x \in]Y[_P, \exists i \in \{1, \ldots, r\}, |g_i(x)| \geq \lambda\}$$

and

$$V_\eta^\lambda = \{x \in [Y]_{P\eta}, \exists i \in \{1, \ldots, r\}, |g_i(x)| \geq \lambda\}.$$

Now, if we write $P = \mathrm{Spf} A$ and $\lambda = |\pi|^{n/m}$, then the ideal

$$(g_1^m, \ldots, g_1^m, \pi^n)$$

is the unit ideal in A_K. It follows that if, for $i = 1, \ldots, r$, we let

$$W_i^\lambda := \{x \in P_K, |g_i(x)| \geq \lambda, |g_1(x)|, \ldots, |g_r(x)|\},$$

we have a (so-called *rational*) affinoid covering

$$P_K := \cup_{i=1}^r W_i^\lambda \cup [Z]_{P\lambda}.$$

We may then consider the coverings induced on $]Y[_P$ and $[Y]_{P\eta}$, respectively. More precisely, as a finite union of affinoid domains, $\cup_{i=1}^r W_i^\lambda$ is an admissible open subset of P_K coming with an admissible (affinoid) covering. It follows that V_λ and V_η^λ are admissible open subsets of $]Y[_P$ and $[Y]_{P\eta}$, respectively, the second one being quasi-compact. Moreover, we have admissible coverings

$$V^\lambda = \cup_{i=1}^r V_i^\lambda \quad \text{and} \quad V_\eta^\lambda = \cup_{i=1}^r V_{\eta,i}^\lambda.$$

And therefore also

$$]Y[_P = V^\lambda \cup [Z]_{P\lambda} \quad \text{and} \quad [Y]_{P\eta} = V_\eta^\lambda \cup ([Y]_{P\eta} \cap [Z]_{P\lambda}).$$

□

Corollary 3.2.3 *Let $X \subset Y \subset P$ be an affine frame and let*

$$Z = V(g_1, \ldots, g_r) \cap P_k \subset P$$

be a closed complement for X in Y. If $\lambda < 1$ and $i = 1, \ldots, r$, let

$$V_i^\lambda := \{x \in]Y[_P, |g_i(x)| \geq \lambda, |g_1(x)|, \ldots, |g_r(x)|\}.$$

Let W be an affinoid open subset of $V_i^{\lambda_0}$ for some $\lambda_0 < 1$ and $\lambda_0 \leq \lambda < 1$. Then, $W \cap V^\lambda$ is affinoid and we have

$$W \cap V^\lambda = W \cap V_i^\lambda.$$

Proof We first show that the first assertion follows from the second. There exists $\eta < 1$ such that $W \subset [Y]_\eta$ and we know from Lemma 3.2.2 that $V_{i\eta}^\lambda := V_i^\lambda \cap [Y]_\eta$ is affinoid. It will follow from the second assertion that

$$W \cap V^\lambda = W \cap V_\eta^\lambda = W \cap V_{i\eta}^\lambda$$

is the intersection of two affinoid open subsets of an affinoid variety. It is therefore also affinoid.

Finally, the second assertion is trivially true for any subset W of $V_i^{\lambda_0}$. □

Proposition 3.2.4 *Let $(X \subset Y \subset P)$ be a quasi-compact frame and Z a closed complement for X in Y. If $\lambda < 1$, then*

$$V^\lambda :=]Y[_P \setminus]Z[_{P\lambda}$$

is an admissible open subset of $]Y[_P$ and we have an admissible covering

$$]Y[_P = V^\lambda \cup [Z]_{P\lambda}.$$

Moreover, if $\eta < 1$, *then*

$$V_\eta^\lambda := [Y]_\eta \cap V^\lambda$$

is a quasi-compact admissible open subset of $]Y[_P$ *and we have an admissible covering*

$$[Y]_{P\eta} = V_\eta^\lambda \cup ([Y]_{P\eta} \cap [Z]_{P\lambda}).$$

Finally, we also have an admissible covering $V^\lambda = \cup V_\eta^\lambda$.

Proof The questions are local on $]Y[_P$. We may therefore assume that P is affine in which case our assertions directly result from the lemma. □

Corollary 3.2.5 *In the situation of the proposition, if* W *is a quasi-compact open subset of* $]Y[_P$, *then* $W \cap V^\lambda$ *is also quasi-compact. Moreover, if* $W \subset]Z[_P$, *there exists* λ *such that* $W \cap V^\lambda = \emptyset$.

Proof Since $]Y[_P = \cup [Y]_{P\eta}$ is an admissible covering, there exists η such that $W \subset [Y]_{P\eta}$ and therefore

$$W \cap V^\lambda = W \cap V_\eta^\lambda$$

is quasi-compact as the intersection of two quasi-compact admissible open subsets inside a quasi-separated rigid analytic variety.

If $W \subset]Z[_P$, then we use the admissible covering $]Z[_P = \cup]Z[_{P\lambda}$ to get $W \subset]Z[_{P\lambda}$ for some $\lambda < 1$ and therefore $W \cap V^\lambda = 0$. □

We list now some properties of V^λ and V_η^λ.

Proposition 3.2.6 *Let* $(X \subset Y \subset P)$ *be a quasi-compact frame and* Z *a closed complement for* X *in* Y. *As usual, we let* $V^\lambda :=]Y[_P \setminus]Z[_{P\lambda}$ *for* $\lambda < 1$ *and* $V_\eta^\lambda := [Y]_\eta \cap V^\lambda$ *for* $\eta < 1$. *Then,*

(i) *We have*

$$]X[_P = \cap_\lambda V^\lambda \quad (\text{and} \quad [X]_{P\eta} = \cap_\lambda V_\eta^\lambda \quad \text{for} \quad \eta < 1)$$

(ii) *If* Z' *is another closed complement (with the same underlying space, though), there exists* $\lambda' < 1$ *such that, with obvious notations,*

$$V'^{\lambda'} \subset V^\lambda \quad (\text{and} \quad V_\eta'^{\lambda'} \subset V_\eta^\lambda \quad \text{for} \quad \eta < 1).$$

(iii) *If* X' *is an open subvariety of* X *and* Z *is a subvariety of some closed complement* Z' *of* X' *in* Y, *then, with obvious notations, we have*

$$V'^\lambda \subset V^\lambda \quad (\text{and} \quad V_\eta'^\lambda \subset V_\eta^\lambda \quad \text{for} \quad \eta < 1).$$

(iv) *If* $X = \cup X_i$ *is a finite union of open subvarieties, and Z is endowed with the scheme structure of the intersection of the closed complements of the* X_i*'s, then, with obvious notations, we have an admissible covering*

$$V^\lambda = \cup V_i^\lambda \quad (\text{and} \quad V_\eta^\lambda = \cup V_{i\eta}^\lambda \quad \text{for} \quad \eta < 1).$$

(v) *If* $X = \cap X_i$ *is a finite intersection of open subvarieties, and Z is endowed with the scheme structure of the union of the closed complements of the* X_i*'s, then, with obvious notations,*

$$\cap V_i^{\lambda_i} \subset V^\lambda \quad (\text{and} \quad \cap V_{i\eta}^{\lambda_i} \subset V_\eta^\lambda \quad \text{for} \quad \eta < 1).$$

with $\lambda = \prod \lambda_i$.

Proof All these properties are automatically derived by reversing the analogous properties of $]Z[_{P\lambda}$ (and intersecting with $[Y]_{P\eta}$). More precisely, the first assertion comes from $]Z[_P = \cup]Z[_{P\lambda}$. The second one from Corollary 2.3.8. For the third one, we may refer for example to Proposition 2.3.9. The last two assertions will result from Propositions 2.3.5 and 2.3.6 respectively. □

Recall that if $g : Y' \to Y$ is a morphism of algebraic k-varieties and Z is a subvariety of Y, then $g^{-1}(Z)$, as a subvariety of Y', is the (isomorphic) image of $Z \times_Y Y'$.

Corollary 3.2.7 *In the situation of the proposition, assume that*

$$X = \coprod X_i$$

is a finite disjoint union of open (and closed) subvarieties. Then, with obvious notations, for $\lambda^2 > \eta$,

$$V_\eta^\lambda = \coprod V_{i\eta}^\lambda$$

is a disjoint admissible covering.

Proof It follows from assertion (iv) of Proposition 3.2.6 that

$$V_\eta^\lambda = \cup V_{i\eta}^\lambda.$$

is an admissible covering. We just have to check that the covering is disjoint and it is sufficient to consider the case of two subsets X_1 and X_2. Since $\eta < \lambda^2$, it follows from assertion (v) that

$$V_{1\eta}^\lambda \cap V_{2\eta}^\lambda \subset V_1^\lambda \cap V_2^\lambda \cap]Y[_{P\lambda^2} = \emptyset.$$

□

Proposition 3.2.8 *Let*

$$\begin{array}{ccccc} X' & \hookrightarrow & Y' & \hookrightarrow & P' \\ \downarrow{\scriptstyle f} & & \downarrow{\scriptstyle g} & & \downarrow{\scriptstyle u} \\ X & \hookrightarrow & Y & \hookrightarrow & P \end{array}$$

be a cartesian morphism of quasi-compact frames (the first square is cartesian).

If Z a closed complement of X in Y, then $Z' := g^{-1}(Z)$ is a closed complement of X' in Y'.

For $\lambda < 1$, we will write

$$V^{\lambda} :=]Y[_P \setminus]Z[_{P\lambda} \quad \text{and} \quad V'^{\lambda} :=]Y'[_{P'} \setminus]Z'[_{P'\lambda},$$

and if $\eta < 1$, we will write

$$V_{\eta}^{\lambda} := [Y]_{P\eta} \cap V^{\lambda} \quad \text{and} \quad V_{\eta}'^{\lambda} := [Y']_{P'\eta} \cap V'^{\lambda}.$$

Then, if $\eta < \lambda$, we have

$$[Y']_{P'\eta} \cap u_K^{-1}(V_{\eta}^{\lambda}) = [Y']_{P'\eta} \cap u_K^{-1}(V^{\lambda}) = V_{\eta}'^{\lambda}.$$

When the morphism is strict (the second square is also cartesian), we always have

$$u_K^{-1}(V^{\lambda}) = V'^{\lambda} \quad \text{and} \quad u_K^{-1}(V_{\eta}^{\lambda}) = V_{\eta}'^{\lambda}.$$

Proof First of all, since our morphism is cartesian, it is clear that Z' is a closed complement of X in Y'.

Now, in order to prove the proposition, we assume first that the morphism is strict. Then, it follows from Proposition 2.2.6 that

$$u_K^{-1}(]Y[_P) =]Y'[_{P'}$$

and from Proposition 2.3.9 that

$$u_K^{-1}(]Z[_{P\lambda}) =]Z'[_{P'\lambda}.$$

It immediately follows from the definitions of V_{λ} and V'^{λ} that

$$u_K^{-1}(V^{\lambda}) = V'^{\lambda}.$$

Proposition 2.3.9 again tells us that

$$u_K^{-1}([Y]_{P\eta}) = [Y']_{P'\eta}$$

and it follows that

$$\begin{aligned} u_K^{-1}(V_\eta^\lambda) &= u_K^{-1}([Y]_{P\eta} \cap V^\lambda) \\ &= u_K^{-1}([Y]_{P\eta}) \cap u_K^{-1}(V'^\lambda) = [Y']_{P'\eta} \cap V'^\lambda = V'^\lambda_\eta . \end{aligned}$$

In order to prove the proposition in general, we may now assume thanks to Proposition 3.1.8 that $P = P'$ and $u = Id_P$. In this situation, Y' is a closed subvariety of Y and we have $X' = X \cap Y'$, $Z' = Z \cap Y'$ with the scheme structure of the intersection. We assume that $\eta < \lambda$ and we want to show that

$$[Y']_{P\eta} \cap V^\lambda = [Y']_{P\eta} \cap V_\eta^\lambda = V'^\lambda_\eta .$$

From Proposition 2.3.5, we know that $]Z'[_{P\lambda} =]Z[_{P\lambda} \cap]Y'[_{P\lambda}$. Taking complements in $]Y'[_{P'}$, we get

$$V'^\lambda = (]Y'[_P \cap V^\lambda) \cup (]Y'[_P \setminus]Y'[_{P\lambda}).$$

Since we assumed that $\eta < \lambda$, we obtain

$$V'^\lambda_\eta = [Y']_{P\eta} \cap V'^\lambda = [Y']_{P\eta} \cap V^\lambda .$$

In particular, since Y' is a subvariety of Y, it follows from Proposition 2.3.4 that $[Y']_\eta \subset [Y]_\eta$ and therefore

$$V'^\lambda_\eta = [Y']_{P\eta} \cap V^\lambda \subset [Y]_{P\eta} \cap V^\lambda = V_\eta^\lambda .$$

Putting this together, we get our equalities. □

Corollary 3.2.9 *With the notations of the proposition, if the morphism is strict and u is a formal blowing up (or more generally a formal modification), we get isomorphisms*

$$V'^\lambda \simeq V^\lambda \quad \text{and} \quad V'^\lambda_\eta \simeq V^\lambda_\eta .$$

Proof In this case, u_K is an isomorphism. □

Corollary 3.2.10 *Let*

$$\begin{array}{ccccc} X' & \hookrightarrow & Y' & \hookrightarrow & P' \\ \downarrow & & \downarrow & & \downarrow \\ X & \hookrightarrow & Y & \hookrightarrow & P \end{array}$$

be a cartesian immersion of quasi-compact frames. Then, with obvious notations, for $\eta < \lambda$, we have

$$[Y']_{P'\eta} \cap V^\lambda_\eta = [Y']_{P'\eta} \cap V^\lambda = V'^\lambda_\eta .$$

When the immersion is strict, we have,

$$V'^{\lambda} = P'_K \cap V^{\lambda} \quad \text{and} \quad V'^{\lambda}_{\eta} = P'_K \cap V^{\lambda}_{\eta}.$$

Proof This is the particular case of the proposition where u_K is an inclusion. □

Corollary 3.2.11 *Let $(X \subset Y \subset P)$ be a quasi-compact frame, $\eta < \lambda < 1$, $u : P' \to P$ a formal model for the open immersion $V^{\lambda}_{\eta} \hookrightarrow P_K$, $Y' := u^{-1}(Y)$ and $X' := u^{-1}(X)$. Then, with obvious notations, we have*

$$P'_K = V'^{\lambda} = V'^{\lambda}_{\eta} \simeq V^{\lambda}_{\eta}$$

Proof Again, we apply the proposition to u. □

Note that this last results applies in particular to the case

$$P = \mathrm{Spf} A, \quad Y = V(f_1, \ldots, f_r) \cap P_k \subset P, \quad Z = V(g) \cap Y.$$

and u comes from the canonical map

$$A \to A\{T_1, \ldots, T_r, T\}/(f_i^n - \pi^m T_i, g^t T - \pi^s).$$

with $\eta = |\pi|^{m/n} < \lambda = |\pi|^{s/t}$.

Proposition 3.2.12 *Let S be a formal $\mathcal{V}$-scheme and $(X \subset Y \subset P)$, $(X' \subset Y' \subset P')$ two quasi-compact S-frames. If Z (resp. Z') is a closed complement for X in Y (resp. X' in Y'), then*

$$Z'' := (Z \times_{S_k} Y') \cup (Y \times_{S_k} Z')$$

is a closed complement for $X'' := X \times_{S_k} X'$ in $Y'' := Y \times_{S_k} Y'$. If we call V^{λ}, V'^{λ} and V''^{λ} the corresponding admissible open subsets of P_K, P'_K and P''_K, respectively, where $P'' := P \times_S P'$, we have for $\lambda > \eta$,

$$V^{\lambda}_{\eta} \times_{S_K} V'^{\lambda}_{\eta} \subset V''^{\lambda^2}_{\eta} \subset V^{\lambda^2}_{\eta} \times_{S_K} V'^{\lambda^2}_{\eta}$$

In the case $X' = Y'$, we even have

$$V''^{\lambda}_{\eta} = V^{\lambda}_{\eta} \times_{S_K} [Y']_{P\eta}.$$

Proof Since

$$X \times_{S_k} X' = \left(X \times_{S_k} Y'\right) \cap \left(X' \times_{S_k} Y\right)$$

and

$$V^{\lambda}_{\eta} \times_{S_K} V'^{\lambda}_{\eta} = \left(V^{\lambda}_{\eta} \times_{S_K} [Y']_{P\eta}\right) \cap \left([Y]_{P\eta} \times_{S_K} V'^{\lambda}_{\eta}\right),$$

we may assume thanks to assertion (v) of Proposition 3.2.6 that $X' = Y'$. In this case, the first projection is a cartesian morphism

$$\begin{array}{ccccc} X'' & \hookrightarrow & Y'' & \hookrightarrow & P'' \\ \downarrow & & \downarrow & & \downarrow p_1 \\ X & \hookrightarrow & Y & \hookrightarrow & P \end{array}$$

and we may use Proposition 3.2.8. □

Proposition 3.2.13 *Let*

$$\begin{array}{ccccc} X_i & \hookrightarrow & Y_i & \hookrightarrow & P_i \\ \downarrow & & \downarrow & & \downarrow \\ X & \hookrightarrow & Y & \hookrightarrow & P \end{array}$$

be a cartesian finite mixed covering of a quasi-compact frame. Then, for any $\eta < 1$, *there exists* $\eta \leq \delta < \lambda_0 < 1$ *such that for all* $\lambda_0 \leq \lambda < 1$, *we have with obvious notations, an admissible covering*

$$V_\eta^\lambda = \cup_i([Y]_{P\eta} \cap V_{i\delta}^\lambda).$$

If the X_i*'s are all disjoint, we may also assume that the* $V_{i\delta}^\lambda$*'s are disjoint.*

Proof Clearly, we may assume that for all i, we have $P_i = P$. It follows from Proposition 2.3.6 that, given $\eta < 1$, there exists $\eta \leq \delta < 1$ such that

$$[Y]_{P\eta} = \cup_i \left([Y]_{P\eta} \cap [Y_i]_{P\delta}\right)$$

is an admissible covering. Using Corollary 3.2.10, we see that for $\eta, \delta < \lambda < 1$, if we intersect with V^λ, we get as expected an admissible covering

$$V_\eta^\lambda = [Y]_{P\eta} \cap V^\lambda = \cup_i([Y]_{P\eta} \cap ([Y_i]_{P\delta} \cap V^\lambda)) = \cup_i([Y]_{P\eta} \cap V_{i\delta}^\lambda).$$

Moreover, it results from Corollary 3.2.7 that it the X_i's are disjoint, and $\lambda^2 > \delta$, the covering will be disjoint. Note however that here again, we have to refer to Corollary 3.2.10 because we consider X_i as an open subset of Y_i and not as an open subset of Y. □

Corollary 3.2.14 *Let*

$$\begin{array}{ccccc} X' & \hookrightarrow & Y' & \hookrightarrow & P' \\ \downarrow & & \downarrow & & \downarrow \\ X & \hookrightarrow & Y & \hookrightarrow & P \end{array}$$

be a cartesian mixed immersion of quasi-compact frames. Assume moreover that X' is open (and closed) in X. Then, for any $\eta < 1$, there exists $\delta < \lambda_0 < 1$ such that for all $\lambda_0 \leq \lambda < 1$, with obvious notations, $[Y]_{P\eta} \cap V_\delta'^\lambda$ is open and closed in V_η^λ.

Proof Since X' is open in X, it is also open in Y and we might call Y'' a closed complement for X' in Y. We see that Y is the union of the closed subvarieties Y' and Y'' (although the scheme structure might be different from the scheme structure of the union). It follows from the proposition that for any $\eta < 1$, there exists $\delta < \lambda_0 < 1$ such that for all $\lambda_0 \leq \lambda < 1$, we have a disjoint admissible covering

$$V_\eta^\lambda = \left([Y]_\eta \cap V_\delta'^\lambda\right) \coprod \left([Y]_\eta \cap V_\delta''^\lambda\right).$$

Of course, the notation V'' corresponds to the inclusion of $X'' := Y'' \cap X$ in Y''. In particular, we see that $[Y]_{P\eta} \cap V_\delta'^\lambda$ is open and closed in V_η^λ as asserted. □

In the literature, one finds sometimes a modified version of V^λ that we will call here $\widetilde{V}^\lambda$. The next proposition shows that, as long as we consider only closed tubes, we can use one or the other.

Proposition 3.2.15† *Let $(X \subset Y \subset P)$ be an affine frame and*

$$Z = V(g_1, \ldots, g_s) \cap Y \subset P$$

a closed complement for X in Y.

For $\lambda < 1$, we let

$$\widetilde{V}^\lambda := \{x \in]Y[_P, \exists i \in \{1, \cdots, s\}, |g_i(x)| \geq \lambda\}.$$

As usual, we write

$$V^\lambda :=]Y[_P \setminus]Z[_{P\lambda} \quad \text{and} \quad V_\eta^\lambda := [Y]_{P\eta} \cap V^\lambda \quad \text{if} \quad \eta \overset{<}{\to} 1.$$

Then, if W is a quasi-compact admissible open subset of $]Y[_P$, there exists $\lambda_0 < 1$ such that if $\lambda_0 \leq \lambda < 1$, we have

$$W \cap \widetilde{V}^\lambda = W \cap V^\lambda.$$

† In Lemma 2.6.2 of [26], Chiarellotto and Tsuzuki, state that $\widetilde{V}^\lambda$ is independent of the generators when the valuation is discrete and $\lambda > |\pi|$. This is clearly wrong as the following example shows: take $P = \widehat{\mathbf{A}}^2_{\mathcal{V}}$ with variables t and s, and let Y (resp. X) be the subset defined modulo $\mathfrak{m}$ by $t = 0$ (resp. $t = 0$ and $s \neq 0$). Then, both s and $s + t$ define $Y \setminus X$ in Y, but if $\lambda^n \leq |\pi|$ and $x := (\pi^{1/n}, 0)$, we have $|s(x)| = 0 < \lambda$ but $|(s+t)(x)| = |\pi|^{1/n} \geq \lambda$.

Actually, if for $\eta < 1$, we write

$$\widetilde{V}^{\lambda}_{\eta} := [Y]_{P\eta} \cap \widetilde{V}^{\lambda},$$

then for $\lambda > \eta$, we have $\widetilde{V}^{\lambda}_{\eta} = V^{\lambda}_{\eta}$.

Proof Note first that, as in Lemma 3.2.2, we can write

$$\widetilde{V}^{\lambda} := \cup_{i=1}^{s} \widetilde{V}^{\lambda}_{i},$$

with

$$\widetilde{V}^{\lambda}_{i} := \{x \in]Y[_P, |g_i(x)| \geq \lambda, |g_1(x)|, \ldots, |g_s(x)|\}.$$

On the other hand, we may write

$$Y = V(g_{s+1}, \ldots, g_r) \cap P_k \subset P.$$

Since W is a quasi-compact admissible open subset of $]Y[_P$, it follows from Proposition 2.3.7 that there exists $\lambda_0 < 1$ such that if $\lambda_0 \leq \lambda < 1$, we have $W \subset]Y[_{P\lambda}$. For both assertions, it is therefore sufficient to check that

$$]Y[_{P\lambda} \cap \widetilde{V}^{\lambda} =]Y[_{P\lambda} \cap V^{\lambda}.$$

And this follows from the fact that

$$]Y[_{P\lambda} \cap \widetilde{V}^{\lambda}_{i} =]Y[_{P\lambda} \cap V^{\lambda}_{i}.$$

for $i \leq s$ and

$$]Y[_{P\lambda} \cap \widetilde{V}^{\lambda}_{i} = \emptyset.$$

for $i > s$ because

$$]Y[_{P\lambda} = \{x \in]Y[_P, \forall j > s, |g_j(x)| < \lambda\}.$$

□

3.3 Strict neighborhoods and tubes

We can use subsets of the form V^{λ} in order to decide if an admissible subset is a strict neighborhood or not. First of all, we see that they give examples of strict neighborhoods:

Proposition 3.3.1 *Let $(X \subset Y \subset P)$ be a quasi-compact frame and Z a closed complement for X in Y. Then, for $\lambda < 1$,*

$$V^{\lambda} :=]Y[_P \setminus]Z[_{P\lambda}$$

is a strict neighborhood of $]X[_P$ in $]Y[_P$.

Proof We saw in Proposition 3.2.4 that the covering

$$]Y[_P = V^\lambda \cup [Z]_{P\lambda}$$

is admissible and this covering refines

$$]Y[_P = V^\lambda \cup]Z[_P$$

which is therefore also admissible. □

Proposition 3.3.2 *Let $(X \subset Y \subset P)$ be a quasi-compact frame and Z a closed complement for X in Y. As usual, for $\lambda, \eta < 1$, let*

$$V^\lambda :=]Y[_P \setminus]Z[_{P\lambda} \quad \text{and} \quad V_\eta^\lambda := [Y]_{P\eta} \cap V^\lambda.$$

If V is an admissible open subset of $]Y[_P$, the following are equivalent:

(i) *V is a strict neighborhood of $]X[_P$.*
(ii) *For any affinoid open subset $W \subset]Y[_P$, there exists $\lambda < 1$ such that*

$$V^\lambda \cap W \subset V.$$

(iii) *For any quasi-compact admissible open subset $W \subset]Y[_P$, there exists $\lambda < 1$ such that*

$$V^\lambda \cap W \subset V.$$

(iv) *For any $\eta < 1$, there exists $\lambda < 1$ such that $V_\eta^\lambda \subset V$.*

Proof Assume first that V is a strict neighborhood of $]X[_P$ in $]Y[_P$ so that the covering of $]Y[_P = V \cup]Z[_P$ is admissible. Since $]Z[_P = \cup_\lambda]Z[_{P\lambda}$ is also an admissible covering, so is

$$]Y[_P = V \cup \cup_\lambda]Z[_{P\lambda}.$$

It follows that if W an affinoid open subset of $]Y[_P$, we have

$$W \subset V \cup]Z[_{P\lambda}$$

for some λ. In other words, we have

$$V^\lambda \cap W \subset V.$$

Clearly, if the condition is satisfied for affinoid open subsets, it is also satisfied for quasi-compact admissible open subsets which are finite unions of affinoid open subsets.

Now, if we apply this condition to $[Y]_\eta$, which is quasi-compact, we get

$$V_\eta^\lambda := V^\lambda \cap [Y]_{P\eta} \subset V.$$

It remains to show that this last condition implies that V is a strict neighborhood of $]X[_P$ in $]Y[_P$. Since the closed tubes $[Y]_{P\eta}$, for $\eta < 1$, form an admissible affinoid covering of $]Y[_P$, it is sufficient to show that we have, for fixed $\eta < 1$, an admissible covering

$$[Y]_{P\eta} = (V \cap [Y]_{P\eta}) \cup (]Z[_P \cap [Y]_{P\eta}).$$

But, since $V_\eta^\lambda \subset V$, it follows from Lemma 3.2.2 that the covering

$$[Y]_{P\eta} = V_\eta^\lambda \cup (]Z[_{P\lambda} \cap [Y]_{P\eta}).$$

is admissible. Since it is a refinement of our covering, we are done. □

Corollary 3.3.3 *Let $(X \subset Y \subset P)$ be a quasi-compact frame with $Y_k \hookrightarrow P$ nilpotent. If Z a closed complement for X in Y, the V^λ's for $\lambda \to 1$ form a cofinal system of (quasi-compact) strict neighborhoods of $]X[_P$ in P_K.*

Proof In this situation, we have $]Y[_P = P_K$ which is quasi-compact. It follows from Proposition 3.3.2 that, if V is a strict neighborhood of $]X[_P$, there exists $\lambda < 1$ such that

$$V^\lambda \subset V.$$

□

This applies in particular to the very important case $Y = P_k$.

Proposition 3.3.4 *Let $(X \subset Y \subset P)$ be a frame and V a strict neighborhood of $]X[_P$ in $]Y[_P$. Then, any Zariski open neighborhood of $]X[_P$ in V is still a strict neighborhood of $]X[_P$ in $]Y[_P$.*

Proof The question is local on P and we may therefore assume that it is a quasi-compact formal $\mathcal{V}$-scheme. We are given a closed analytic subset T of V such that $T \cap]X[_P = \emptyset$ and we want to show that the open complement of T in V is a strict neighborhood of $]X[_P$ in $]Y[_P$. Note first that, if Z is a closed complement of X in Y, our hypothesis just means that $T \subset]Z[_P$.

Now, if W is an affinoid domain in $]Y[_P$, by Proposition 3.3.2, there exists $\mu < 1$ such that $W \cap V^\mu \subset V$. Moreover, $W \cap V^\mu \cap T$ is an analytic subset of $W \cap V^\mu$ which is quasi-compact by Proposition 3.2.5. It follows that $W \cap V^\mu \cap T$ is also quasi-compact. Since it is contained in $]Z[_P$, Proposition 3.2.5 again tells us that there exists $\lambda < 1$ such that

$$V^\lambda \cap (W \cap V^\mu \cap T) = \emptyset$$

If we take $\lambda > \mu$, we see that

$$(W \cap V^\lambda) \cap T = \emptyset$$

or in other words, that

$$W \cap V^\lambda \subset V \setminus T.$$

It follows from Proposition 3.3.2 that $V \setminus T$ is a strict neighborhood of $]X[_P$ in $]Y[_P$. □

Note however that there are in general analytic subsets T of $]Z[_P$ that meet all strict neighborhoods of $]X[_P[$ in $]Y[_P$. Take for example

$$P := \mathbf{A}^2_{\mathcal{V}}, \quad Y := \mathbf{A}^2_k, \quad P = \mathbf{A}^2_k \setminus 0 \quad \text{and} \quad T = \{(x, x), |x| < 1\}.$$

It is therefore essential to assume from the beginning that T is a closed analytic subset of some strict neighborhood.

Definition 3.3.5 *A morphism of frames*

$$\begin{array}{ccccc} X' & \hookrightarrow & Y' & \hookrightarrow & P' \\ \downarrow f & & \downarrow g & & \downarrow u \\ X & \hookrightarrow & Y & \hookrightarrow & P \end{array}$$

is flat *(resp.* smooth, étale*) if u is flat (resp. smooth, étale) in a neighborhood of X'.*

An S-frame $(X \subset Y \subset P)$ is flat *(resp.* smooth, étale*) over S if the morphism of frames*

$$\begin{array}{ccccc} X & \hookrightarrow & Y & \hookrightarrow & P \\ \downarrow & & \downarrow & & \downarrow \\ S_k & = & S_k & \hookrightarrow & S \end{array}$$

is flat (resp. smooth, étale).

Of course, the latter means that P is smooth (or étale) over S in a neighborhood of X. We may now state a corollary to the previous proposition.

Corollary 3.3.6 *Let*

$$\begin{array}{ccccc} X' & \hookrightarrow & Y' & \hookrightarrow & P' \\ \downarrow f & & \downarrow g & & \downarrow u \\ X & \hookrightarrow & Y & \hookrightarrow & P \end{array}$$

be a flat (resp. smooth, étale) morphism of frames. Then, there exists a strict neighborhood V' of $]X'[_{P'}$ in $]Y'[_{P'}$ such that the induced map $V' \to]Y[_P$ is flat (resp. smooth, étale).

Proof We saw in Proposition 2.2.17 that the induced map

$$]X'[_P \to]X[_P$$

is flat (resp. smooth, étale). Moreover, we know that the flat (resp. smooth, étale) locus is Zariski open. This assertion therefore follows from Proposition 3.3.4. □

As a very important example, we will consider the Monsky–Washnitzer situation.

Proposition 3.3.7 *Let P be a formal subscheme of $\widehat{\mathbf{P}}^N_{\mathcal{V}}$ and*

$$X := \widehat{\mathbf{A}}^N_k \cap P.$$

Then, the family of

$$V_\rho := \mathbf{B}^N(0, \rho^+) \cap P_K$$

is a cofinal family of strict neighborhoods of $]X[_P$ in P_K when $\rho > 1$.

Proof This follows from Corollary 3.3.3 once we have proven Lemma 3.3.8 below. □

Lemma 3.3.8 *With the hypothesis and notations of the proposition, if Z denotes the intersection of P_k with the hyperplane at infinity and $\lambda := 1/\rho$, we have for $\rho > 1$,*

$$V_\rho = V^\lambda := P_K \setminus]Z[_{P\lambda}.$$

Proof Note that, if $P' \subset P$ is a formal subscheme and we define V'_ρ and V'^λ in the same way, then

$$V_\rho \cap P'_K = V'_\rho$$

by definition and

$$V^\lambda \cap P'_K = V'^\lambda$$

by Proposition 3.2.8. It is therefore sufficient to consider the case $P = \widehat{\mathbf{P}}^N_{\mathcal{V}}$. Moreover, the question is local. Thus, if $T_0, \ldots, T_N$ denote the projective coordinates in $\mathbf{P}^N$ (and $\mathbf{A}^N$ is the open subset defined by $T_0 \neq 0$), it is sufficient to prove the lemma when P is the open subset of $\widehat{\mathbf{P}}^N_{\mathcal{V}}$ defined by $T_i \neq 0$ with

$i = 0, \dots, N$. In the case $i = 0$, we simply have $P = \widehat{\mathbf{A}}^N_{\mathcal{V}}$ and the assertion is trivial. We may therefore assume that $i \neq 0$. We will denote by

$$T_0/T_i, \dots, T_{i-1}/T_i, T_{i+1}/T_i, \dots, T_N/T_i$$

the affine coordinates on P. We have to show that

$$P_K \cap \mathbf{B}^N(0, \rho^+) = \{x \in P_K, |T_0/T_i(x)| \geq \lambda\}.$$

This is an easy exercise in rigid analytic geometry but we can work it out. Inclusion is clear because, if $x \in \mathbf{B}^N(0, \rho^+)$, then necessarily $|T_i/T_0(x)| \leq \rho$ and therefore

$$|T_0/T_i(x)| \geq \frac{1}{\rho} = \lambda.$$

Conversely, if we assume that $x \in P_K$, then $|T_j/T_i(x)| \leq 1$ for all $j = 0, \dots, n$. If moreover, we have $|T_0/T_i(x)| \geq \lambda$, then we see that

$$|T_0/T_i(x)| \geq \lambda |T_j/T_i(x)|$$

for all $j = 0, \dots, n$ and therefore $|T_j/T_0|(x) \leq \rho$ so that $x \in \mathbf{B}^N(0, \rho^+)$. □

Corollary 3.3.9 *Let X be an affine variety over $\mathcal{V}$, $X \hookrightarrow \mathbf{A}^N_{\mathcal{V}}$ a closed embedding into an affine space and Y the closure of the image of X in $\mathbf{P}^N_{\mathcal{V}}$. Then, the family of*

$$V_\rho := \mathbf{B}^N(0, \rho^+) \cap X_K^{\mathrm{rig}}$$

is a cofinal family of strict neighborhoods of $\widehat{X}_K$ in Y_K^{rig} when $\rho > 1$.

Proof We have $X_K^{\mathrm{rig}} = \mathbf{A}_K^{N,\mathrm{rig}} \cap Y_K^{\mathrm{rig}}$. Moreover, since Y is proper, we have $\widehat{Y}_K = Y_K^{\mathrm{rig}}$. It follows that

$$\mathbf{B}^N(0, \rho^+) \cap X_K^{\mathrm{rig}} = \mathbf{B}^N(0, \rho^+) \cap \mathbf{A}_K^{N,\mathrm{rig}} \cap Y_K^{\mathrm{rig}} = \mathbf{B}^N(0, \rho^+) \cap \widehat{Y}_K.$$

□

The next result is fundamental. As we will see later, it implies that as long as we are only interested in strict neighborhoods we can make étale changes at the level of P and proper changes at the level of Y. But it is also needed in order to give a local criterion for the overconvergence of a connection.

We first introduce the following definition.

Definition 3.3.10 *A morphism of frames*

$$\begin{array}{ccccc} X' & \hookrightarrow & Y' & \hookrightarrow & P' \\ \downarrow f & & \downarrow g & & \downarrow u \\ X & \hookrightarrow & Y & \hookrightarrow & P \end{array}$$

is proper (*resp.* projective, *resp.* finite) *if* g *is proper (resp. projective, resp. finite).*

An S-frame $(X \subset Y \subset P/S)$ *is* proper (*resp.* projective, finite) *over* S *if the morphism of frames*

$$\begin{array}{ccccc} X & \hookrightarrow & Y & \hookrightarrow & P \\ \downarrow & & \downarrow & & \downarrow \\ S_k & = & S_k & \hookrightarrow & S \end{array}$$

is proper (resp. projective, finite).

Of course, the latter means that Y is proper (*resp.* projective, finite) over S_k.

Proposition 3.3.11 *Let*

$$\begin{array}{ccccc} & \nearrow & Y' & \hookrightarrow & P' \\ X & & \downarrow g & & \downarrow u \\ & \searrow & Y & \hookrightarrow & P \end{array}$$

be a proper étale cartesian morphism of quasi-compact frames. As usual, let Z *be a closed complement of* X *in* Y, $Z' := g^{-1}(Z)$ *and for* $\eta, \lambda < 1$,

$$V^\lambda :=]Y[_P \setminus]Z[_{P\lambda}, \quad V_\eta^\lambda := [Y]_{P\eta} \cap V^\lambda,$$

$$V'^\lambda :=]Y'[_{P'} \setminus]Z'[_{P'\lambda} \quad \text{and} \quad V_\eta'^\lambda := [Y']_{P'\eta} \cap V'^\lambda.$$

Then, given $\eta < 1$, *there exists* $\eta \leq \delta < \lambda_0 < 1$ *such that for* $\lambda_0 \leq \lambda < 1$, *the morphism* u_K *induces an isomorphism*

$$u_K^{-1}([Y]_{P\eta}) \cap V_\delta'^\lambda = [Y']_{P'\delta} \cap u_K^{-1}(V_\eta^\lambda) \simeq V_\eta^\lambda.$$

Proof We fix $\eta < 1$. Since u is étale in a neighborhood of X, the canonical immersion of X in $X'' := u^{-1}(X)$ has an étale retraction and it follows that X is open and closed in X''. Let us write

$$Y'' := u^{-1}(Y) \quad \text{and} \quad Z'' := u^{-1}(Z)$$

so that we can split our morphism of frames as follows (cartesian mixed immersion followed by strict morphism)

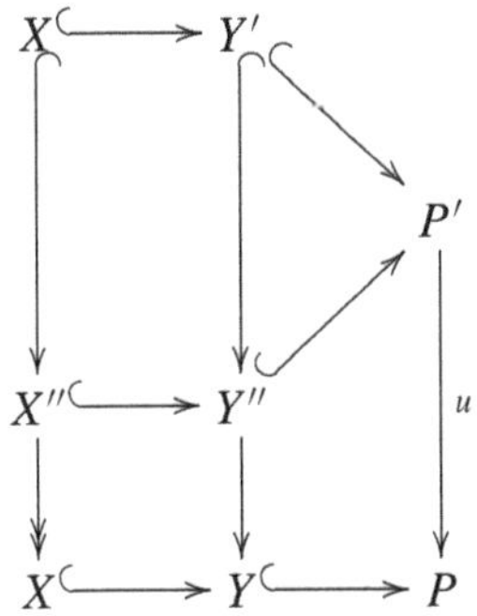

and we may define as usual

$$V''^{\lambda} :=]Y''[_{P'} \backslash]Z''[_{P'\lambda} \quad \text{and} \quad V''^{\lambda}_{\eta} := [Y'']_{P'\eta} \cap V''^{\lambda}$$

for $\lambda < 1$. It follows from Corollary 3.2.14, that there exists

$$\eta \leq \delta < \lambda_0 < 1$$

such that for all $\lambda_0 \leq \lambda < 1$, we have a closed (and also open) immersion

$$[Y'']_{P'\eta} \cap V'^{\lambda}_{\delta} \hookrightarrow V''^{\lambda}_{\eta}.$$

We also know from Proposition 2.3.9 that $[Y'']_{P'\eta} = u_K^{-1}([Y]_{P\eta})$ and from Proposition 3.2.8 that $V''^{\lambda}_{\eta} = u_K^{-1}(V^{\lambda}_{\eta})$. In other words, we have

$$u_K^{-1}([Y]_{P\eta}) \cap V'^{\lambda}_{\delta} \subset u_K^{-1}(V^{\lambda}_{\eta}).$$

Note also that, since the morphism of frames is cartesian and $\eta \leq \delta < \lambda$, it results from Proposition 3.2.8 that

$$\begin{aligned} u_K^{-1}([Y]_{P\eta}) \cap V'^{\lambda}_{\delta} &= u_K^{-1}([Y]_{P\eta}) \cap [Y']_{P'\delta} \cap u_K^{-1}(V^{\lambda}_{\delta}) \\ &= [Y']_{P'\delta} \cap u_K^{-1}([Y]_{P\eta} \cap V^{\lambda}_{\delta}) = [Y']_{P'\delta} \cap u_K^{-1}(V^{\lambda}_{\eta}), \end{aligned}$$

which therefore proves our first equality.

We can now choose a formal model for the open immersion $V^{\lambda_0}_{\eta} \hookrightarrow P_K$ and pull everything back through this morphism. In other words, we may assume that

$$P_K = [Y]_{P\eta} = V^{\lambda_0}_{\eta} = V^{\lambda_0}$$

and we know that $V'^{\lambda_0}_{\delta}$ is a closed (and open) analytic subset of P'_K. Let P'' be the schematic closure of $V'^{\lambda_0}_{\delta}$ in P'_K. This is a flat formal closed subscheme of P' with $P''_K = V'^{\lambda_0}_{\delta}$. Since P'' is flat, specialization is surjective on closed points. But we also have $P''_K \subset]Y'[_{P'}$ and it follows that the underlying space of P''_k is a (closed) subset of Y'. Since g is proper, this implies that $P''_k \to P_k$ is proper too. It follows from Corollary 3.2.10 that we may replace P' by P'' without changing V'^{λ}_{δ} and we may thus assume that u is proper. It follows that u_K too is proper. Note also that now, we have

$$P'_K = [Y']_{P'\delta} = V'^{\lambda_0}_{\delta} = V'^{\lambda_0}.$$

Recall from Corollary 3.3.6 that u_K is étale on some strict neighborhood of $]X[_{P'}$ in $]Y'[_{P'}$. It then follows from Proposition 3.3.2 that if we choose a bigger $\lambda_0 < 1$, then u_K induces an étale morphism

$$V'^{\lambda} \to V^{\lambda}$$

for $\lambda_0 \leq \lambda < 1$ and we want to show that it is an isomorphism it we choose λ_0 close enough to 1. Being étale and proper, it is necessarily finite. It follows that the canonical map

$$\mathcal{O}_{P_K} \to u_{K*}\mathcal{O}_{P'_K}$$

is a morphism of coherent modules on V^{λ} for $\lambda_0 \leq \lambda < 1$. We know from Proposition 2.3.14 that this is an isomorphism on $]X[_P$. On the other hand, as any morphism of coherent modules, it is an isomorphism on a Zariski open subset of P_K. But we know from Proposition 3.3.4 that such a neighborhood is strict. Since we have $[Y]_{P\eta} = P_K$, the immersion $Y \hookrightarrow P_k$ is nilpotent. It follows from Corollary 3.3.3 that if we take a bigger λ_0, the induced map $V'^{\lambda} \to V^{\lambda}$ is an isomorphism for $\lambda_0 \leq \lambda < 1$. The proposition is proved. □

Note that we have a sequence of open immersions

$$V'^{\lambda}_{\eta} \subset u_K^{-1}(V^{\lambda}_{\eta}) \subset V'^{\lambda}_{\delta}$$

for λ close enough to 1, but there is no reason to believe that one can take $\delta = \eta$ in general. This is however often the case in practice.

Corollary 3.3.12 *In the theorem, assume that $Y \hookrightarrow P$ is nilpotent or that $Y' = Y$, $g = \mathrm{Id}_Y$ and u is étale in the neighborhood of Y. Then, there exists η_0 such that for $\eta_0 \leq \eta < 1$, there exists $\lambda_0 < 1$ such that for $\lambda_0 \leq \lambda < 1$, the morphism u_K induces an isomorphism*

$$V'^{\lambda}_{\eta} \simeq V^{\lambda}_{\eta}.$$

Proof In the first case, we have $[Y]_{P\eta} =]Y[_P = P_K$ so that $V^{\lambda}_{\eta} = V^{\lambda}$ and the theorem reads

$$V'^{\lambda}_{\delta} = [Y']_{P'\delta} \cap u_K^{-1}(V^{\lambda}) \simeq V^{\lambda}.$$

We may then choose $\eta_0 = \delta$.

In the second case, we may assume from Proposition 2.3.15 that u_K induces an isomorphism $[Y]_{P\eta} \simeq [Y]_{P'\eta}$. It follows that

$$V_\eta'^\lambda = [Y]_{P'\eta} \cap V_\delta'^\lambda = u_K^{-1}([Y]_{P\eta}) \cap V_\delta'^\lambda \simeq V_\eta^\lambda.$$

□

In order to extend this proposition to the case of a smooth morphism, we need some preliminary results.

Proposition 3.3.13 *Let*

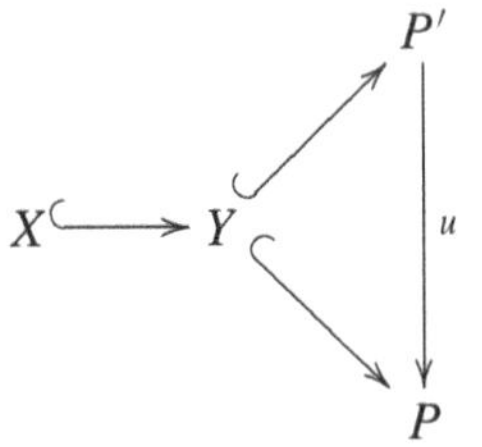

be a morphism of affine frames. Let I' be the ideal of Y in P', $X' := u^{-1}(X)$ and $\omega_{X'/X}$ the conormal sheaf of X in X'.

(i) *Assume that $t_1, \ldots, t_d \in I'$ induce a basis $(\bar{t}_1, \ldots, \bar{t}_d)$ of $\omega_{X'/X}$. If we embed Y into $\widehat{\mathbf{A}}^d_P$ using the zero section, then $(t_1, \ldots, t_d)$ induce an étale morphism of affine frames*

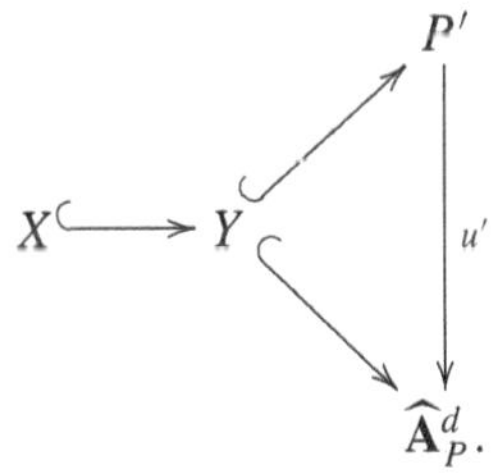

(ii) *Assume that the complement of X in Y is a hypersurface and that $\omega_{X'/X}$ is free. Then, there exist $t_1, \ldots, t_d \in I'$ inducing a basis $(\bar{t}_1, \ldots, \bar{t}_d)$ of $\omega_{X'/X}$.*

Proof Since we chose our affine coordinates $(t_1, \ldots, t_d)$ in I', the diagram is commutative. The first assertion then follows from Lemma 2.3.18.

In order to prove the second assertion, we write,

$$P = \mathrm{Spf} A, \quad P' = \mathrm{Spf} A', \quad X = Y \cap D(g)$$

and we denote by I the ideal of Y in P. Then we have

$$\Gamma(X, \omega_{X'/X}) := (I'/I'^2 + IA')_{\bar{g}}$$

and this is a free $(A/I)_{\overline{g}}$-module. Thus, we can find a basis of the form $(\overline{t}_1, \ldots, \overline{t}_d)$ with $t_1, \ldots, t_d \in I'$. $\square$

Note that, if

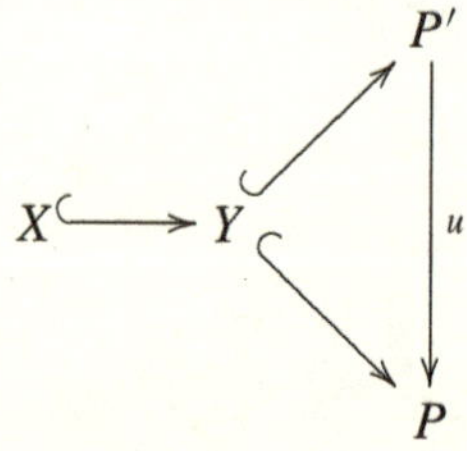

is a smooth morphism of frames, then, thanks to the second part of the proposition, the condition of the first part is always satisfied locally on $(X \subset Y \subset P')$.

Proposition 3.3.14 *Let*

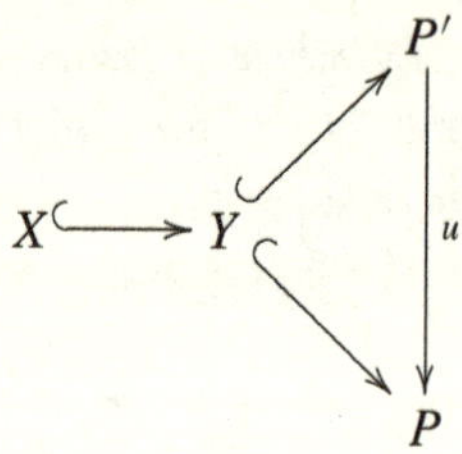

be a smooth morphism of affine frames.

Let I' be the ideal of Y in P' and $X' := u^{-1}(X)$. Assume that there exist $t_1, \ldots, t_d \in I'$ inducing a basis $(\overline{t}_1, \ldots, \overline{t}_d)$ of $\omega_{X'/X}$.

Then, given $\eta < 1$, there exists $\eta \leq \delta < \lambda_0 < 1$ such that for $\lambda_0 \leq \lambda < 1$, with our usual notations, the morphism

$$(t_1, \ldots, t_d) : P' \to \widehat{\mathbf{A}}^d_P$$

induces an isomorphism

$$u_K^{-1}([Y]_{P\eta}) \cap V'^{\lambda}_{\delta} = [Y]_{P'\delta} \cap u_K^{-1}(V^{\lambda}_{\eta}) \simeq V^{\lambda}_{\eta} \times \mathbf{B}^d(0, \eta^+)$$

when Y is embedded in $\widehat{\mathbf{A}}^d_P$ using the zero section.

Proof On one hand, we have a morphism of frames

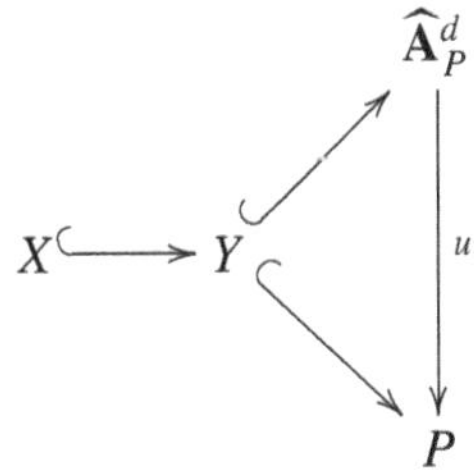

On the other hand, we have an étale morphism

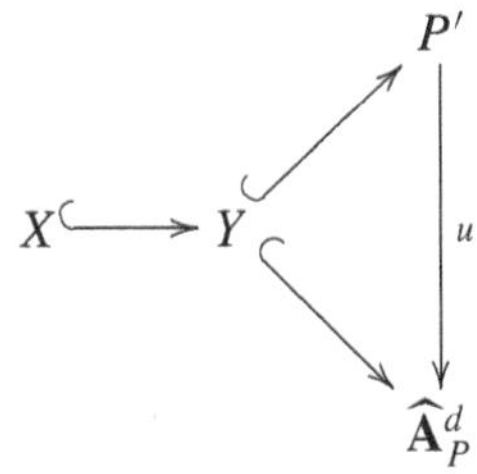

of affine frames. We may then apply Proposition 3.3.11 and conclude with the last assertion of Proposition 3.2.12. □

Corollary 3.3.15 *In the proposition, assume that either $Y \hookrightarrow P$ is nilpotent or else that $g = \mathrm{Id}_Y$, u is smooth in the neighborhood of Y and $t_1, \ldots, t_d$ induces a basis of $\omega_{Y'/Y}$ with $Y' = u^{-1}(Y)$. Then, given $\eta < 1$, there exists $\lambda_0 < 1$ such that for $\lambda_0 \leq \lambda < 1$, the morphism u_K induces an isomorphism*

$$V'^{\lambda}_{\eta} \simeq V^{\lambda}_{\eta} \times \mathbf{B}^d(0, \eta^+).$$

Proof This follows from Corollary 3.3.12. □

3.4 Standard neighborhoods

In this section, we will see how to build strict neighborhoods from scratch and use them to prove the strong fibration theorem.

Recall that if $(X \subset Y \subset P)$ is a quasi-compact frame, Z is a closed complement of X in Y and $\lambda, \eta < 1$, we write

$$V^{\lambda} :=]Y[_P \backslash]Z[_{P\lambda} \quad \text{and} \quad V^{\lambda}_{\eta} := [Y]_{\eta} \cap V^{\lambda}.$$

Proposition 3.4.1 *Let $(X \subset Y \subset P)$ be a quasi-compact frame. If we are given two sequences $\eta_n \overset{<}{\to} 1$ and $\lambda_n \overset{<}{\to} 1$, then*

$$V_{\underline{\eta}}^{\underline{\lambda}} := \cup_n V_{\eta_n}^{\lambda_n}$$

is a strict neighborhood of $]X[_P$ in $]Y[_P$ and the covering is admissible. Moreover, for fixed $\underline{\eta}$, the subsets $V_{\underline{\eta}}^{\underline{\lambda}}$ form a fundamental system of strict neighborhoods of $]X[_P$ in $]Y[_P$.

Proof Since this question is local, it follows from Lemma 3.4.2 below, that $V_{\underline{\eta}}^{\underline{\lambda}}$ is an admissible open subset and that the covering is admissible.

Now, we want to show that $V_{\underline{\eta}}^{\underline{\lambda}}$ is a strict neighborhood of $]X[_P$ in $]Y[_P$. We use the last criterion of Proposition 3.3.2. Given $\eta < 1$, we can find $\eta \le \eta_n < 1$. We let $\lambda := \lambda_n$ and then

$$V_\eta^\lambda \subset V_{\eta_n}^{\lambda_n} \subset V_{\underline{\eta}}^{\underline{\lambda}}.$$

Finally, we want to show that for fixed $\eta_n \overset{<}{\to} 1$, the strict neighborhoods $V_{\underline{\eta}}^{\underline{\lambda}}$ are cofinal. Let V be a strict neighborhood of $]X[_P$ in $]Y[_P$. We use again the last criterion of Proposition 3.3.2: For each n, there exists $\lambda_n < 1$ such that $V_{\eta_n}^{\lambda_n} \subset V$ and it follows that $V_{\underline{\eta}}^{\underline{\lambda}} \subset V$. □

Lemma 3.4.2 *Let $(X \subset Y \subset P)$ be an affine frame and*

$$Z = V(g_1, \ldots, g_r) \cap P_k \subset P$$

a closed complement for X in Y.

As before, if $\eta, \lambda < 1$, we write for $i = 1, \ldots, r$,

$$V_{\eta,i}^\lambda := \{x \in [Y]_{P\eta}, |g_i(x)| \ge \lambda, |g_1(x)|, \ldots, |g_r(x)|\}.$$

If $\eta_n \overset{<}{\to} 1, \lambda_n \overset{<}{\to} 1$, we define for $i = 1, \ldots, r$,

$$V_{\underline{\eta},i}^{\underline{\lambda}} := \cup_n V_{\eta_n,i}^{\lambda_n}.$$

Then, $V_{\underline{\eta}}^{\underline{\lambda}}$ as well as the $V_{\underline{\eta},i}^{\underline{\lambda}}$ are admissible open subsets and the coverings

$$V_{\underline{\eta}}^{\underline{\lambda}} = \cup_i V_{\underline{\eta},i}^{\underline{\lambda}}, \quad V_{\underline{\eta}}^{\underline{\lambda}} := \cup_n V_{\eta_n}^{\lambda_n} \quad \text{and} \quad V_{\underline{\eta},i}^{\underline{\lambda}} := \cup_n V_{\eta_n,i}^{\lambda_n}.$$

are all admissible.

Proof The question is local on $]Y[_P$ and we know that $]Y[_P = \cup_m [Y]_{P\eta_m}$ is an admissible covering. For each $m \in \mathbf{N}$ and each i, we have a finite affinoid covering

$$[Y]_{P\eta_m} \cap V_{\underline{\eta},i}^{\underline{\lambda}} := \cup_{n \le m} V_{\eta_n,i}^{\lambda_n}.$$

It follows that for each i, $V_{\underline{\eta},i}^{\underline{\lambda}}$ is an admissible open subset and that the covering

$$V_{\underline{\eta},i}^{\underline{\lambda}} := \cup_n V_{\eta_n,i}^{\lambda_n}$$

is admissible. We also have a finite covering

$$[Y]_{P\eta_m} \cap V_{\underline{\eta}}^{\underline{\lambda}} := \cup_{n \leq m} V_{\eta_n}^{\lambda_n}.$$

Since, for each n, we have a finite affinoid covering

$$V_{\eta_n}^{\lambda_n} = \cup_{i=1}^{r} V_{\eta_n,i}^{\lambda_n}$$

we see that

$$[Y]_{P\eta_m} \cap V_{\underline{\eta}}^{\underline{\lambda}} := \cup_{\substack{n \leq m \\ 1 \leq i \leq r}} V_{\eta_n,i}^{\lambda_n}$$

is also a finite affinoid covering. It follows that $V_{\underline{\eta}}^{\underline{\lambda}}$ is an admissible open subset and that the coverings

$$V_{\underline{\eta}}^{\underline{\lambda}} = \cup_{i=1}^{r} \underline{V_{\underline{\eta},i}^{\underline{\lambda}}} \quad \text{and} \quad V_{\underline{\eta}}^{\underline{\lambda}} := \cup_n V_{\eta_n}^{\lambda_n}$$

are both admissible. □

Definition 3.4.3 *With the notations of the proposition, the strict neighborhood*

$$V_{\underline{\eta}}^{\underline{\lambda}} := \cup_n V_{\eta_n}^{\lambda_n}$$

is said to be standard.

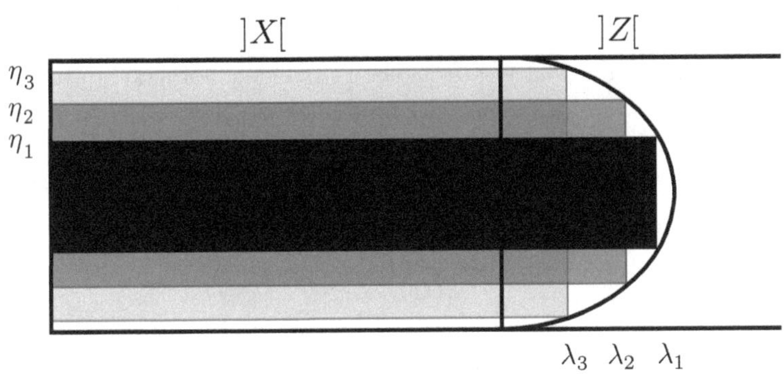

Fig. 3.2 $V_{\underline{\eta}}^{\underline{\lambda}}$

Proposition 3.4.4 *Let*

$$\begin{array}{ccccc} X' & \hookrightarrow & Y' & \hookrightarrow & P' \\ \downarrow f & & \downarrow g & & \downarrow u \\ X & \hookrightarrow & Y & \hookrightarrow & P \end{array}$$

be a strict morphism of quasi-compact frames, Z a closed complement for X in Y and $Z' := g^{-1}(Z)$. Let $\eta_n \overset{<}{\to} 1$ and $\lambda_n \overset{<}{\to} 1$. Denote by $V_{\underline{\eta}}^{\underline{\lambda}}$ and $V'^{\underline{\lambda}}_{\underline{\eta}}$ the corresponding standard strict neighborhoods of $]X[_P$ in $]Y[_P$, and $]X'[_{P'}$ in $]Y'[_{P'}$ respectively. Then,

$$u_K^{-1}(V_{\underline{\eta}}^{\underline{\lambda}}) = V'^{\underline{\lambda}}_{\underline{\eta}}.$$

Proof Follows from Proposition 3.2.8. □

Note that it is absolutely necessary to assume that our morphism is strict and not only cartesian. Consider for example the following: let $P = \widehat{\mathbf{A}}^1_{\mathcal{V}}$ with variable s, $Y = P_k$ and X defined by $s \neq 0$. Take $P' = \widehat{\mathbf{A}}^2_{\mathcal{V}}$ with variables t and s, and let Y' (resp. X') be the subset defined in P_k by $t = 0$ (resp. $t = 0$ and $s \neq 0$). Of course, u is the projection and both f and g are the identity. We assume that for all n, we have $\eta_n < \lambda_n$. Then $V_{\underline{\eta}}^{\underline{\lambda}}$ is simply the annulus $\mathbf{A}(0, \lambda_0^+, 1^+)$ and

$$]Y'[_{P'} \cap u_K^{-1}(V_{\underline{\eta}}^{\underline{\lambda}}) = \{(t, s) \in \mathbf{B}^2(0, 1^+), |t| < 1 \quad \text{and} \quad \lambda_0 \leq |s|\}.$$

On the other hand, we have

$$V'^{\underline{\lambda}}_{\underline{\eta}} = \{(t, s) \in \mathbf{B}^2(0, 1^+), \exists n \in \mathbf{N}, |t| \leq \eta_n < \lambda_n \leq |s|\}.$$

Proposition 3.4.5 *Let $\sigma : K \hookrightarrow K'$ be an isometric embedding, $(X \subset Y \subset P)$ be a quasi-compact frame and V' a strict neighborhood of $]X^\sigma[_{P^\sigma}$ in $]Y^\sigma[_{P^\sigma}$. Then, there exists a strict neighborhood V of $]X[_P$ in $]Y[_P$ with $V^\sigma \subset V'$.*

Proof We may assume that V' is a standard strict neighborhood in which case we even have $V^\sigma = V'$. Of course, we need to choose a presentation

$$Z := V(f_1, \ldots, f_r) \cap P_k$$

of the complement of X in Y and work with the presentation

$$Z^\sigma := V(f_1^\sigma, \ldots, f_r^\sigma) \cap P_k.$$

□

Using standard neighborhoods, we can prove some other results on the local nature of strict neighborhoods.

Proposition 3.4.6 *Let $(X \subset Y \subset P)$ be a quasi-compact frame and $(X' \subset Y' \subset P')$ a strict open subframe. If V' is a strict neighborhood of $]X'[_{P'}$ in $]Y'[_{P'}$, there exists a strict neighborhood V of $]X[_P$ in $]Y[_P$ with $V \cap P'_K \subset V'$.*

Proof Shrinking it if necessary, we may assume that V' is the standard strict neighborhood of $]X \cap P'[_{P'}$ in $]Y \cap P'[_{P'}$ associated to two sequences $\eta_n \overset{<}{\to} 1$ and $\lambda_n \overset{<}{\to} 1$. It is then sufficient to take for V the standard strict neighborhood of $]X[_P$ in $]Y[_P$ associated to the same sequences. It follows from Proposition 3.4.4 that $V' = V \cap P'_K$. □

It is sometimes necessary, as we will see in the Monsky–Washnitzer setting for example, to consider morphisms of strict neighborhoods that do not come from morphisms of frames.

Definition 3.4.7 *Let $(X \subset Y \subset P)$ (resp. $(X' \subset Y' \subset P')$) be a frame and V (resp. V') be an admissible open subset of P_K (resp. $P'_{K'}$). A morphism $u : V' \to V$ is* compatible to *a morphism $f : X' \to X$ if*

$$\forall x \in V' \cap]X'[_{P'}, \quad f(\overline{x}) = \overline{u(x)}.$$

Proposition 3.4.8 *Let $(X \subset Y \subset P)$ be a quasi-compact frame, $(X' \subset Y' \subset P')$ any frame and V' a strict neighborhood of $]X'[_{P'}$ in $]Y'[_{P'}$. If $f : X' \to X$ and $u : V' \to P_K$ are two compatible morphisms, there exists a strict neighborhood V'' of $]X'[_{P'}$ in V' with $u(V'') \subset]Y[_P$.*

Proof Given η, there exists λ such that, with our usual notations, $V'^{\lambda}_{\eta} \subset V'$. Since P is quasi-compact and Y closed in P, then $P_K \backslash]Y[_P$ is also quasi-compact and it follows that

$$V'^{\lambda}_{\eta} \cap u^{-1}(P_K \backslash]Y[_P)$$

too is quasi-compact. On the other hand, since u is compatible with f, we have

$$u(V' \cap]X'[_{P'}) \subset]X[_P$$

and it follows that $V'^{\lambda}_{\eta} \cap u^{-1}(P_K \backslash]Y[_P)$ does not meet $]X'[_{P'}$. Now, Corollary 3.2.5 tells us that there exists $\mu \geq \lambda$ such that

$$V'^{\lambda}_{\eta} \cap u^{-1}(P_K \backslash]Y[_P) \cap V'^{\mu} = \emptyset.$$

In other words, we have $V'^{\mu}_{\eta} \subset V'$ and $u(V'^{\mu}_{\eta}) \subset]Y[_P$. It follows that if we are given a sequence $\eta_n \overset{<}{\to} 1$, we can find a sequence $\mu_n \overset{<}{\to} 1$ such that the strict neighborhood $V'' := V'^{\underline{\mu}}_{\underline{\eta}}$ of $]X'[_{P'}$ in $]Y'[_{P'}$ satisfies $V'' \subset V'$ and $u(V'') \subset]Y[_P$. □

Proposition 3.4.9 *Let $(X \subset Y \subset P)$ be a quasi-compact frame and $\{X_i\}_i$ a finite set of open subsets of Y such that $X = \cap_i X_i$. If V is a strict neighborhood of $]X[_P$ in $]Y[_P$, there exists for each i, a strict neighborhood V_i of $]X_i[_P$ in $]Y[_P$ such that*

$$\cap_i V_i \subset V.$$

Proof By induction, it is sufficient to consider the case $X = X_1 \cap X_2$. Shrinking it if necessary, we may assume that V is the standard strict neighborhood associated to two sequences $\eta_n \overset{<}{\to} 1$ and $\lambda_n \overset{<}{\to} 1$. We choose for V_1 and V_2 the standard strict neighborhoods associated to the sequences $\eta_n \overset{<}{\to} 1$ and $\lambda'_n \overset{<}{\to} 1$ with $\lambda'_n = \sqrt{\lambda_n}$. Thus, with obvious notations, we have

$$\begin{aligned} V_1 \cap V_2 &= \left(\cup_n([Y]_{P\eta_n} \cap V_1^{\lambda'_n})\right) \cap \left(\cup_m(Y]_{P\eta_m} \cap V_2^{\lambda'_m})\right) \\ &= \cup_{n,m}\left([Y]_{P\eta_n} \cap V_1^{\lambda'_n} \cap [Y]_{P\eta_m} \cap V_2^{\lambda'_m}\right). \end{aligned}$$

If $n \leq m$, we have

$$[Y]_{P\eta_n} \cap V_1^{\lambda'_n} \cap [Y]_{P\eta_m} \cap V_2^{\lambda'_m} \subset [Y]_{P\eta_n} \cap V_1^{\lambda'_n} \cap V_2^{\lambda'_n} \subset [Y]_{P\eta_n} \cap V^{\lambda_n},$$

the last equality coming from part (v) of Proposition 3.2.6. Thus we see that

$$V_1 \cap V_2 \subset V$$

as expected. □

Proposition 3.4.10 *Let $(X \subset Y \subset P)$ be a quasi-compact frame and $X = \cup_i X_i$ a finite open covering. If, for each i, V_i is a strict neighborhood of $]X_i[_P$ in $]Y[_P$, there exists a strict neighborhood V of $]X[_P$ in $]Y[_P$ and an admissible covering $V = \cup_i V'_i$ where V'_i is a strict neighborhood of $]X_i[_P$ in V_i.*

Proof By induction, it is sufficient to consider the case $X = X_1 \cup X_2$. Shrinking them if necessary, we may assume that V_1 and V_2 are the standard strict neighborhoods associated to two sequences $\eta_n \overset{<}{\to} 1$ and $\lambda_n \overset{<}{\to} 1$. Using assertion (iv) of Proposition 3.2.6, we see that the standard strict neighborhood of $]X[_P$ in $]Y[_P$ associated to the same sequences will work. □

Proposition 3.4.11 *Let $(X \subset Y \subset P)$ be a quasi-compact frame, V a strict neighborhood of $]X[_P$ in $]Y[_P$ and W a quasi-compact admissible open subset of $]Y[_P$. If $W \cap]X[_P = \emptyset$, there exists a strict neighborhood V' of $]X[_P$ in V such that $W \cap V' = \emptyset$.*

Proof It is sufficient to consider the case $V =]Y[_P$ since we may always intersect with V in the end. It follows from Proposition 3.3.3 that there exists λ_0 such that $W \cap V^{\lambda_0} = \emptyset$. We may therefore choose two sequences $\eta_n \overset{<}{\to} 1$ and $\lambda_0 \leq \lambda_n \overset{<}{\to} 1$ and take $V' := V_{\underline{\eta}}^{\underline{\lambda}}$. □

Theorem 3.4.12 *Let*

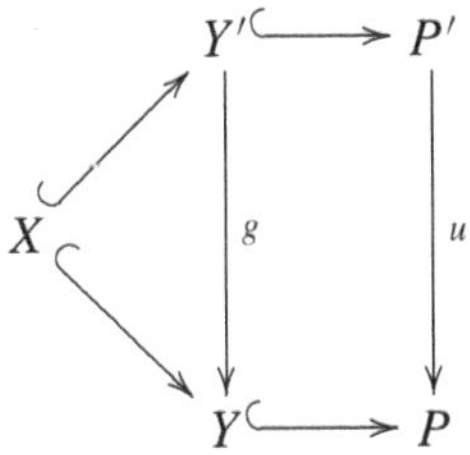

be a proper étale quasi-compact morphism of frames. Then, u_K induces an isomorphism between a strict neighborhood V' of $]X[_{P'}$ in $]Y'[_{P'}$ and a strict neighborhood V of $]X[_P$ in $]Y[_P$.

Proof The question is local on P that we may assume affine, and in particular quasi-compact. Also, using Proposition 3.1.10, we may assume that X is dense in Y'. Since u is étale in a neighborhood of X, the immersion $X \hookrightarrow u^{-1}(X)$ has an étale section. It is therefore an open (and closed) immersion. Since X is dense in Y, it follows that

$$g^{-1}(X) = Y \cap u^{-1}(X) - X$$

and our morphism is therefore cartesian.

Now, we know from Proposition 3.3.11 that given $\eta_n \stackrel{<}{\to} 1$, there exists $\delta_n \stackrel{<}{\to} 1$ and $\lambda_n \stackrel{<}{\to} 1$ such that, for each $n \in \mathbf{N}$, the morphism u_K induces an isomorphism

$$V'_n := u_K^{-1}([Y]_{P\eta_n}) \cap V'^{\lambda_n}_{\eta_n} = [Y']_{P'\delta_n} \cap u_K^{-1}(V^{\lambda_n}_{\eta_n}) \simeq V^{\lambda_n}_{\eta_n}.$$

Thus, if we let $V := V^{\underline{\lambda}}_{\underline{\eta}}$ and $V' := \cup_n V'_n$, we see that u_K induces an isomorphism $V' \simeq V$. And it remains to show that V' is a strict neighborhood of $]X[_{P'}$ in $]Y[_{P'}$.

Let us first show that $V' := \cup_n V'_n$ is an admissible open subset and that the covering is admissible. To start with, note that

$$]Y'[_{P'} = \cup_m \left([Y']_{P'\delta_m} \cap u_K^{-1}([Y]_{P\eta_m})\right)$$

is an admissible covering. It is therefore sufficient to show that for each m, the covering

$$\left([Y']_{P'\delta_m} \cap u_K^{-1}([Y]_{P\eta_m})\right) \cap V' = \cup_{n \leq m} \left([Y']_{P'\delta_m} \cap u_K^{-1}(V^{\lambda_m}_{\eta_n})\right)$$

is admissible. And we are done.

In order to show that V' is a strict neighborhood of $]X[_{P'}$ in $]Y[_{P'}$, we consider a quasi-compact admissible open subset W of $]Y'[_{P'}$. Since W is quasi-compact, there exists $N \in \mathbf{N}$ such that

$$W \subset [Y']_{P'\delta_N} \cap u_K^{-1}([Y]_{P\eta_N})$$

and it follows that

$$W \cap V'^{\lambda_N} = W \cap \left(u_K^{-1}([Y]_{P\eta_N}) \cap V'^{\lambda_N}_{\eta_N}\right) = W \cap V'_N \subset V'.$$

□

We can now state and prove the local form of the strong fibration theorem.

Corollary 3.4.13 *Let*

$$\begin{array}{ccccc} & & & & P' \\ & & & \nearrow & \downarrow \\ X & \hookrightarrow & Y & & \downarrow u \\ & & & \searrow & \downarrow \\ & & & & P \end{array}$$

be a smooth morphism of affine frames.

Let I' be the ideal of Y in P' and $X' := u^{-1}(X)$. Assume that there exist $t_1, \ldots, t_d \in I'$ inducing a basis $(\bar{t}_1, \ldots, \bar{t}_d)$ of the conormal sheaf of X in X'.

Then, the morphism

$$(t_1, \ldots, t_d) : P' \to \widehat{\mathbf{A}}^d_P$$

induces an isomorphism between a strict neighborhood V' of $]X[_{P'}$ in $]Y'[_{P'}$ and a strict neighborhood V'' of $]X[_P \times \mathbf{B}^d(0, 1^-)$ in $]Y[_P \times \mathbf{B}^d(0, 1^-)$.

Proof As we saw in Proposition 3.3.13, we have a (proper) étale morphism

$$\begin{array}{ccccc} & & & & P' \\ & & & \nearrow & \downarrow \\ X & \hookrightarrow & Y & & \downarrow u \\ & & & \searrow & \downarrow \\ & & & & \widehat{\mathbf{A}}^d_P \end{array}$$

of affine frames. We may then apply the theorem. □

We can even be a little more precise:

Corollary 3.4.14 *In the situation of the previous corollary, with our usual notations, we have:*

For all $\eta < 1$, there exists $\eta \leq \delta < \lambda_0 < 1$ such that for all $\lambda_0 \leq \lambda < 1$, u induces an isomorphism between an admissible open subset W_η^λ of in $]Y'[_{P'}$ and $V_\eta^\lambda \times B^d(0, \eta^+)$ with

$$V_\eta'^\lambda \subset W_\eta^\lambda \subset V_\delta'^\lambda.$$

In particular, if $\eta_n, \lambda_n \overset{<}{\to} 1$, the subsets

$$W_{\underline{\eta}}^{\underline{\lambda}} := \cup_n W_{\eta_n}^{\lambda_n}$$

are strict neighborhoods of $]X[_{P'}$ in $]Y'[_{P'}$ and for fixed $\underline{\eta}$, they form a cofinal system of strict neighborhoods.

Proof If we forget about δ, the first assertion follows from Proposition 3.3.2 which tells us that, with the notations of the corollary, for all $\eta < 1$, there exists $\lambda_0 < 1$ such that for all $\lambda_0 \leq \lambda < 1$, we have

$$V_\eta^\lambda \times B^d(0, \eta^+) \subset V''.$$

The fact that

$$V_\eta'^\lambda \subset W_\eta^\lambda \subset V_\delta'^\lambda$$

is a consequence of Proposition 3.3.14. Finally, the last assertion follows immediately from the analog assertion for the $V_\eta'^\lambda$'s. □

And we finish with the global form of the strong fibration theorem.

Corollary 3.4.15 *(Strong fibration theorem) Let*

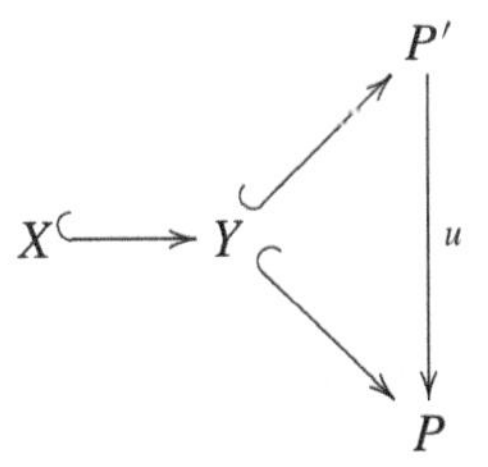

be a smooth morphism of frames. Then, locally on $(X \subset Y \subset P')$, there is an isomorphism between a strict neighborhood V' of $]X[_{P'}$ in $]Y'[_{P'}$ and a strict neighborhood V'' of $]X[_P \times \mathbf{B}^d(0, 1^-)$ in $]Y[_P \times \mathbf{B}^d(0, 1^-)$.

Proof Follows from Proposition 3.3.13 since locally on $(X \subset Y \subset P')$, we may assume that P' is affine and X is the complement of a hypersurface in Y. □

4

Calculus

We fix a complete ultrametric field K with $\mathcal{V}$, k and π as usual.

4.1 Calculus in rigid analytic geometry

Before doing anything else, we need to recall some basic notions from differential calculus on rigid analytic varieties. Everything we say below has an exact analog in complex analytic geometry as well as in algebraic geometry.

There are four different ways to describe the main object of our study: as a crystal, as a module with a stratification, as a differential module, and finally, as a module with an integrable connection.

We fix a morphism of rigid analytic varieties $p : V \to T$.

4.1.1 Crystals and stratifications

From a theoretical point of view, the crystalline approach is the best way to introduce calculus. A *(finitely presented) infinitesimal crystal on* V/T is a family of (coherent) $\mathcal{O}_{V'}$-modules $\mathcal{E}_{V'}$ for each nilpotent immersion $V_0' \hookrightarrow V'$ (also called a *thickening*) over T, where V_0' is a rigid analytic variety over V:

$$\begin{array}{ccc} V_0' & \hookrightarrow & V' \\ \downarrow & & \downarrow \\ V & \longrightarrow & T \end{array}$$

Moreover, with obvious notations, for any morphism $u : V'' \to V$ over T that induces a morphism $V_0'' \to V_0'$ over V, we want an isomorphism $\varphi_u : u^*\mathcal{E}_{V'} \simeq \mathcal{E}_{V''}$. These isomorphisms should be compatible in the sense that

$$\varphi_{v\circ u} = \varphi_v \circ v^*(\varphi_u)$$

whenever this has a meaning. This is called the *cocycle condition*. A morphism of crystals is simply a compatible family of morphisms of $\mathcal{O}_{V'}$-modules $\mathcal{E}_{V'} \to \mathcal{F}_{V'}$. We get an abelian category with exact fibers $\mathcal{E} \mapsto \mathcal{E}_{V'}$ which is functorial in V/T and also with respect to isometric extensions of K. Note that we get an equivalent category if we consider only thickenings V' where V'_0 is an affinoid open subset of V. Finally, if $\mathcal{E}$ and $\mathcal{F}$ are two crystals on V/T, the family of

$$\mathcal{E}_{V'} \otimes_{\mathcal{O}_{V'}} \mathcal{F}_{V'},$$

also define crystal. The same is true with the family of

$$\mathcal{H}om_{\mathcal{O}_{V'}}(\mathcal{E}_{V'}, \mathcal{F}_{V'})$$

when the crystal $\mathcal{E}$ is finitely presented.

We now turn to stratifications. Let

$$\delta : V \hookrightarrow V \times_T V$$

be the diagonal embedding. We will write $V^{(n)}$ for the n-th infinitesimal neighborhood of V in $V \times_T V$ (defined by $\mathcal{I}^{n+1}$ if V is defined by $\mathcal{I}$) and

$$p_1^{(n)}, p_2^{(n)} : V^{(n)} \to V$$

for the projections (which are homeomorphisms). We will always consider $V \times_T V$ as well as $V^{(n)}$ as rigid analytic varieties over V using the first projection.

A *stratification* on an $\mathcal{O}_V$-module $\mathcal{E}$ is a compatible sequence of linear isomorphisms called the *Taylor isomorphisms*

$$\{\epsilon^{(n)} : p_2^{(n)*}\mathcal{E} \simeq p_1^{(n)*}\mathcal{E}\}_{n\in\mathbf{N}}$$

on $V^{(n)}$ with $\epsilon^{(0)} = Id_{\mathcal{E}}$ that satisfy a cocycle condition on triple products. A morphism of stratified modules is a morphism of $\mathcal{O}_V$-modules that is compatible with the data. Stratified $\mathcal{O}_V$-modules form an abelian category with exact and faithful forgetful functor. Moreover, this construction is functorial in V/T and also with respect to isometric extensions of K. The tensor product of two stratified modules has a canonical stratification and the same is true for the internal Hom if the first module is coherent.

If $\mathcal{E}$ is a stratified module, we will also consider the *Taylor morphisms*

$$\theta^{(n)} := \epsilon^{(n)} \circ p_2^{(n)*} : \mathcal{E} \to p_1^{(n)*}\mathcal{E}$$

which are semi-linear with respect to the second projection $p_2^{(n)*}$.

It will also be necessary (and convenient) to take inverse limits and look at

$$\widehat{\epsilon} := \varprojlim \epsilon^{(n)} : \varprojlim p_2^{(n)*}\mathcal{E} \simeq \varprojlim p_1^{(n)*}\mathcal{E}$$

as well as

$$\widehat{\theta} := \varprojlim \theta^{(n)} : \mathcal{E} \to \varprojlim p_1^{(n)*}\mathcal{E}$$

which is semi-linear with respect to

$$\widehat{p}_2^* : \mathcal{O}_V \to \varprojlim \mathcal{O}_{V^{(n)}}.$$

Any crystal $\mathcal{E}$ defines a stratified module $\mathcal{E} = \mathcal{E}_V$ with Taylor isomorphisms given by the composite map

$$p_2^{(n)*}\mathcal{E} \simeq \mathcal{E}_{V^{(n)}} \simeq p_1^{(n)*}\mathcal{E}.$$

We obtain a functor from the category of crystals on V/T to the category of stratified modules on V/T.

4.1.2 Connections and differential modules

With the same notations as above, the sheaf of differential operators on V/T can be defined as

$$\mathcal{D}_{V/T} := \varinjlim \mathcal{D}_{n,V/T}.$$

with

$$\mathcal{D}_{n,V/T} = \mathcal{H}om_{\mathcal{O}_V}(\mathcal{O}_{V^{(n)}}, \mathcal{O}_V).$$

Also, by definition, there is an exact sequence

$$0 \to \Omega^1_{V/T} \to \mathcal{O}_{V^{(1)}} \to \mathcal{O}_V \to 0$$

and

$$d := p_2^{(1)*} - p_1^{(1)*} : \mathcal{O}_V \to \Omega^1_{V/T} \subset \mathcal{O}_{V^{(1)}}$$

makes $\Omega^1_{V/T}$ universal for T-derivations into coherent modules. A *connection* on an $\mathcal{O}_V$-module $\mathcal{E}$ is an $\mathcal{O}_T$-linear map

$$\nabla : \mathcal{E} \to \mathcal{E} \otimes_{\mathcal{O}_V} \Omega^1_{V/T}$$

satisfying the *Leibnitz rule*

$$\nabla(fs) = f\nabla(s) + s \otimes df.$$

A *horizontal* map is an $\mathcal{O}_V$-linear map compatible with the connections. Modules with a connection form an abelian category with exact and faithful forgetful functor. Moreover, this construction is functorial in V/T as well as in K. Also, if $\mathcal{E}, \mathcal{F}$, are two $\mathcal{O}_V$-modules with a connection, then $\mathcal{E} \otimes_{\mathcal{O}_V} \mathcal{F}$ is endowed with the connection

$$\nabla(s \otimes t) = \nabla(s) \otimes t + s \otimes \nabla(t).$$

Also, when $\mathcal{E}$ is coherent, $\mathcal{H}om_{j_X^\dagger \mathcal{O}_{]Y[_P}}(\mathcal{E}, \mathcal{F})$ is endowed with

$$\nabla(\varphi)(s) = \nabla(\varphi(s)) - (\varphi \otimes \mathrm{Id}_{\Omega_{V/T}})\nabla(s).$$

Any stratification ϵ on a coherent $\mathcal{O}_V$-module $\mathcal{E}$ induces a connection

$$\nabla = \theta^{(1)} - p_1^{(1)*} : \mathcal{E} \to \mathcal{E} \otimes_{\mathcal{O}_V} \Omega^1_{V/T} \subset p_1^{(1)*}\mathcal{E}$$

and this is functorial again.

In order to define the notion of integrability, it is necessary to introduce $\Omega^2_{V/T} = \Lambda^2 \Omega^1_{V/T}$ and describe $d : \Omega^1_{V/T} \to \Omega^2_{V/T}$. The more natural way to do that is to consider for $i = 1, 2, 3$, the projection

$$p_i : V \times_T V \times_T V \to V \times_T V$$

that forgets the i-th factor. Using the identification

$$V \times_T V \times_T V \simeq (V \times_T V) \times_V (V \times_T V),$$

we get maps

$$p_i^{(1)} : V^{(1)} \times_V V^{(1)} \to V^{(1)}$$

and d is the composition of

$$p_3^{(1)*} - p_2^{(1)*} + p_1^{(1)*} : \Omega^1_{V/T} \to \Omega^1_{V/T} \otimes_{\mathcal{O}_V} \Omega^1_{V/T}$$

with the projection

$$\Omega^1_{V/T} \otimes_{\mathcal{O}_V} \Omega^1_{V/T} \to \Omega^2_{V/T}.$$

Then we can define *integrability* in the usual way. More precisely, using Leibnitz rule again, a connection ∇ on a module $\mathcal{E}$ extends to a morphism

$$\nabla : \mathcal{E} \otimes_{\mathcal{O}_V} \Omega^1_{V/T} \to \mathcal{E} \otimes_{\mathcal{O}_V} \Omega^2_{V/T}$$

and the connection is *integrable* if the *curvature*

$$\nabla^2 : \mathcal{E} \to \mathcal{E} \otimes_{\mathcal{O}_V} \Omega^2_{V/T}$$

is zero. This is automatic if the connection comes from a stratification or if $\dim V/T \leq 1$.

4.1.3 The smooth case

Assume now that V is smooth over T.

Then, the above functor from crystals to stratified modules is an equivalence of categories. More precisely, given a stratified module $\mathcal{E}$, it is sufficient to define $\mathcal{E}_{V'}$ (and the isomorphisms φ_u) when V' is an affinoid variety. But since V is smooth over T, the morphism $V'_0 \to V$ lifts to a morphism $h : V' \to V$ and we may set $\mathcal{E}_V := h^*\mathcal{E}$. If we are given $u : V'' \to V'$ and some lifting $h' : V'' \to V$ of the structural map $V''_0 \to V$, we may consider the morphism

$$(h \circ u, h') : V'' \to V \times_T V.$$

It factors through some $V^{(n)}$ and we may pull back $\epsilon^{(n)}$ in order to get

$$\varphi_u : u^*(h^*\mathcal{E}) \simeq h'^*\mathcal{E}.$$

Note also that, since V is smooth over T, $\mathcal{D}_{V/T}$ has a natural structure of $\mathcal{O}_V$-algebra. It is not difficult to check that a stratification on an $\mathcal{O}_V$-module is equivalent to a left $\mathcal{D}_{V/T}$-module structure. And it follows that the category of crystals which is equivalent to the category of modules with stratifications is also equivalent to the category of $\mathcal{D}_{V/T}$-modules. As an example, if P is any differential operator on V, then

$$\mathcal{E} := \mathcal{D}_{V/T}/\mathcal{D}_{V/T}P$$

is a stratified module and inherits a integrable connection.

If V is still smooth and moreover, $\operatorname{Char} K = 0$, then $\mathcal{D}_{V/T}$ acts faithfully on $\mathcal{O}_V$ and one can inductively identify $\mathcal{D}_{n,V/T}$ with

$$\{P \in \mathcal{E}nd_{\mathcal{O}_T}(\mathcal{O}_V), \forall f \in \mathcal{O}_V, [P, f] \in \mathcal{D}_{n-1,V/T}\}.$$

In this situation, it is equivalent to give a $\mathcal{D}_{V/T}$-module structure (or a stratification) or an integrable connection on an $\mathcal{O}_V$-module $\mathcal{E}$. More precisely the four categories are all equivalent.

For further reference, we give a local description of these equivalences (assuming V smooth and $\operatorname{Char} K = 0$). Assume that we have étale coordinates $t_1, \ldots, t_d \in \Gamma(V, \mathcal{O}_V)$. It means that the induced morphism

$$(t_1, \ldots, t_d) : V \to \mathbf{A}_T^{d,\mathrm{rig}}$$

is étale, or equivalently, since V is smooth, that $\Omega^1_{V/T}$ is free with basis $dt_1, \ldots, dt_d$. For each $i = 1, \ldots, n$, we set

$$\tau_i := p_2^* t_i - p_1^* t_i \in \mathcal{O}_{V \times_T V}.$$

Then, if we use standard multi-index notations, the morphism

$$(\underline{\tau}) = (\tau_1, \ldots, \tau_d) : V \times_T V \to V \times_T \mathbf{A}_T^{d,\mathrm{rig}} = V \times_K \mathbf{A}_K^{d,\mathrm{rig}}$$

is also étale and induces an isomorphism

$$V^{(n)} \simeq V \times_K \mathrm{Spm} K[\underline{\tau}]/(\underline{\tau})^n.$$

We will implicitly identify these two analytic varieties. In particular, we see that

$$\mathcal{O}_{V^{(n)}} = \mathcal{O}_V[\underline{\tau}]/(\underline{\tau})^n \quad \text{and} \quad \varprojlim \mathcal{O}_{V^{(n)}} = \mathcal{O}_V[[\underline{\tau}]].$$

Note that, since $\Omega^1_{V/T}$ is free on the images $dt_1, \ldots, dt_d$ of $\tau_1, \ldots, \tau_d$, the ring $\mathcal{D}_{V/T}$ is the subalgebra of $\mathcal{E}nd_{\mathcal{O}_T}(\mathcal{O}_V)$ generated by the dual basis $\partial/\partial t_1, \ldots, \partial/\partial t_d$. Finally, if for all $\underline{k}$, we write

$$\underline{\partial}^{[\underline{k}]} := \frac{1}{\underline{k}!}\underline{\partial}^{\underline{k}}/\underline{\partial t}^{\underline{k}} \in \mathcal{D}_{V/T},$$

then the inverse image with respect to the second projection is

$$\begin{array}{cccc} \widehat{p}_2^* : & \mathcal{O}_V & \longrightarrow & \mathcal{O}_V[[\tau]] \\ & f & \longmapsto & \sum \underline{\partial}^{[\underline{k}]}(f)\underline{\tau}^{\underline{k}} \end{array}.$$

Now, if $\mathcal{E}$ is a module with an integrable connection, we have

$$\nabla(s) = \sum_{i=1}^{n} \partial/\partial t_i(s) \otimes dt_i.$$

Moreover, if $\mathcal{E}$ is coherent so that

$$\varprojlim p_1^{(n)*}\mathcal{E} \simeq \mathcal{E} \otimes_{\mathcal{O}_V} \mathcal{O}_V[[\underline{\tau}]].$$

then, we have

$$\theta(s) = \sum \underline{\partial}^{[\underline{k}]}(s) \otimes \underline{\tau}^{\underline{k}} \quad \text{and} \quad \epsilon(f \otimes s) = \sum f\underline{\partial}^{[\underline{k}]}(s) \otimes \underline{\tau}^{\underline{k}}$$

and by symmetry,

$$\epsilon^{-1}(s \otimes f) = \sum (-1)^{|\underline{k}|}\underline{\tau}^{\underline{k}} \otimes f\underline{\partial}^{[\underline{k}]}(s).$$

In practice, we will stick to modules with integrable connection, using the other interpretations whenever it is convenient. We will denote by $\mathrm{MIC}(V/T)$ the category of coherent $\mathcal{O}_V$-modules with an integrable connection. This is an abelian category with unit, tensor product and internal Hom. We will generally assume that V is smooth and $\mathrm{Char} K = 0$.

4.1.4 Cohomology

We will always use the vocabulary of derived categories and derived functors. In general, the reader can easily translate into usual cohomological language by writing

$$R^i FM := H^i(RFM)$$

and replacing any exact triangle

$$M' \to M \to M'' \to$$

by a long exact sequence

$$\cdots \to H^i(M') \to H^i(M) \to H^i(M'') \to H^{i+1}(M) \to \cdots.$$

To be more precise, if $M \to I$ is an injective resolution of a complex (bounded below) and F a left exact functor, then $RFM := FI$ (dually, one can also define LG for a right exact functor). Of course, this is only defined up to a quasi-isomorphism. We should also recall that any exact sequence of complexes gives rise to an exact triangle, that any exact triangle induces a long exact sequence as above and that derived functors preserve exact triangles.

We now come back to differential calculus. For each approach of the problem – crystals, stratifications, $\mathcal{D}$-modules, or connections – there is a different way to describe the cohomology of our objects.

To start with, a crystal is a particular case of a sheaf on the infinitesimal site. More precisely, the infinitesimal site $\mathrm{INF}(V/T)$ is the category of thickenings $V_0' \hookrightarrow V'$ over T with V_0 a rigid analytic variety over V. A family of morphisms $\{V_i' \to V'\}$ in $\mathrm{INF}(V/T)$ is called a covering if each map is an open immersion and they define an admissible covering of V'. If $u : V \to W$ is a morphism of rigid analytic varieties over T and $\mathrm{rig}(W)$ denotes the rigid site of W (admissible open subsets), there is a canonical morphism of sites

$$u_{INF} : \mathrm{INF}(V/T) \to \mathrm{rig}(W)$$

and the relative cohomology of a crystal $\mathcal{E}$ is simply $Ru_{INF*}\mathcal{E}$. We shall not use this notion in the future.

We now turn to the stratification case. I do not want to go into the details, but for each $n \in \mathbf{N}$, we can build the *Cech–Alexander complex* of $\mathcal{E}$

$$\mathrm{CA}^{(n)}(\mathcal{E}) := [\mathcal{E} \to p_1^{(n)*}\mathcal{E} \to \cdots]$$

with

$$d^0 = \epsilon_n \circ p_2^{(n)*} - p_1^{(n)*}, \quad \text{etc.}$$

and compute $Ru_*\mathrm{CA}^{(n)}(\mathcal{E})$. Again, we shall not use this approach in the future.

When V is smooth over T, for a (complex of) $\mathcal{D}_{V/T}$-module $\mathcal{E}$, the relative cohomology is simply given by

$$Ru_*R\mathcal{H}om_{\mathcal{D}_{V/T}}(\mathcal{O}_V, \mathcal{E}).$$

We finish with the most important case for us, the cohomology of a module with an integrable connection, also called *de Rham cohomology*. There is no need to assume that V is smooth over T. First of all, using the same method as above, we may always define the *de Rham complex*

$$\mathcal{O}_V \to \Omega^1_{V/T} \to \Omega^2_{V/T} \to \cdots.$$

If $\mathcal{E}$ is a module with an integrable connection on V, one may use the Leibnitz rule in order to get the de Rham complex

$$\mathcal{E} \to \mathcal{E} \otimes_{\mathcal{O}_V} \Omega^1_{V/T} \to \mathcal{E} \otimes_{\mathcal{O}_V} \Omega^2_{V/T} \to \cdots$$

of $\mathcal{E}$ and compute its *relative de Rham cohomology*

$$Ru_{\mathrm{dR}}\mathcal{E} := Ru_*(\mathcal{E} \otimes_{\mathcal{O}_V} \Omega^\bullet_{V/T}).$$

If $p : V \to T$ denotes the structural map, we will sometimes write

$$\mathcal{H}^i_{\mathrm{dR}}(V/T, \mathcal{E}) := R^i p_{\mathrm{dR}}\mathcal{E}.$$

Also, in the case $\mathcal{E} = \mathcal{O}_V$ with the trivial connection, we will simply write

$$\mathcal{H}^i_{\mathrm{dR}}(V/T) := R^i p_{\mathrm{dR}}\mathcal{O}_V.$$

Actually, when $W = T = \mathrm{Spm} K$, we will write $R\Gamma_{\mathrm{dR}}(V/K, \mathcal{E})$ and for each $i \in \mathbf{N}$,

$$H^i_{\mathrm{dR}}(V/K, \mathcal{E}) := R^i\Gamma_{\mathrm{dR}}(V/K, \mathcal{E}).$$

Again, we will drop $\mathcal{E}$ from the notation when $\mathcal{E}$ is just the trivial module with the trivial connection.

These different cohomology theories coincide as long as $\mathcal{E}$ is coherent, V is smooth over T and $\mathrm{Char} K = 0$. We can make this more precise. When V is smooth over T, then the trivial thickening V is a covering of the final object of $\mathrm{INF}(V/T)$. And its self product is ind-representable by $\varinjlim V^{(n)}$. If $\mathcal{E}$ is coherent, inverse limits behave well, and we have

$$Ru_{INF*}\mathcal{E} = Ru_* \varprojlim \mathrm{CA}^{(n)}(\mathcal{E}).$$

It should be possible to extend this to more general crystals (and even arbitrary sheaves) by introducing $R\varprojlim$. The analog problem in the algebraic case is dealt with in the appendix of [9].

On the other hand, using the *Spencer complex*, one can show as in Proposition 1.1.2 of [24], that if V is smooth over T, then

$$Ru_{\mathrm{dR}}\mathcal{E} = Ru_*R\mathcal{H}om_{\mathcal{D}_{V/T}}(\mathcal{O}_V, \mathcal{E})$$

Using Grothendieck's method of linearization of differential operators, one can also show that the first approach (cohomology of crystals or Cech–Alexander cohomology) and the second one (de Rham cohomology of cohomology of $\mathcal{D}$-modules) also coincide. This is beyond the scope of this course and will not be used. Actually, we will mainly stick to the de Rham approach.

It is important to note that de Rham cohomology is functorial in the sense that if

$$\begin{array}{ccccc} V' & \xrightarrow{u'} & W' & \longrightarrow & T' \\ \downarrow{\scriptstyle v'} & & \downarrow{\scriptstyle v} & & \downarrow{\scriptstyle u} \\ V & \xrightarrow{u} & W & \longrightarrow & T \end{array}$$

is a commutative diagram, there is a canonical morphism

$$Lv^*Ru_{\mathrm{dR}}\mathcal{E} \to Ru'_{\mathrm{dR}}v'^*\mathcal{E}.$$

This morphism is built by using the composite

$$v^{-1}Ru_*(\mathcal{E} \otimes_{\mathcal{O}_V} \Omega^\bullet_{V/T}) \to Ru'_*v'^{-1}(\mathcal{E} \otimes_{\mathcal{O}_V} \Omega^\bullet_{V/T})$$

of and

$$Ru'_*v'^{-1}(\mathcal{E} \otimes_{\mathcal{O}_V} \Omega^\bullet_{V/T}) \to Ru'_*(v'^*\mathcal{E} \otimes_{\mathcal{O}_{V'}} \Omega^\bullet_{V'/T'}),$$

and extending it with the functor $\mathcal{O}_{W'} \otimes^L_{v^{-1}\mathcal{O}_W} -$.

Note also that de Rham cohomology is also functorial with respect to isometric embeddings $\sigma : K \hookrightarrow K'$ in the sense that there is a canonical map

$$(Ru_{\mathrm{dR}}\mathcal{E})^\sigma \to Ru^\sigma_{\mathrm{dR}}\mathcal{E}^\sigma.$$

Finally, we should say a few words about the *Gauss–Manin* construction. We use filtered derived categories but the reader who wishes so can translate into spectral sequences. If $v : V' \to V$ is a smooth morphism of smooth rigid analytic varieties over T, there is a short exact sequence of locally free sheaves

$$0 \to v^*\Omega^1_{V/T} \to \Omega^1_{V'/T} \to \Omega^1_{V'/V} \to 0$$

and it follows that for each $i \in \mathbf{N}$, $\Omega^i_{V'/T}$ has a filtration by locally free subsheaves

$$Fil^k\Omega^i_{V'/T} := \mathrm{Im}\left(\Omega^{i-k}_{V'/T} \otimes_{\mathcal{O}_{V'}} v^*\Omega^k_{V/T} \to \Omega^i_{V'/T}\right)$$

and that

$$Gr^k\Omega^i_{V'/T} = \Omega^i_{V'/V} \otimes_{\mathcal{O}_{V'}} v^*\Omega^k_{V/T}.$$

It follows that, if $\mathcal{E}'$ is an $\mathcal{O}_{V'}$-module with an overconvergent integrable connection, the de Rham complex

$$K^\bullet := \mathcal{E}' \otimes_{\mathcal{O}_{V'}} \Omega^\bullet_{V'/T}$$

comes with its *Gauss–Manin* filtration

$$Fil^k K^\bullet = \mathrm{Im}\left(K^\bullet[-k] \otimes_{\mathcal{O}_{V'}} v^*\Omega^k_{V/T} \to K^\bullet\right)$$

such that

$$Gr^k K^\bullet = L^\bullet[-k] \otimes_{\mathcal{O}_{V'}} v^*\Omega^k_{V/T}$$

with

$$L^\bullet := \mathcal{E}' \otimes_{\mathcal{O}_{V'}} \Omega^\bullet_{V'/V}$$

(de Rham complex of $\mathcal{E}'$ relative to V). Applying Rv_*, we see that $Rv_{\mathrm{dR}}\mathcal{E}'$ has a canonical filtration with

$$\begin{aligned} Gr^k Rv_{\mathrm{dR}}\mathcal{E}' &= Rv_*(L^\bullet[-k] \otimes_{\mathcal{O}_{V'}} v^*\Omega^k_{V/T}) \\ &= Rv_*L^\bullet[-k] \otimes_{\mathcal{O}_{V'}} \Omega^k_{V/T} = Rv_{\mathrm{dR}}\mathcal{E}'_{|V'}[-k] \otimes_{\mathcal{O}_{V'}} \Omega^k_{V/T}. \end{aligned}$$

It follows in particular that there is a spectral sequence

$$E_1^{p,q} = \mathcal{H}^q_{\mathrm{dR}}(V'/V, \mathcal{E}') \otimes_{\mathcal{O}_{V'}} \Omega^p_{V/T} \Rightarrow R^{p+q} v_{\mathrm{dR}}\mathcal{E}'$$

inducing a connection on $\mathcal{H}^q_{\mathrm{dR}}(V'/V, \mathcal{E}')$. If we translate into the language of differential modules, we obtain the so-called *Picard–Fuchs differential equations*.

4.2 Examples

We will consider here some very simple examples of low rank modules with connection on low dimension rigid analytic varieties and compute their cohomology. We assume that $\mathcal{V}$ has mixed characteristic p. In other words, we assume $\mathrm{Char}K = 0$ and $\mathrm{Char}k = p > 0$.

4.2.1 The Dwork module $\mathcal{L}$

We start with a first example coming from Dwork theory, and giving rise to the so-called *Dwork crystal*. It is also called the *Artin–Schreier* or *exponential* or *additive module*. We may consider for $\alpha \in K$, the free rank one module $\mathcal{L}_\alpha$ on a disk $\mathbf{D}(0, \rho^\pm)$ on some generator θ whose connection is given by

$$\nabla(\theta) = -\alpha\theta \otimes dt.$$

In general, we assume $\rho \in |K^*| \otimes \mathbf{Q} \subset \mathbf{R}_{>0}$ and $\rho \geq 0$ but we will also allow $\mathbf{D}(0, \infty^-) := \mathbf{A}_K^{1,\mathrm{rig}}$.

We have $\partial(\theta) = -\alpha\theta$, so that $\partial^k(\theta) = (-\alpha)^k\theta$ and $\partial^{[k]}(\theta) = (-\alpha)^{[k]}\theta$ where we use the divided power notation

$$x^{[k]} = \frac{x^k}{k!}.$$

Thus, we see that

$$\begin{aligned}\epsilon(1 \otimes \theta) &= \sum \partial^{[k]}(\theta) \otimes \tau^k = \sum (-\alpha)^{[k]}\theta \otimes \tau^k = \theta \otimes \sum (-\alpha)^{[k]}\tau^k \\ &= \theta \otimes \sum (-\alpha\tau)^{[k]} = \theta \otimes \exp -\alpha\tau = (\exp -\alpha\tau)(\theta \otimes 1).\end{aligned}$$

Note also that there is a short exact sequence of $\mathcal{D}$-modules

$$\begin{array}{ccccccccc} 0 & \longrightarrow & \mathcal{D} & \xrightarrow{\partial+\alpha} & \mathcal{D} & \longrightarrow & \mathcal{L}_\alpha & \longrightarrow & 0 \\ & & & & 1 & \longmapsto & \theta & & \end{array}$$

giving an isomorphism $\mathcal{L}_\alpha \simeq \mathcal{D}/(\partial + \alpha)$.

We want to compute the de Rham cohomology of $\mathcal{L}_\alpha$ on $\mathbf{D}(0, \rho^\pm)$. If we write

$$A := \Gamma(\mathbf{D}(0, \rho^\pm), \mathcal{O}) \subset K[[t]],$$

we have to compute the cohomology of the de Rham complex

$$[A\theta \xrightarrow{\nabla} A\theta \otimes dt].$$

If $f := \sum a_k t^{[k]} \in A$, we have

$$\begin{aligned}\nabla(f\theta) &= \partial(f)\theta \otimes dt + f\nabla(\theta) = \sum a_k t^{[k-1]}\theta \otimes dt - \alpha f\theta \otimes dt \\ &= (\sum a_{k+1}t^{[k]} - \alpha \sum a_k t^{[k]})\theta \otimes dt = \sum (a_{k+1} - \alpha a_k)t^{[k]}\theta \otimes dt.\end{aligned}$$

We may therefore identify $H^0_{\mathrm{dR}}(\mathbf{D}(0, \rho^\pm), \mathcal{L}_\alpha)$ and $H^1_{\mathrm{dR}}(\mathbf{D}(0, \rho^\pm), \mathcal{L}_\alpha)$ with the kernel and the cokernel of the morphism

$$\begin{array}{ccc} A & \xrightarrow{D} & A \\ \sum a_k t^{[k]} & \longmapsto & \sum (a_{k+1} - \alpha a_k)t^{[k]} \end{array}.$$

We see that $f \in \ker D$ if and only if for each k, we have $a_{k+1} = \alpha a_k$ which means that $a_k = \alpha^k a_0$. Thus we have

$$f = \sum \alpha^k a_0 t^{[k]} = a_0 \sum (\alpha t)^{[k]} = a_0 \exp \alpha t.$$

It follows that $H^0_{\mathrm{dR}}(\mathbf{D}(0, \rho^\pm), \mathcal{L}_\alpha) = 0$ unless $\exp \alpha t$ converges on $\mathbf{D}(0, \rho^\pm)$. Since for all $k \in \mathbf{N}$, we have

$$|p|^{\frac{1}{p-1}} \le |k!|^{1/k} \le |p|^{\frac{1}{p-1} - \frac{\log_p k}{k}},$$

we see that the convergence of $\exp \alpha t$ is satisfied if and only if

$$\frac{|\alpha|\rho}{|p|^{\frac{1}{p-1}}} < 1$$

in the case of the closed disk and

$$\frac{|\alpha|\rho}{|p|^{\frac{1}{p-1}}} \le 1$$

in the open disk case. And when $\exp \alpha t$ converges, we get an isomorphism with the trivial module with connection:

$$\mathcal{O}_{\mathbf{D}(0,\rho^\pm)} \simeq \mathcal{L}_\alpha, \, 1 \mapsto (\exp \alpha t)\theta.$$

In other words, in this case, we may assume that $\alpha = 0$. Anyway, at this point we are done with the computation of H^0 and we have

$$\dim H^0_{\mathrm{dR}}(\mathbf{D}(0, \rho^\pm), \mathcal{L}_\alpha) = \begin{cases} 1 & \text{if} \quad |\alpha|\rho \le |p|^{\frac{1}{p-1}} \quad (\text{or } <, + \text{case}) \\ 0 & \text{otherwise} \end{cases}$$

Let us turn now to the computation of $H^1_{\mathrm{dR}}(\mathbf{D}(0, \rho^\pm), \mathcal{L}_\alpha)$.

Note first that, when $|\alpha|\rho < |p|^{\frac{1}{p-1}}$, then $H^1_{\mathrm{dR}}(\mathbf{D}(0, \rho^+), \mathcal{L}_\alpha)$ is infinite dimensional. Of course, it is sufficient to consider the case of the trivial module with the trivial connection on the unit disk. Then, the functions

$$g_n := \sum_{i \ge n} p^i t^{p^i - 1}$$

form an infinite family of linearly independent functions that cannot be integrated.

Now, in order to compute $H^1_{\mathrm{dR}}(\mathbf{D}(0, \rho^-), \mathcal{L}_\alpha)$ when $|\alpha|\rho \le |p|^{\frac{1}{p-1}}$, it is sufficient to look at the case $\alpha = 0$. If $g := \sum_k b_k t^{[k]} \in A$, then

$$I(g) := \int g dt := \sum_k b_{k-1} t^{[k]} \in A$$

and we get a section I to our map D. Thus, we see that

$$H^1_{\mathrm{dR}}(\mathbf{D}(0, \rho^-), \mathcal{L}_\alpha) = 0.$$

Of course, if we do not want to assume that $\alpha = 0$, we may use

$$I(g) := \int g \exp(\alpha t) dt.$$

How about the remaining cases. Actually, we can show that

$$H^1_{\mathrm{dR}}(\mathbf{D}(0, \rho^-), \mathcal{L}_\alpha) = 0$$

also when $|\alpha|\rho > |p|^{\frac{1}{p-1}}$. Of course, it is sufficient to describe an inverse I to D. Reversing

$$D(t^{[k]}) = t^{[k-1]} - \alpha t^{[k]},$$

gives

$$I(t^{[k]}) = -\frac{1}{\alpha}(t^{[k]} - I(t^{[k-1]})),$$

and, by induction,

$$I(t^{[k]}) = -\frac{1}{\alpha}\sum_{i \leq k} \frac{1}{\alpha^{k-i}} t^{[i]}.$$

We can compute the norm of $I(t^{[k]})$ on $\mathbf{D}(0, \rho'^+)$ when

$$\frac{1}{|\alpha|}|p|^{\frac{1}{p-1}} \leq \rho' < \rho.$$

We have

$$\|I(t^{[k]})\|' = \frac{1}{|\alpha|} \sup_{i \leq k} \frac{1}{|\alpha|^{k-i}} \frac{1}{|i!|} \rho'^i$$

and since

$$\frac{1}{|\alpha|^{k-i}} \leq \frac{\rho'^{k-i}}{|p|^{\frac{k-i}{p-1}}}$$

we obtain

$$\|I(t^{[k]})\|' = \leq \frac{1}{|\alpha|} \sup_{i \leq k} \frac{\rho'^{k-i}}{|p|^{\frac{k-i}{p-1}}} \frac{1}{|p|^{\frac{i}{p-1}}} \rho'^i = \frac{1}{|\alpha|} \frac{\rho'^k}{|p|^{\frac{k}{p-1}}}.$$

If $g := \sum_k b_k t^{[k]} \in A$, then for $\rho'' < \rho$, we have

$$\frac{|b_k|\rho''^k}{|k!|} \to 0.$$

Our estimates for factorials give

$$\frac{1}{|p|^{\frac{k}{p-1}}}|p|^{\log_p(k)} \leq \frac{1}{|k!|}$$

and it follows that, if we choose $\rho'' > \rho'$, we have

$$\frac{1}{|\alpha|}\frac{|b_k|\rho''^k}{|p|^{\frac{k}{p-1}}} \to 0$$

also. Thus we see that

$$I(g) := \sum_k b_k I(t^{[k]}) \in A$$

and we are done.

To summarize, we see that we almost always have

$$H^i_{\mathrm{dR}}(\mathbf{D}(0, \rho^{\pm}), \mathcal{L}_\alpha) = 0.$$

The only exceptions are the following:

$$\begin{cases} \dim_K H^0_{\mathrm{dR}}(\mathbf{D}(0, \rho^+), \mathcal{L}_\alpha) = 1 \\ \dim_K H^1_{\mathrm{dR}}(\mathbf{D}(0, \rho^+), \mathcal{L}_\alpha) = \infty \end{cases} \quad \text{if} \quad |\alpha|\rho < |p|^{\frac{1}{p-1}},$$

$$\dim_K H^0_{\mathrm{dR}}(\mathbf{D}(0, \rho^-), \mathcal{L}_\alpha) = 1 \quad \text{if} \quad |\alpha|\rho \leq |p|^{\frac{1}{p-1}}$$

and

$$\dim_K H^1_{\mathrm{dR}}(\mathbf{D}(0, \rho^+), \mathcal{L}_\alpha) = ? \quad \text{if} \quad |\alpha|\rho \geq |p|^{\frac{1}{p-1}}.$$

Note that we did not actually check the case of $H^1_{\mathrm{dR}}(\mathbf{D}(0, \rho^+), \mathcal{L}_\alpha)$ when $|\alpha|\rho = |p|^{\frac{1}{p-1}}$.

This first approach shows that de Rham cohomology looks fine on "open" varieties but that it can be bad on affinoid varieties. The next example will teach us that life is not even that simple.

4.2.2 The Kummer module $\mathcal{K}$

We turn now to another example coming from Dwork theory, the one that gives rise to *Kummer* or *multiplicative crystals*. First of all, recall that

$$\mathbf{A}(0, \eta^{\pm}, \rho^{\pm}) := \mathbf{D}(0, \rho^{\pm}) \setminus \mathbf{D}(0, \eta^{\mp})$$

denotes the annulus of radii $0 < \eta < \rho$. We slightly extend the definition in order to include the following

$$\mathbf{A}(0, \rho^-, \rho^+) := \mathbf{D}(0, \rho^+) \setminus D(0, \rho^-), \quad \mathbf{A}(0, 0^-, \rho^{\pm}) := \mathbf{D}(0, \rho^{\pm}) \setminus 0,$$
$$\mathbf{A}(0, \eta^{\pm}, \infty^-) := \mathbf{A}_K^{1,\mathrm{rig}} \setminus \mathbf{D}(0, \eta^{\mp}) \quad \text{and} \quad \mathbf{A}(0, 0^-, \infty^-) := \mathbf{A}_K^{1,\mathrm{rig}} \setminus 0.$$

We will consider for $\beta \in K$, the free rank one module $\mathcal{K}_\beta$ on the annulus $\mathbf{A}(0, \eta^\pm, \rho^\pm)$ on some generator θ whose connection is given by

$$\nabla(\theta) = \beta\theta \otimes \frac{dt}{t}.$$

We have $\partial(\theta) = \frac{\beta}{t}\theta$, so that

$$\partial^k(\theta) = \frac{\beta(\beta-1)\cdots(\beta-k+1)}{t^k}\theta$$

and

$$\partial^{[k]}(\theta) = \binom{\beta}{k}\frac{1}{t^k}\theta.$$

Thus, we see that

$$\epsilon(1\otimes\theta) = \sum\binom{\beta}{k}\frac{1}{t^k}\theta\otimes\tau^k = \theta\otimes\sum\binom{\beta}{k}(\frac{\tau}{t})^k = (1+\frac{\tau}{t})^\beta(\theta\otimes 1).$$

Note also that there is a short exact sequence of $\mathcal{D}$-modules

$$\begin{array}{ccccccccc} 0 & \longrightarrow & \mathcal{D} & \xrightarrow{t\partial-\beta} & \mathcal{D} & \longrightarrow & \mathcal{K}_\beta & \longrightarrow & 0 \\ & & & & 1 & \longmapsto & \theta & & \end{array}$$

giving an isomorphism $\mathcal{K}_\beta \simeq \mathcal{D}/(t\partial - \beta)$.

We want to compute the de Rham cohomology of $\mathcal{K}_\beta$. We have

$$\nabla(t^k\theta) = kt^{k-1}\theta\otimes dt + t^k\beta\theta\otimes\frac{dt}{t} = (k+\beta)t^{k-1}\theta\otimes dt.$$

If we write

$$A := \Gamma(\mathbf{A}(0, \eta^\pm, \rho^\pm), \mathcal{O}),$$

we may identify $H^0_{\mathrm{dR}}(\mathbf{A}(0, \eta^\pm, \rho^\pm), \mathcal{K}_\beta)$ and $H^1_{\mathrm{dR}}(\mathbf{A}(0, \eta^\pm, \rho^\pm), \mathcal{K}_\beta)$ with the kernel and the cokernel, respectively, of the morphism

$$\begin{array}{ccc} A & \xrightarrow{D} & A \\ t^k & \longmapsto & (k+\beta)t^{k-1}. \end{array}$$

We first see that $\ker D = 0$ unless $\beta \in \mathbf{Z}$ in which case, we get an isomorphism with the trivial crystal

$$\begin{array}{ccc} \mathcal{O} & \xrightarrow{\sim} & \mathcal{K}_\beta \\ 1 & \longmapsto & t^{-\beta}\theta. \end{array}$$

In this case, we see that H^0 is always one dimensional and it is not difficult to verify that H^1 is infinite dimensional in the case of a closed or semi-open annulus. In order to finish the case $\beta \in \mathbf{Z}$, it only remains to describe the first cohomology space of the trivial module with trivial connection on an open annulus. As it is well known, the residue map

$$\begin{array}{ccc} Adt & \longrightarrow & K \\ f := \sum_k a_k t^k dt & \longmapsto & \mathrm{res}(f) = a_{-1} \end{array}$$

induces an isomorphism of H^1 with K. Of course, in the case of $\mathcal{K}_\beta$ with $\beta \in \mathbf{Z}$, we should use $\mathrm{res}(t^\beta \omega)$.

We turn now to the more interesting case $\beta \notin \mathbf{Z}$. For simplicity, we only consider the case of an open annulus. Otherwise, we already know from our previous experience that the cohomology behaves badly in general.

Remember that β is a *Liouville number* (see for example [76], Definition 11.3.3) if it is arbitrarily closed to an integer in the following sense:

$$\text{either} \quad \underline{\lim}_{k\in\mathbf{N}} |\beta + k|^{1/k} < 1 \quad \text{or} \quad \underline{\lim}_{k\in\mathbf{N}} |\beta - k|^{1/k} < 1.$$

If this is not the case, then the map

$$\begin{array}{ccc} A & \xrightarrow{\ I\ } & A \\ t^k & \longmapsto & \frac{1}{k+1+\beta} t^{k+1} \end{array}$$

is a well-defined inverse for D and all cohomology spaces vanish. This is just because, then, we have $\frac{1}{|k+1+\beta|} \leq 1$ whenever $k \in \mathbf{Z}$. On the contrary, if β is a Liouville number, one can show that H^1 is infinite dimensional. Assume for example that

$$\underline{\lim}_{k\in\mathbf{N}} |\beta + k|^{1/k} < 1.$$

Then, there exists $\lambda > 1$ and an increasing sequence k_n such that

$$\left|\frac{1}{\beta + k_n}\right| \geq \lambda^{k_n}.$$

If $\rho = |\pi|^{r/s}$ and e_n is the integral part of $k_n r/s$, one defines

$$g_N := \sum_{n\geq N} \pi^{-e_n} t^{k_n - 1}.$$

This is an infinite family of linearly independent functions that cannot be integrated because

$$\left|\frac{\pi^{-e_n}}{\beta + k_n}\right| \rho^{k_n} \geq |\pi|^{k_n \frac{r}{s} - e_n} \lambda^{k_n} \geq |\pi| \lambda^{k_n} \to \infty.$$

To summarize, and if we stick to open annuli for simplicity, we see that we almost always have $H^i_{\mathrm{dR}}(\mathbf{A}(0,\eta^-,\rho^-),\mathcal{K}_\beta)=0$ with the exceptions

$$\dim_K H^0_{\mathrm{dR}}(\mathbf{A}(0,\eta^-,\rho^-),\mathcal{K}_\beta)=1 \quad \text{if}\beta\in\mathbf{Z}$$

and

$$\dim_K H^1_{\mathrm{dR}}(\mathbf{A}(0,\eta^-,\rho^-),\mathcal{K}_\beta)=\infty \quad \text{if} \quad \beta \quad \text{is} \quad \text{Liouville.}$$

In particular, we see that there exists rank one modules with connection on open annuli with infinite dimensional de Rham cohomology.

4.2.3 Playing with Dwork and Kummer modules

We can extend our differential modules $\mathcal{L}$ and $\mathcal{K}$ in the following way. If V is any smooth rigid analytic variety on K and $f\in\Gamma(V,\mathcal{O}_V)$, then we may consider f as a morphism

$$f: V\to\mathbf{A}^{1,\mathrm{rig}}_K$$

and let $\mathcal{L}_{\alpha,f} =: f^*\mathcal{L}_\alpha$. This is the free rank one module with connection

$$\nabla(\theta)=-\alpha\theta\otimes df.$$

In the same way, if $h\in\Gamma(V,\mathcal{O}_V^\times)$, then we may consider h as a morphism

$$h: V\to\mathbf{A}^{1,\mathrm{rig}}_K\setminus 0$$

and let $\mathcal{K}_{\beta,h} =: h^*\mathcal{K}_\beta$. This is the free rank one module with connection

$$\nabla(\theta)=-\beta\theta\otimes\frac{df}{f}.$$

We can say that $\mathcal{L}$ is *additive* in the sense that

$$\mathcal{L}_{\alpha,f_1+f_2}=\mathcal{L}_{\alpha,f_1}\otimes\mathcal{L}_{\alpha,f_2}$$

and that $\mathcal{K}$ is *multiplicative* in the sense that

$$\mathcal{K}_{\beta,h_1h_2}=\mathcal{K}_{\beta,h_1}\otimes\mathcal{K}_{\beta,h_2}.$$

More generally *exponential sums* come from mixing additive and multiplicative modules:

$$\mathcal{L}:=\otimes_i\mathcal{L}_{\alpha_i,f_i}\otimes_j\mathcal{K}_{\beta_j,h_j}$$

We shall come back later to these objects.

Actually, $\mathcal{K}$ and $\mathcal{L}$ modules appear in many other situations. For example, related to Kedlaya's algorithm for hyperelliptic curves ([52] and [42]), we can consider the following. First, we fix a monic separable polynomial $Q \in \mathcal{V}[X]$ of degree $d > 0$. In order to visualize things better, we assume that it splits over $\mathcal{V}$ as

$$Q = \prod_{i=1}^{d} (X - c_i).$$

Having coefficients in $\mathcal{V}$ means that for each i, $|c_i| \leq 1$ and the separability assumption means that for each $i \neq j$, we have $|c_i - c_j| = 1$ (they lie in different residual classes). We let

$$V = \{x \in \mathbf{D}(0, \rho^-), |Q(x)| > \eta\} = \mathbf{D}(0, \rho^-) \setminus \coprod_i \mathbf{D}(c_i, \eta^+)$$

with some $\eta < 1 < \rho$ (it is an easy exercise to check the equality). As we shall see shortly, we have

$$K \simeq H^0_{\mathrm{dR}}(V) \quad \text{and} \quad \{\frac{P(x)}{Q(x)} dx, \deg P < d\} \simeq H^1_{\mathrm{dR}}(V).$$

More precisely, any element of the ring of function on V can be uniquely written in the form

$$f = \sum_{k \geq 0} a_k x^k + \sum_i \sum_{k<0} a_{ik}(x - c_i)^k.$$

Moreover, we can write

$$\sum_i \frac{a_{i,-1}}{x - c_i} = \frac{P(x)}{Q(x)}$$

with $\deg P < \deg Q$, and conversely. Thus, we see that f can be uniquely written as

$$f = \sum_{k \geq 0} a_k x^k + \frac{P(x)}{Q(x)} + \sum_i \sum_{k<1} a_{ik}(x - c_i)^k$$

with $\deg P < \deg Q$. And integration works as usual. The details are left to the reader.

We assume now that $p \neq 2$ and we consider the module with connection $\mathcal{K}_{\frac{1}{2},Q}$ on V. It is the free module of rank one whose connection is given by

$$\nabla(\theta) = \frac{1}{2}\theta \otimes \frac{dQ}{Q} = \sum_i \frac{1}{2}\theta \otimes \frac{dx}{x - c_i}.$$

As above, one can show that

$$0 = H^0_{\mathrm{dR}}(V, \mathcal{K}_{\frac{1}{2},Q}) \quad \text{and} \quad \{\frac{P(x)}{Q(x)}\theta \otimes dx, \deg P < d - 1\} \simeq H^1_{\mathrm{dR}}(V, \mathcal{K}_{\frac{1}{2},Q}).$$

Actually, it might be convenient to note that Q induces a finite morphism

$$Q : V \to \mathbf{A}(0, \eta^-, \rho'^-)$$

with $\rho' = \rho^{1/d}$. This implies that any function f on V can be written uniquely as

$$f = \sum_{k \in \mathbf{Z}} P_k(x) Q(x)^k$$

with $\deg P_k < d$. And a few more computations give the above results. This is left to the reader.

It is quite easy to deduce the cohomology of hyperelliptic curves from this discussion. We consider the plane analytic curve U defined by

$$y^2 = Q(x) \quad \text{with} \quad y \in \mathbf{A}(0, \eta^-, \rho^-).$$

If we denote by

$$V := \{x \in \mathbf{D}(0, (\rho^{2/d})^-), \quad |Q(x)| > \eta^2\},$$

the first projection

$$\begin{array}{ccc} U & \xrightarrow{p} & V \\ (x, y) & \longmapsto & x \end{array}$$

is finite étale and we have

$$p_* \mathcal{O}_V \simeq \mathcal{O}_V \oplus \mathcal{K}_{\frac{1}{2}, Q}.$$

More precisely, $p_* \mathcal{O}_V$ is clearly free on the images of 1 and y and we have

$$2y dy = dQ \mod y^2 - Q$$

which gives

$$dy = \frac{1}{2y} dQ = \frac{y}{2y^2} dQ = \frac{1}{2} y \frac{dQ}{Q} \mod y^2 - Q.$$

It follows that

$$H^i_{\mathrm{dR}}(U) = H^i_{\mathrm{dR}}(V) \oplus H^i_{\mathrm{dR}}(V, \mathcal{K}_{\frac{1}{2}, Q}).$$

In the same way, if we compute the cohomology of $K_{i/r, Q}$ for all $i = 0, \ldots, r-1$ as we did above in the case $r = 2$, we get the cohomology of the superelliptic curve $y^r = Q(x)$ for $p \nmid r$.

4.2.4 The Legendre family

In order to give other examples, it might be useful to come back to the notion of *Gauss–Manin connection* and the corresponding *Picard–Fuchs differential equation*. We will only consider here the case of a smooth quasi-Stein family of curves

$$p : V \to U$$

where U is some smooth curve over K. The exact sequence of locally free sheaves of finite rank

$$0 \to p^*\Omega^1_{U/K} \to \Omega^1_{V/K} \to \Omega^1_{V/U} \to 0$$

gives rise to an isomorphism

$$\Omega^2_{V/K} \simeq \Omega^1_{U/K} \otimes_{\mathcal{O}_U} \Omega^1_{V/U}.$$

On the other hand, there is an exact sequence on U

$$p_*\mathcal{O}_V \xrightarrow{d} p_*\Omega^1_{V/U} \to \mathcal{H}^1_{\mathrm{dR}}(V/U) \to 0.$$

Thus, we get a surjective map

$$p_*\Omega^2_{V/K} \simeq \Omega^1_{U/K} \otimes p_*\Omega^1_{V/U} \to \Omega^1_{U/K} \otimes \mathcal{H}^1_{\mathrm{dR}}(V/U) \simeq \mathcal{H}^1_{\mathrm{dR}}(V/U) \otimes \Omega^1_{U/K}.$$

Now, any section θ of $\mathcal{H}^1_{\mathrm{dR}}(V/U)$ can be lifted to $p_*\Omega^1_{V/U}$ and therefore also to some $\omega \in p_*\Omega^1_{V/K}$. We have $d\omega \in p_*\Omega^2_{V/K}$ and we may consider its image

$$\nabla(\theta) \in \mathcal{H}^1_{\mathrm{dR}}(V/U) \otimes \Omega^2_{U/K}.$$

In other words, we have $\nabla(\overline{\omega}) = \overline{d\omega}$.

We will apply these considerations to the Legendre family of elliptic curves (see for example §7 of [83]) when $p \neq 2$. More precisely, we consider the relative affine curve E of equation

$$y^2 = x(x-1)(x-t), \quad y \neq 0$$

with t in some admissible open subset U of $\mathbf{A}^{1,\mathrm{rig}}_K \setminus \{0, 1\}$. Let V be the Zariski open subset of the relative line over U defined by $x \neq 0, 1, t$ and

$$\mathcal{E} := \mathcal{H}^1_{\mathrm{dR}}(V/U, \mathcal{K}_{\frac{1}{2}, x(x-1)(x-t)}).$$

With the same arguments as above, it is not difficult to see that $\mathcal{E}$ is the free rank two module generated by the images θ and ϕ of dx/y and xdx/y, respectively, and that

$$\mathcal{H}^1_{\mathrm{dR}}(E/U) \simeq \mathcal{H}^1_{\mathrm{dR}}(V/U) \oplus \mathcal{E}.$$

Alternatively, one can see $\mathcal{E}$ as the relative de Rham cohomology of the Zariski closure of E in $\mathbf{A}_U^{2,\mathrm{rig}}$ or even in $\mathbf{P}_U^{2,\mathrm{rig}}$.

If we want to describe the connection on $\mathcal{E}$, we should first remark that

$$dy = \frac{-x^2+x}{2y}dt + \frac{3x^2-2(t+1)x+t}{2y}dx$$

It follows that

$$d(\frac{dx}{2y}) = -\frac{1}{2y^2}dy \wedge dx$$

$$= -\frac{-x^2+x}{2y}dt \wedge \frac{1}{2x(x-1)(x-t)}dx = \frac{1}{2(x-t)}dt \wedge \frac{dx}{2y}$$

and we want to write it as a linear combination of dx/y and xdx/y modulo exact forms. We use the fact that

$$dt \wedge d(\frac{y}{x-t}) = dt \wedge \left(\frac{dy}{x-t} - \frac{ydx}{(x-t)^2}\right)$$

$$= dt \wedge \left(\frac{(3x^2-2(t+1)x+t)dx}{2y(x-t)} - \frac{x(x-1)dx}{y(x-t)}\right)$$

$$= \left(\frac{x^2-2tx+t}{x-t}\right)dt \wedge \frac{dx}{2y} = -\left(-x+t+\frac{t(t-1)}{x-t}\right)dt \wedge \frac{dx}{2y}.$$

Dividing by $2t(t-1)$, we see that

$$d(\frac{dx}{2y}) = \frac{1}{2(x-t)}dt \wedge \frac{dx}{2y}$$

$$= \frac{x-t}{2t(t-1)}dt \wedge \frac{dx}{2y} - dt \wedge d(\frac{y}{2t(t-1)(x-t)})$$

$$= -\frac{1}{2(t-1)}dt \wedge \frac{dx}{2y} + \frac{1}{2t(t-1)}dt \wedge x\frac{dx}{2y} - dt \wedge d(\text{something}).$$

It follows that

$$\nabla(\theta) = -\frac{1}{2(t-1)}\theta dt + \frac{1}{2t(t-1)}\phi dt.$$

Now, we have

$$d(\frac{xdx}{2y}) = xd(\frac{dx}{2y}) = \frac{x}{2(x-t)}dt \wedge \frac{dx}{2y}$$

$$= (\frac{1}{2} + \frac{t}{2(x-t)})dt \wedge \frac{dx}{2y} = \frac{1}{2}dt \wedge \frac{dx}{2y} + td(\frac{dx}{2y})$$

and it follows that

$$\nabla(\phi) = \frac{1}{2}\theta dt + t\nabla(\theta)$$

and therefore,

$$\nabla(\phi) = -\frac{1}{2(t-1)}\theta dt + \frac{1}{2(t-1)}\phi dt.$$

In other words, the matrix of the connection of $\mathcal{E}$ with respect to (θ, ϕ) is

$$\begin{bmatrix} -\frac{1}{2(t-1)} & -\frac{1}{2(t-1)} \\ \frac{1}{2t(t-1)} & \frac{1}{2(t-1)} \end{bmatrix}.$$

Finally, using

$$\partial/\partial t(\theta) = -\frac{1}{2(t-1)}\theta + \frac{1}{2t(t-1)}\phi \quad \text{and} \quad \partial/\partial t(\phi) = \frac{1}{2}\theta + t\partial/\partial t(\theta),$$

one easily sees that θ satisfies the Picard–Fuchs equation

$$t(1-t)\partial^2/\partial t^2(\theta) + (1-2t)\partial/\partial t(\theta) - \frac{1}{4}\theta = 0.$$

More precisely, the first relation gives

$$\frac{\phi}{2} = t(t-1)\partial/\partial t(\theta) + \frac{t}{2}\theta.$$

Applying $\partial/\partial t$ on both sides provides us with

$$\frac{1}{4}\theta + \frac{t}{2}\partial/\partial t(\theta) = [t(t-1)\partial^2/\partial t^2(\theta) + (2t-1)\partial/\partial t(\theta)] + [\frac{t}{2}\partial/\partial t(\theta) + \frac{1}{2}\theta].$$

It is then sufficient to rearrange the terms.

Since, clearly θ and $\theta' := \partial/\partial t(\theta)$ form a basis of $\mathcal{E}$, we get a presentation

$$\begin{array}{ccccccccc} 0 & \longrightarrow & \mathcal{D} & \xrightarrow{L} & \mathcal{D} & \longrightarrow & \mathcal{E} & \longrightarrow & 0 \\ & & & & 1 & \longmapsto & \theta & & \end{array}$$

with

$$L := t(1-t)\partial^2/\partial t^2 + (1-2t)\partial/\partial t - \frac{1}{4}.$$

4.2.5 Hypergeometric differential equations

The above Picard–Fuchs equation is a particular case of *hypergeometric differential equation* (see for example [40]). Given parameters a, b, c, the *hypergeometric series*

$${}_2F_1(a, b, c; t) := \sum_{k=0}^{\infty} \frac{(a)_k (b)_k}{(c)_k} t^{[k]},$$

where

$$(x)_k := x(x+1)\cdots(x+k-1)$$

is the *Pochhammer symbol*, is a formal solution of the hypergeometric differential equation associated to the hypergeometric differential operator

$$L := t(1-t)\partial^2 + (c-(a+b+1)t)\partial - ab.$$

The case $a := b := \frac{1}{2}$ and $c := 1$ is the Picard–Fuchs equation associated to the Legendre family of elliptic curve.

More generally, given $a_1, \ldots, a_n, b_1, \ldots, b_m, \alpha \in K$, one considers the *hypergeometric differential operator*

$$L := \partial(t\partial + b_1 - 1)\cdots(t\partial + b_m - 1) + \alpha(t\partial + a_1)\cdots(t\partial + a_n).$$

Then, the *generalized hypergeometric series*

$${}_nF_m(a_1, \ldots, a_n, b_1, \ldots, b_m; -\alpha t) := \sum_{k=0}^{\infty} \frac{\prod_{i=1}^{n}(a_i)_k}{\prod_{i=1}^{m}(b_i)_k}(-\alpha t)^{[k]}$$

is a formal solution of the differential equation associated to L.

There is a lot of literature on these equations. For example, the article [5] shows how the study of ${}_3F_2$ can be reduced to the study of ${}_2F_1$. The later case being extensively studied by Dwork in [41].

Anyway, let us consider what happens for small values of m and n. First of all, for $m = n = 0$, we have

$$L = \partial + \alpha$$

and we fall back on Dwork crystal $\mathcal{L}_\alpha$. In the case $n = 1$ and $m = 0$, we get

$$L = \alpha[(1/\alpha + t)\partial + a]$$

which is isomorphic to the Kummer crystal $\mathcal{K}_{-a}$ after a translation by $1/\alpha$.

But still, the case $n = 0$ and $m = 1$ is new since we obtain

$$L = t\partial^2 + b\partial + \alpha.$$

As is well known, this is closely related to the *Bessel operator*

$$P := t^2\partial^2 + t\partial - (\alpha t^2 + d^2)$$

which is worked out in detail in [3] and defines a rank two module with an integrable connection on $\mathbf{A}_K^{1,\mathrm{rig}} \setminus 0$ whose behavior is a lot more chaotic than what we have seen so far. A particular case is done by Tsuzuki in [81].

4.3 Calculus on strict neighborhoods

Starting at Theorem 4.3.9 below and until the end of the section, we will assume that $\mathrm{Char} K = 0$. Also, throughout this section, we fix a formal $\mathcal{V}$-scheme S.

Recall that the generic fiber functor $P \mapsto P_K$ behaves as well as we may expect with respect to products and immersions, and in particular with respect to infinitesimal neighborhoods. More precisely, if S is a formal $\mathcal{V}$-scheme, P a formal S-scheme, and $P^{(n)} \subset P \times_S P$ denotes the n-th infinitesimal neighborhood of P over S, we have $(P_K)^{(n)} = P_K^{(n)}$.

Proposition 4.3.1 *Let $X \hookrightarrow P$ be an S-immersion of an algebraic S_k-variety into a formal S-scheme. If $X \hookrightarrow P \hookrightarrow P^{(n)}$ is the composite immersion, we have an isomorphism*

$$]X[_{P^{(n)}} \simeq]X[_P^{(n)}$$

and, when X is quasi-compact, for $\eta < 1$, an isomorphism

$$[X]_{P^{(n)}\eta} \simeq [X]_{P\eta}^{(n)} \quad (\text{resp.}]X[_{P^{(n)}\eta} \simeq]X[_{P\eta}^{(n)}).$$

Proof As usual, it is sufficient to do the case of a closed tube of radius η. By functoriality, the open immersion $[X]_{P\eta} \hookrightarrow P_K$ induces an open immersion $[X]_{P\eta}^{(n)} \hookrightarrow P_K^{(n)}$. The assertion is now purely set-theoretic and follows from Proposition 2.3.9 since the the map

$$\delta^{(n)} : P_K \hookrightarrow P_K^{(n)}$$

is a homeomorphism. □

It is very important to recall however that, in general, if we embed X diagonally into $P \times_S P$, then

$$]X[_{P\times_S P} \neq]X[_P \times_{S_K}]X[_P.$$

This is immediate if we simply consider the embedding of $\mathbf{A}_k^1$ in $\widehat{\mathbf{A}}_{\mathcal{V}}^1$. It is clear that the tube of $\mathbf{A}_k^1$ in $\widehat{\mathbf{A}}_{\mathcal{V}}^2$ which is defined by $|x - y| < 1$ in $\mathbf{B}^2(0, 1)$ is not equal to the full ball.

If $(X \subset Y \subset P)$ is an S-frame, we may consider the diagonal embedding

$$\delta : (X \subset Y \subset P) \hookrightarrow (X \subset Y \subset P \times_S P)$$

as well as the projections

$$p_1, p_2 : (X \subset Y \subset P \times_S P) \to (X \subset Y \subset P).$$

Of course, we can also consider

$$\delta^{(n)} : (X \subset Y \subset P) \hookrightarrow (X \subset Y \subset P^{(n)})$$

and

$$p_1^{(n)}, p_2^{(n)} : (X \subset Y \subset P^{(n)}) \to (X \subset Y \subset P).$$

Proposition 4.3.2 *Let $(X \subset Y \subset P)$ be an S-frame and V a strict neighborhood of $]X[_P$ in $]Y[_P$. Then $V^{(n)}$ is a strict neighborhood of $]X[_{P^{(n)}}$ in $]Y[_{P^{(n)}}$. Actually,*

$$V' := (V \times_{S_K} V) \cap]Y[_{P\times_S P}$$

is a strict neighborhood of $]X[_{P\times_S P}$ in $]Y[_{P\times_S P}$ and

$$V^{(n)} = V' \cap]Y[_P^{(n)}.$$

Proof We have

$$\begin{aligned} V' &:= (V \times_{S_K} V) \cap]Y[_{P\times_S P} \\ &= (p_1^{-1}(V) \cap]Y[_{P\times_S P}) \cap (p_2^{-1}(V) \cap]Y[_{P\times_S P}). \end{aligned}$$

Since the inverse image of a strict neighborhood is a strict neighborhood and the intersection of two strict neighborhoods is also a strict neighborhood, we see that V' is a strict neighborhood of $]X[_{P\times_S P}$ in $]Y[_{P\times_S P}$.

In particular, if we prove that $V' \cap]Y[_P^{(n)} = V^{(n)}$, the latter will be a strict neighborhood of $]X[_{P^{(n)}}$ in $]Y[_{P^{(n)}}$ as the inverse image of V' by the morphism of S-frames

$$(X \subset Y \subset P^{(n)}) \hookrightarrow (X \subset Y \subset P \times_S P).$$

By functoriality, the open immersion $V \hookrightarrow]Y[_P$ induces an open immersion

$$V^{(n)} \hookrightarrow]Y[_P^{(n)}.$$

By restriction, we also have an open immersion

$$V' \cap]Y[_P^{(n)} \hookrightarrow]Y[_P^{(n)}.$$

The assertion is therefore set-theoretic and we may assume that $n = 0$ in which case, we clearly have

$$V' \cap]Y[_P = (V \times_{S_K} V) \cap P_K = V.$$

□

Proposition 4.3.3 *Let $(X \subset Y \subset P)$ be a quasi-compact S-frame and Z a closed complement for X in Y. As usual, for $\lambda, \eta < 1$, we will write*

$$V^\lambda :=]Y[_P \setminus]Z[_{P\lambda} \quad \text{and} \quad V_\eta^\lambda := [Y]_{P\eta} \cap V^\lambda.$$

We will also set

$$W^\lambda :=]Y[_{P\times_S P} \setminus]Z[_{P\times_S P\lambda} \quad \text{and} \quad W_\eta^\lambda := [Y]_{P\times_S P\eta} \cap W^\lambda.$$

Then, for $\eta < \lambda$, we have

$$W_\eta^\lambda = [Y]_{P\times_S P\eta} \cap (V^\lambda \times_{S_K} V^\lambda) = [Y]_{P\times_S P\eta} \cap (V_\eta^\lambda \times_{S_K} V_\eta^\lambda)$$

Proof It follows from Proposition 3.2.8 that

$$[Y]_{P\times_S P\eta} \cap p_1^{-1}(V^\lambda) = W_\eta^\lambda = [Y]_{P\times_S P\eta} \cap p_2^{-1}(V^\lambda),$$

and

$$[Y]_{P\times_S P\eta} \cap p_1^{-1}(V_\eta^\lambda) = W_\eta^\lambda = [Y]_{P\times_S P\eta} \cap p_2^{-1}(V_\eta^\lambda).$$

It is therefore sufficient to consider the intersection in both cases. □

Definition 4.3.4 *Let $(X \subset Y \subset P)$ be an S-frame, V a strict neighborhood of $]X[_P$ in $]Y[_P$ and $\mathcal{E}$ an $\mathcal{O}_V$-module. A stratification on $\mathcal{E}$ is said to be* overconvergent *if there exists a strict neighborhood*

$$V' \subset (V \times_{S_K} V) \cap]Y[_{P\times_S P}$$

of $]X[_{P\times_S P}$ in $]Y[_{P\times_S P}$ and an isomorphism

$$\epsilon : p_2^*\mathcal{E}_{|V'} \simeq p_1^*\mathcal{E}_{|V'}$$

such that the Taylor isomorphism of $\mathcal{E}$ is induced on $V' \cap]Y[_P^{(n)}$ by ϵ for each n. We will also say that the connection of $\mathcal{E}$ is overconvergent. *In the case $Y = X$, we simply say* convergent.

Note that in the convergent case, there is only one strict neighborhood, namely $]X[_P$ itself. And we see that a stratified $\mathcal{O}_{]X[_P}$-module $\mathcal{E}$ is convergent if the Taylor isomorphisms of $\mathcal{E}$ comes from an isomorphism $\epsilon : p_2^*\mathcal{E} \simeq p_1^*\mathcal{E}$ on $]X[_{P^2}$.

When V is smooth and $\mathrm{Char} K = 0$, will denote by $MIC^\dagger(V/S_K)$ the full subcategory of $MIC(V/S_K)$ of coherent $\mathcal{O}_V$-modules with an overconvergent integrable connection. Note that by definition, morphisms in $MIC^\dagger(V/S_K)$ are just morphisms of modules with connection and that we do not require any compatibility a priori with the Taylor isomorphisms at the level of strict neighborhoods. This is an important issue and we will come back to this question in the future.

Proposition 4.3.5 *Let*

$$\begin{array}{ccccc} X' & \hookrightarrow & Y' & \hookrightarrow & P' \\ \downarrow f & & \downarrow g & & \downarrow u \\ X & \hookrightarrow & Y & \hookrightarrow & P \end{array}$$

be a morphism from an S'-frame to an S-frame over some morphism of formal $\mathcal{V}$-schemes $v : S' \to S$. Let V be a strict neighborhood of $]X[_P$ in $]Y[_P$ and $\mathcal{E}$ a module with an overconvergent stratification on V. Then, the stratification of $u_K^*\mathcal{E}$ is overconvergent on $u_K^{-1}(V) \cap]Y'[_{P'}$.

Proof This is completely formal. Our morphism extends to a morphism

$$\begin{array}{ccccc} X' & \hookrightarrow & Y' & \hookrightarrow & P' \times_{S'} P' \\ \downarrow{\scriptstyle f} & & \downarrow{\scriptstyle g} & & \downarrow{\scriptstyle u \times_S u} \\ X & \hookrightarrow & Y & \hookrightarrow & P \times_S P \end{array}$$

and we know that if V' is a strict neighborhood of $]X[_{P\times_S P}$ in $]Y[_{P\times_S P}$, then

$$(u \times_S u)^{-1}(V') \cap]Y'[_{P'\times_{S'} P'}$$

will be a strict neighborhood of $]X'[_{P'\times_{S'} P'}$ in $]Y'[_{P'\times_{S'} P'}$. □

In order to understand the true nature of the overconvergence condition, we need a local description. Before that, we prove a result from rigid analytic geometry.

Proposition 4.3.6 *Let V be a rigid analytic variety and $\eta < 1$. Denote by*

$$p_\eta : V \times_K B^d(0, \eta^+) \to V$$

and

$$p^{(n)} : V \times_K \mathrm{Spm} K[\underline{t}]/(\underline{t})^n \to V,$$

the projections. If $\mathcal{F}$ is a coherent module on V, the canonical map

$$p_{\eta*} p_\eta^* \mathcal{F} \to \varprojlim_n p^{(n)*} \mathcal{F}$$

is injective.

Proof It is sufficient to show that if V is affinoid, the map

$$\Gamma(V, p_{\eta*} p_\eta^* \mathcal{F}) \to \Gamma(V, \varprojlim_n p^{(n)*} \mathcal{F})$$

is injective. If we let $A := \Gamma(V, \mathcal{O}_V)$ and $M := \Gamma(V, \mathcal{F})$, this map is the canonical map

$$M \otimes_A A\{\underline{t}/\eta\} \to M \otimes_A A[[\underline{t}]].$$

Note that we do not need to complete the tensor product because M is finitely presented. By induction on the number of generators of M, we may assume that M is a quotient of A. It is therefore an affinoid algebra and we are reduced to the case $M = A$ which is part of the definition of the ring of convergent power series. □

In the future, we will write

$$\mathcal{F} \otimes_{\mathcal{O}_V} \mathcal{O}_V\{\underline{t}/\eta\} := p_{\eta *} p_\eta^* \mathcal{F} \subset \varprojlim_n p^{(n)*}\mathcal{F} =: \mathcal{F} \otimes_{\mathcal{O}_V} \mathcal{O}_V[[\underline{t}]].$$

Definition 4.3.7 *A* local frame *on S is a (smooth) affine frame $(X \subset Y \subset P)$ with coordinates $t_1, \ldots, t_m$ on P over S étale in the neighborhood of X. We will simply say that $(t_1, \ldots, t_m)$ is a* set of coordinates *on the local frame. The local frame is said to be* strictly local *if the closed complement Z of X in Y is a hypersurface.*

Proposition 4.3.8 *Any smooth S-frame $(X \subset Y \subset P)$ has an open covering*

$$\begin{array}{ccccc} X_i & \hookrightarrow & Y_i & \hookrightarrow & P_i \\ \downarrow & & \downarrow & & \downarrow \\ X & \hookrightarrow & Y & \hookrightarrow & P \end{array}$$

by strictly local frames.

Proof We may first localize on P and assume that $P = \mathrm{Spf} A$. Then, we may find $t_1, \ldots, t_N$ such that $\Omega^1_{P/S}$ is generated by $dt_1, \cdots, dt_N$. It follows, that if $x \in X$, then their images

$$dt_1(x), \ldots, dt_N(x) \in \Omega^1_{P/S} \otimes_{\mathcal{O}_P} k(x)$$

form a set of generators and we may extract a basis $dt_1(x), \ldots, dt_m(x)$. It follows from Nakayama's lemma that there exists a neighborhood P' of x in P such that $dt_1, \ldots, dt_m$ form a basis of $\Omega^1_{P'/S}$. We may therefore localize on X and assume that $dt_1, \ldots, dt_m$ form a basis of $\Omega^1_{P/S}$ in the neighborhood of X. Of course, we may then localize again on X and assume that its complement in Y is a hypersurface. □

From now on, in this section, we assume that $\mathrm{Char} K = 0$.

Theorem 4.3.9† *Let $(X \subset Y \subset P)$ be a strictly local S-frame with coordinates $t_1, \ldots, t_m$ and corresponding derivations $\partial_1, \ldots, \partial_m$.*

If Z is denotes a closed complement of X in Y, we write as usual for $\lambda, \eta < 1$,

$$V^\lambda :=]Y[_P \setminus]Z[_{P\lambda} \quad \text{and} \quad V^\lambda_\eta := [Y]_{P\eta} \cap V^\lambda.$$

† The proof of Proposition 2.2.13 in [14] is not correct and there is no evidence either that the assertion itself is true in full generality.

Let V be a smooth strict neighborhood of $]X[_P$ in $]Y[_P$ and $\mathcal{E}$ a coherent $\mathcal{O}_V$-module with an integrable connection ∇.

Then, ∇ is overconvergent if and only if for each $\eta < 1$, there exists $\eta \leq \delta < \lambda_0 < 1$ such that for all $\lambda_0 \leq \lambda < 1$, we have

$$\forall s \in \Gamma(V_\delta^\lambda, \mathcal{E}), \quad \|\underline{\partial}^{[\underline{k}]}(s)\|\eta^{|\underline{k}|} \to 0$$

with $\| - \|$ a Banach norm on $\Gamma(V_\eta^\lambda, \mathcal{E})$. And it is actually sufficient to check this property for generators of $\Gamma(V_\delta^\lambda, \mathcal{E})$.

If $t_1, \ldots, t_m$ are actually étale in a neighborhood of Y, we may choose $\delta = \eta$.

Before giving the proof to this theorem, we will need some intermediate results.

Note that this theorem is generally stated with $\delta = \eta$. I do not know how to prove this in general or even whether it is true or not. This is obviously the case when $Y = P_k$ and therefore in particular in the "Monsky–Washnitzer setting". This is also the case when the local coordinates are étale in the neighborhood of Y and therefore in particular in the quasi-projective situation (when X is a subvariety of some projective space $\mathbf{P}_k^N$ and P is an open formal subscheme of $\widehat{\mathbf{P}}_\mathcal{V}^N$).

Lemma 4.3.10 *Let $(X \subset Y \subset P)$ be a local S-frame with coordinates $t_1, \ldots, t_d$. Then, the morphism*

$$(\tau_1, \ldots, \tau_d) : P \times_S P \to \widehat{\mathbf{A}}_P^d$$

with

$$\tau_i := p_2^*(t_i) - p_1^*(t_i),$$

induces an isomorphism between a strict neighborhood of $]X[_{P\times_S P}$ in $]Y[_{P\times_S P}$ and a strict neighborhood of $]X[_P \times \mathbf{B}^d(0, 1^-)$ in $]Y[_P \times \mathbf{B}^d(0, 1^-)$.

More precisely, if we set for $\lambda, \eta < 1$,

$$W^\lambda :=]Y[_{P\times_S P} \setminus]Z[_{P\times_S P\lambda} \quad \text{and} \quad W_\eta^\lambda := [Y]_{P\times_S P\eta} \cap W^\lambda$$

then, for each $\eta < 1$, there exists $\eta \leq \delta < \lambda_0 < 1$ such that for all $\lambda_0 \leq \lambda < 1$, we have

$$W_\eta^\lambda \subset V_\eta^\lambda \times_K B^d(0, \eta^+) \subset W_\delta^\lambda.$$

If $t_1, \ldots, t_d$ are actually étale in a neighborhood of Y, we may choose $\delta = \eta$.

Proof Follows from Corollary 3.4.14, Proposition 3.3.14 and Corollary 3.3.15. □

Lemma 4.3.11 *Let $(X \subset Y \subset P)$ be a local S-frame. With the notations of the previous lemma, denote by*

$$p : V_\eta^\lambda \times \mathbf{B}^d(0, \eta^+) \to V_\eta^\lambda,$$

as well as

$$p_1^{(n)}, p_2^{(n)} : V_\eta^{\lambda(n)} \to V_\eta^\lambda,$$

the projections, and by

$$q : V_\eta^\lambda \times \mathbf{B}^d(0, \eta^+) \hookrightarrow W_\delta^\lambda \overset{p_2}{\to} V_\delta^\lambda,$$

the composite map.

If $\mathcal{E}$ (resp. $\mathcal{F}$) is a coherent module on V_δ^λ (resp. V_η^λ), the canonical map

$$p_*\mathcal{H}om(q^*\mathcal{E}, p^*\mathcal{F}) \to \varprojlim_n \mathcal{H}om(p_2^{(n)*}\mathcal{E}_{|V_\eta^\lambda}, p_1^{(n)*}\mathcal{F})$$

is injective.

Proof Locally, $\mathcal{E}$ has a finite presentation. Since our functors are left exact, it is therefore sufficient to consider the case $\mathcal{E} := \mathcal{O}_{V_\delta^\lambda}$. In other words, we are reduced to show that the map

$$p_*p^*\mathcal{F} \to \varprojlim_n p_1^{(n)*}\mathcal{F}$$

is injective. But this has been proved in Proposition 4.3.6. □

In the next lemma, we have to assume that the frame is strictly local because we need V_η^λ and V_δ^λ to be affine.

Lemma 4.3.12 *Let $(X \subset Y \subset P)$ be a strictly local S-frame and use the notations of the previous lemma. Let $\mathcal{E}$ be a coherent module with an integrable connection on V_δ^λ and $\| - \|$ a Banach norm on $\Gamma(V_\eta^\lambda, \mathcal{E})$.*

Then, the following conditions are equivalent:

(i) *There exists a morphism*

$$\epsilon : q^*\mathcal{E} \to p^*\mathcal{E}_{|V_\eta^\lambda}$$

such that the Taylor isomorphism of $\mathcal{E}$ is induced on $V_\eta^{\lambda(n)}$ by ϵ for each $n \in \mathbf{N}$.

(ii) *We have*

$$\forall s \in \Gamma(V_\delta^\lambda, \mathcal{E}), \quad \|\underline{\partial}^{[\underline{k}]}(s)\|\eta^{|\underline{k}|} \to 0.$$

And it is actually sufficient to check this property for generators of $\Gamma(V_\delta^\lambda, \mathcal{E})$.

Moreover, the morphism ϵ is unique with this property.

Proof Of course, the uniqueness of ϵ comes from Lemma 4.3.11.

We will write

$$A := \Gamma(V_\delta^\lambda, \mathcal{O}_{P_K}), \quad B := \Gamma(V_\eta^\lambda, \mathcal{O}_{P_K}) \quad \text{and} \quad M := \Gamma(V_\delta^\lambda, \mathcal{E}).$$

Assume first that the Taylor isomorphisms are induced by

$$\epsilon : q^*\mathcal{E} \to p^*\mathcal{E}_{|V_\eta^\lambda}.$$

Then, if $s \in M$, by definition, the element

$$\epsilon(q^*(s)) \in \Gamma(V_\eta^\lambda \times_K B^d(0, \eta^+), p^*\mathcal{E}) = M \otimes_A B\{\underline{\tau}/\eta\}$$

has image

$$\theta(s) = \sum \underline{\partial}^{[\underline{k}]}(s) \otimes \underline{\tau}^{\underline{k}} \in M \otimes_A B[[\underline{\tau}]].$$

It follows from Proposition 4.3.6 that

$$\|\underline{\partial}^{[\underline{k}]}(s)\|\eta^{|\underline{k}|} \to 0.$$

This shows that (i) implies (ii).

Note that this applies in particular to the trivial connection on the trivial module in which case $\epsilon = \mathrm{Id}$. This means that if $f \in A$, the image of $q^*(f)$ in $B\{\underline{\tau}/\eta\}$ is $\theta(f) = \sum \underline{\partial}^{[\underline{k}]}(f)\underline{\tau}^{\underline{k}}$. And this implies that it is actually sufficient to check the condition for generators of $\Gamma(V_\delta^\lambda, \mathcal{E})$.

Conversely, if we assume that for all $s \in M$, we have

$$\|\underline{\partial}^{[\underline{k}]}(s)\|\eta^{|\underline{k}|} \to 0,$$

then, we may consider the morphism

$$\begin{aligned} M &\xrightarrow{\varphi} M \otimes_A B\{\underline{\tau}/\eta\} \\ s &\longmapsto \sum \underline{\partial}^{[\underline{k}]}(s) \otimes \underline{\tau}^{\underline{k}}. \end{aligned}$$

If $f \in A$ and $s \in M$, then, by definition of the Taylor morphisms, we have $\theta(fs) = \theta(f)\theta(s) \in B[[\underline{\tau}]]$. It follows from Proposition 4.3.6 that $\varphi(fs) = q^*(f)\varphi(s)$. In other words, φ is linear with respect to $q^* : A \to B\{\underline{\tau}/\eta\}$. Thus, it extends to a morphism

$$\Gamma(V_\eta^\lambda \times \mathbf{B}^d(0, \eta^+), q^*\mathcal{E}) \to \Gamma(V_\eta^\lambda \times \mathbf{B}^d(0, \eta^+), p^*\mathcal{E})$$

which corresponds to a morphism

$$\epsilon : q^*\mathcal{E} \to p^*\mathcal{E}_{|V_\eta^\lambda}.$$

This shows that (ii) implies (i).

Of course, by construction, the Taylor isomorphism of $\mathcal{E}$ is induced on $V_\eta^{\lambda(n)}$ by ϵ for each n. □

Proof (of the theorem) Assume first that ∇ is overconvergent. By definition, there exists a strict neighborhood

$$V' \subset (V \times_{S_K} V) \cap]Y[_{P \times_S P}$$

of $]X[_{P \times_S P}$ in $]Y[_{P \times_S P}$ and an isomorphism

$$\epsilon : p_2^* \mathcal{E}_{|V'} \simeq p_1^* \mathcal{E}_{|V'}$$

such that the Taylor isomorphism of $\mathcal{E}$ is induced on $V' \cap]Y[_P^{(n)}$ by ϵ. Lemma 4.3.10 tells us that, with our usual notations, given any $\eta < 1$, there exists $\eta \leq \delta < \lambda_0 < 1$ such that for all $\lambda_0 \leq \lambda < 1$, we have

$$W_\eta^\lambda \subset V_\eta^\lambda \times_K B^d(0, \eta^+) \subset W_\delta^\lambda.$$

Moreover, raising λ_0 if necessary, we may assume that $W_\delta^\lambda \subset V'$. We may then consider the restriction of $\mathcal{E}$ to V_δ^λ and apply Lemma 4.3.12.

Conversely, assume that the condition is satisfied and fix a sequence $\eta_i \overset{<}{\to} 1$. Then, there exists a sequence $\lambda_i \overset{<}{\to} 1$ with $V_i := V_{\eta_i}^{\lambda_i} \subset V$ and, thanks to Lemma 4.3.12, a unique morphism

$$\epsilon_i : p_2^* \mathcal{E}_{|W_i} \to p_1^* \mathcal{E}_{|W_i},$$

with $W_i := V_i \times \mathbf{B}^d(0, \eta_i^+)$, such that the Taylor isomorphism of $\mathcal{E}$ is induced on V_i by ϵ_i for each n. And it follows from Lemma 4.3.11 that ϵ_i and ϵ_{i+1} coincide on $W_i \cap W_{i+1} := W_{\eta_i}^{\lambda_{i+1}}$. Thus, if we set $W := \cup W_i$, we get a morphism

$$\epsilon : p_2^* \mathcal{E}_{|W} \to p_1^* \mathcal{E}_{|W},$$

which induces the Taylor isomorphisms on infinitesimal neighborhoods.

Exchanging the two factors in $P \times_S P$ simply turns τ_i into $-\tau_i$ and it follows that there also exists a morphism

$$\epsilon' : p_1^* \mathcal{E}_{|W'} \to p_2^* \mathcal{E}_{|W'},$$

such that the inverse of the Taylor isomorphisms of $\mathcal{E}$ is induced by ϵ'. Now, Lemma 4.3.11 again tells us that $\epsilon' \circ \epsilon = \mathrm{Id}$ and $\epsilon \circ \epsilon' = \mathrm{Id}$ on some strict neighborhood contained in $W \cap W'$. In other words, if we shrink W a little bit, we may assume that ϵ is an isomorphism. □

Corollary 4.3.13 *Let P be a smooth affine formal S-scheme with some étale coordinates in the neighborhood of a closed subvariety X of P_k.*

An integrable connection ∇ on a coherent $\mathcal{O}_{]X[_P}$-module $\mathcal{E}$ is convergent if and only if for each $\eta < 1$, we have

$$\forall s \in \Gamma([X]_\eta, \mathcal{E}), \quad \|\underline{\partial}^{[\underline{k}]}(s)\|\eta^{|\underline{k}|} \to 0.$$

Proof This is simply the particular case $X = Y$. □

Another very important application of Lemma 4.3.11 concerns the local nature of overconvergence. We prove an intermediate result.

Lemma 4.3.14 *Let $(X \subset Y \subset P)$ be a smooth S-frame, V a strict neighborhood of $]X[_P$ in $]Y[_P$ and $\mathcal{E}$ a coherent $\mathcal{O}_V$-module with an overconvergent integrable connection. Assume that, for $i = 1, 2$,*

$$\epsilon_i : p_2^*\mathcal{E}_{|V_i'} \simeq p_1^*\mathcal{E}_{|V_i'}$$

is a Taylor isomorphism on a strict neighborhood of $]X[_{P\times_S P}$ in $]Y[_{P\times_S P}$. Then, there exists a strict neighborhood $V' \subset V_1 \cap V_2'$ of $]X[_{P\times_S P}$ in $]Y[_{P\times_S P}$ such that $\epsilon_{2|V'} = \epsilon_{1|V'}$.

Proof Using Proposition 3.1.12, we see that the assertion is local on P. We may therefore assume that the frame is local. In particular, P is quasi-compact and we may therefore also assume that $V_1' = V_2'$ is a standard strict neighborhood associated to some sequences $\eta_n \overset{<}{\to} 1$ and $\lambda_n \overset{<}{\to} 1$. Refining the sequence λ_n if necessary, it follows from Lemma 4.3.11 that, with our usual notations, ϵ_1 and ϵ_2 coincide on $W_{\eta_n}^{\lambda_n}$ for each n. And therefore, they coincide on the standard strict neighborhood $W_{\underline{\eta}}^{\underline{\lambda}}$. □

Proposition 4.3.15 *Let $(X \subset Y \subset P)$ be a smooth S-frame, V a strict neighborhood of $]X[_P$ in $]Y[_P$ and $\mathcal{E}$ a coherent $\mathcal{O}_V$-module with an integrable connection. Let*

$$\begin{array}{ccccc} X_i & \hookrightarrow & Y_i & \hookrightarrow & P_i \\ \downarrow & & \downarrow & & \downarrow \\ X & \hookrightarrow & Y & \hookrightarrow & P \end{array}$$

be a finite open covering or a finite mixed covering, and for each $i \in I$, let V_i a be strict neighborhood of $]X_i[_{P_i}$ in $]Y_i[_{P_i} \cap V$, then the connection of $\mathcal{E}$ is overconvergent if and only if for each i, the connection of $\mathcal{E}_{|V_i}$ is overconvergent (with respect to $(X_i \subset Y_i \subset P_i)$).

Proof This is an immediate consequence of the uniqueness of the Taylor isomorphism in Lemma 4.3.14, using again Proposition 3.1.12 but also Proposition 3.4.10. □

4.4 Radius of convergence

We still assume in this section that $\mathrm{Char}K = 0$.

Definition 4.4.1 *Let $V \to T$ be a (smooth) morphism of rigid analytic varieties with étale coordinates $t_1, \ldots, t_d$, $\mathcal{E}$ a coherent $\mathcal{O}_V$-module with an integrable connection and*

$$\theta : \mathcal{E} \to \mathcal{E} \otimes_{\mathcal{O}_V} \mathcal{O}_V[[\underline{\tau}]])$$

its Taylor morphism. Let $\eta < 1$.

(i) *A section $s \in \Gamma(V, \mathcal{E})$ is* η-convergent *with respect to $t_1, \ldots, t_d$ if*

$$\theta(s) \in \Gamma(V, \mathcal{E} \otimes_{\mathcal{O}_V} \mathcal{O}_V\{\underline{\tau}/\eta\}).$$

A function $f \in \Gamma(V, \mathcal{O}_V)$ is η-convergent *if it is η-convergent when $\mathcal{O}_V$ is endowed with the trivial connection.*

(ii) *The module $\mathcal{E}$ is* η-convergent *with respect to $t_1, \ldots, t_d$ if any section of $\mathcal{E}$ on V is η-convergent. If $V = \mathrm{Spf} A$ is affinoid, we will also say that $M := \Gamma(V, \mathcal{E})$ is* η-convergent *with respect to $t_1, \ldots, t_d$.*

(iii) *Finally, V is* of η-type *with respect to $t_1, \ldots, t_d$ if the trivial module with the trivial connection is η-convergent.*

In other words, $\mathcal{E}$ is η-convergent on V if and only if the Taylor morphism

$$\theta : \Gamma(V, \mathcal{E}) \to \Gamma(V, \mathcal{E} \otimes_{\mathcal{O}_V} \mathcal{O}_V[[\tau]])$$

factors through $\Gamma(V, \mathcal{E} \otimes_{\mathcal{O}_V} \mathcal{O}_V\{\underline{\tau}/\eta\})$. This is a *global* condition.

These definitions depend heavily on the choice of the étale coordinates (even multiplying by a constant changes convergence). It should also be noticed that it is not $\mathcal{O}_V$-linear in general: if s is η-convergent and f is some function, then fs is not necessarily η-convergent (unless f itself is).

Finally, the notion overconvergence for a module is not stable by restriction to an open subset. For example, the unit disc $\mathbf{D}(0, 1^+)$ is of η type for any η with respect to the canonical coordinate, but the punctured disk $\mathbf{D}(0, 1^+) \setminus \{0\}$ is not of any η type since

$$\theta(1/t) = \sum \frac{(-1)^i}{t^{i+1}} \tau^i$$

is nowhere convergent.

However, we have the following results.

Proposition 4.4.2 *Let $V \to T$ be a smooth morphism of rigid analytic varieties with étale coordinates $t_1, \ldots, t_d$ and $\mathcal{E}$ a coherent $\mathcal{O}_V$-module with an integrable connection.*

(i) *Let $V = \cup V_i$ be an admissible covering. Then, a section $s \in \Gamma(V, \mathcal{E})$ is η-convergent on V if and only if $s_{|V_i}$ is η-convergent for each i.*
(ii) *Assume V is of η-type. Then, $\mathcal{E}$ is η-convergent if and only if s is η-convergent when s runs through a family of generators of $\Gamma(V, \mathcal{E})$.*

Proof The first assertion is an immediate consequence of the definition of a sheaf. More precisely, if

$$\mathcal{Q}_\eta := \operatorname{coker}(\mathcal{O}_V\{\underline{\tau}/\eta\} \to \mathcal{O}_V[[\underline{\tau}]]),$$

and

$$\mathcal{E}_\eta := \ker(\mathcal{E} \to \mathcal{E} \otimes_{\mathcal{O}_V} \mathcal{O}_V[[\underline{\tau}]] \to \mathcal{E} \otimes_{\mathcal{O}_V} \mathcal{Q}_\eta)$$

then, $s \in \Gamma(V, \mathcal{E})$ is η-convergent if and only if $s \in \Gamma(V, \mathcal{E}_\eta)$.

The second assertion follows from the fact that η-convergence is both an additive and a multiplicative notion. □

Corollary 4.4.3 *Let $V \to T$ be a smooth morphism of rigid analytic varieties with étale coordinates $t_1, \ldots, t_d$ and $\mathcal{E}$ a coherent $\mathcal{O}_V$-module with an integrable connection. Let $V = \cup V_i$ be an admissible covering. If $\mathcal{E}_{|V_i}$ is η-convergent for all i, then $\mathcal{E}$ is η-convergent and the converse is true if V is affinoid and each V_i is of η-type.*

Proof The direct implication follows from the first assertion of Proposition 4.4.2. The converse will hold thanks to Proposition 4.4.2 because then $\mathcal{E}$ is generated by its global sections. □

Proposition 4.4.4 *Let $V \to T$ be a smooth morphism of rigid analytic varieties with étale coordinates $t_1, \ldots, t_d$ and $\mathcal{E}$ a coherent $\mathcal{O}_V$-module with an integrable connection. Assume that $V = \operatorname{Spm} A$ is affinoid.*

(i) *Let $\| - \|$ be a Banach norm on $M := \Gamma(V, \mathcal{E})$. Then a section $s \in M$ is η-convergent if and only if*

$$\|\underline{\partial}^{[\underline{k}]}(s)\|\eta^{|\underline{k}|} \to 0.$$

(ii) *Let $t'_1, \ldots, t'_d$ be another set of étale coordinates on V and $\| - \|$ a Banach norm on A. Assume that*

$$\forall i = 1, \ldots, d, \quad \forall \underline{k} = (k_1, \ldots, k_d) \neq 0, \quad \|\underline{\partial}^{[\underline{k}]}(t'_i)\|\eta^{|\underline{k}|} \leq \eta'.$$

Then, if a section s of $\mathcal{E}$ is η'-convergent with respect to $t'_1, \dots, t'_d$, it is η-convergent with respect to $t_1, \dots, t_d$.

Proof Since $\theta(s) = \sum_{\underline{k}} \underline{\partial}^{[\underline{k}]}(s)\underline{\tau}^{\underline{k}}$, the first assertion results from the fact that, by definition, a formal power series

$$\sum_{\underline{k}=0}^{\infty} s_{\underline{k}}\underline{\tau}^{\underline{k}} \in M \otimes_A A[[\underline{\tau}]]$$

falls in

$$M \otimes_A A\{\underline{\tau}/\eta\}$$

if and only if

$$\|s_{\underline{k}}\|\eta^{|\underline{k}|} \to 0.$$

Now, the condition of the second assertion means that

$$\tau'_i = \theta(t'_i) - t'_i = \sum_{\underline{k}>0} \underline{\partial}^{[\underline{k}]}(t'_i)\underline{\tau}^{\underline{k}}$$

belongs to $A\{\underline{\tau}/\eta\}$ and that $\|\tau'_i\| \leq \eta'$. In other words, it means that $A\{\underline{\tau}'/\eta'\} \subset A\{\underline{\tau}/\eta\}$. Thus, clearly, if $\theta(s) \in A\{\underline{\tau}'/\eta'\}$, then *a fortiori*, $\theta(s) \in A\{\underline{\tau}/\eta\}$. □

The situation is very nice in the good reduction case as we will see shortly. In this case, we consider the following definition of η-convergence that does not depend on the choice of étale coordinates.

Definition 4.4.5 *Let $u : P \to S$ be a smooth morphism of formal $\mathcal{V}$-schemes and $\eta < 1$. Let $\mathcal{E}$ be a coherent $\mathcal{O}_{P_K}$-module with an integrable connection. Then, a section s of $\mathcal{E}$ is* η-convergent *if and only if there exists an open affine covering $P = \cup P_i$ with étale coordinates on each P_i which makes $s_{|P_{iK}}$ η-convergent (with respect to these coordinates). The module $\mathcal{E}$ itself is* η-convergent *if any section is so.*

Proposition 4.4.6 *Let $u : P \to S$ be a smooth morphism of formal $\mathcal{V}$-schemes, $\mathcal{E}$ a coherent $\mathcal{O}_{P_K}$-module with an integrable connection and $\eta < 1$. Then*

(i) *If we have étale coordinates $t_1, \dots, t_d$ on P, then a section s of $\mathcal{E}$ is η-convergent if and only if it is η-convergent with respect to $t_1, \dots, t_d$. In particular, $\mathcal{E}$ is η-convergent if and only if it is η-convergent with respect to $t_1, \dots, t_d$.*

(ii) *The module $\mathcal{E}$ is η-convergent if and only $\Gamma(P_K, \mathcal{E})$ is generated by η-convergent sections. In particular, the trivial module is always η-convergent.*

(iii) *If $P = \cup P_i$ is an open covering, then $\mathcal{E}$ is η-convergent if and only if $\mathcal{E}_{|P_{iK}}$ is also η-convergent for each i.*

Proof Before doing anything, assume that $P = \mathrm{Spf} A$ is affine with étale coordinates $t_1, \ldots, t_d$. If $f \in A$, then

$$\theta(f) = \sum_{\underline{k}} \underline{\partial}^{[\underline{k}]}(f)\underline{\tau}^{\underline{k}} \in A[[\underline{\tau}]]$$

and therefore, for each $k = (k_1, \ldots, k_d)$, we have $\underline{\partial}^{[\underline{k}]}(f) \in A$. On the other hand, if $\| - \|$ is a Banach norm on A_K and $f \in A$, then $\|f\| \leq 1$. It follows that if $k = (k_1, \ldots, k_d)$, then $\|\underline{\partial}^{[\underline{k}]}(f)\| \leq 1$. Thus, if $t'_1, \ldots, t'_d$ is another set of étale coordinates on A, the condition

$$\forall i = 1, \ldots, d, \quad \forall \underline{k} = (k_1, \ldots, k_d) \neq 0, \quad \|\underline{\partial}^{[\underline{k}]}(t'_i)\|\eta^{|\underline{k}|} \leq \eta.$$

is satisfied. It follows from the second part of Proposition 4.4.4 that a section s of $\mathcal{E}$ which is η-convergent with respect to $t'_1, \ldots, t'_d$ is also η-convergent with respect to $t_1, \ldots, t_d$. And conversely.

We consider now the first assertion of the proposition. We know from the first part of Proposition 4.4.2 that the condition is sufficient and we want to show that it is necessary. Thus, we assume that s is η-convergent, which means that there exists an open affine covering $P = \cup P_i$ and étale coordinates on each P_i such that $s_{|P_{iK}}$ is η-convergent with respect to this set of étale coordinate. But then, as we saw at the beginning of this proof, it is also η-convergent with respect to $t_1, \ldots, t_d$. It follows from the first part of Proposition 4.4.2, that s is η-convergent with respect to $t_1, \ldots, t_d$.

We prove now a special case of the second assertion, namely the case of the trivial module when $P = \mathrm{Spf} A$ is affine with étale coordinates $t_1, \ldots, t_d$. We show that P_K is of η-type with respect to these coordinates. If $f \in A_K$, there exists N such that $\pi^N f \in A$. Since our coordinates are defined on P, it follows that for all $k_1, \ldots, k_d \in \mathbf{N}$, we have $\underline{\partial}^{[\underline{k}]}(\pi^N f) \in A$ so that $\|\underline{\partial}^{[\underline{k}]}(f)\| \leq |\pi|^{-N}$. Thus, for each $\eta < 1$,

$$\|\underline{\partial}^{[\underline{k}]}(f)\|\eta^{\underline{k}} \to 0$$

and we can use the first assertion of Proposition 4.4.4.

Now, we jump to the last assertion. It follows from the second part of Proposition 4.4.2 that the condition is sufficient. For the converse, we may clearly assume that we are working with affine formal schemes and that we have étale coordinates on P. Any section of $\mathcal{E}$ on P_{iK} can be written $s = \sum f_k s_k$ with $f_k \in \mathcal{O}_{P_{iK}}$ and $s_k \in \mathcal{E}$. It follows from the second part of the proposition that each f_k is η-convergent and we know that each s_k is also η-convergent. It follows that s is η-convergent.

In order to prove the second assertion, we may therefore assume P affine with étale coordinates. Moreover, thanks to the second part of Proposition 4.4.2, it is sufficient to consider the trivial module. This has already be done. □

This last result shows that the notion of η-convergence is local with respect to the formal topology. We introduce now the notion of radius of convergence.

Definition 4.4.7 *Let $V \to T$ be a smooth morphism of rigid analytic varieties with étale coordinates $t_1, \ldots, t_d$. If $\mathcal{E}$ is a coherent $\mathcal{O}_V$-module with an integrable connection and $s \in \Gamma(V, \mathcal{E})$, the* radius of convergence *of s on V is*

$$R(s) = \sup\{\eta, s \text{ is } \eta\text{-convergent on } V\}.$$

And the radius of convergence *of $\mathcal{E}$ is*

$$R(\mathcal{E}, V) = \inf_{s \in \Gamma(V,\mathcal{E})} R(s).$$

In other words, $R(s)$ is characterized by the fact that s is η-convergent when $\eta < R(s)$ and that it is not η-convergent if $\eta > R(s)$. And we have the analog statement for $R(\mathcal{E}, V)$: $\mathcal{E}$ is η-convergent when $\eta < R(\mathcal{E}, V)$ and it is not η-convergent if $\eta > R(\mathcal{E}, V)$.

Proposition 4.4.8 *Let $V \to T$ be a smooth morphism of rigid analytic varieties with étale coordinates $t_1, \ldots, t_d$, $\mathcal{E}$ a coherent $\mathcal{O}_V$-module with an integrable connection and $s \in \Gamma(V, \mathcal{E})$.*

(i) *If $V = \cup V_i$ is an admissible covering, then,*

$$R(s) = \inf_i R(s_{|V_i}).$$

(ii) *If $V = \operatorname{Spm} A$ is affine and $\| - \|$ is a Banach norm on $\Gamma(V, \mathcal{E})$, then,*

$$R(s) = \inf\{1, \underline{\lim}_k \|\partial^{[\underline{k}]}(s)\|^{-1/|\underline{k}|}\}.$$

Proof The first assertion directly follows from Proposition 4.4.2.

We consider now the second assertion. Assume that s is η-convergent. We saw in Proposition 4.4.4 that

$$\|\partial^{[\underline{k}]}(s)\|\eta^{|\underline{k}|} \to 0.$$

Thus, there exists N such that, for $|k| \geq N$, we have $\|\partial^{[\underline{k}]}(s)\|\eta^{|\underline{k}|} \leq 1$ and consequently, $\eta \leq \|\partial^{[k]}(s)\|^{-1/|\underline{k}|}$. In other words, we have

$$\eta \leq \underline{\lim}_k \|\partial^{[\underline{k}]}(s)\|^{-1/|\underline{k}|}.$$

And since this is true whenever s is η-convergent, we get

$$R(s) \leq \underline{\lim}_k \|\partial^{[\underline{k}]}(s)\|^{-1/|\underline{k}|}.$$

Conversely, assume that

$$\eta < \delta := \underline{\lim}_k \|\partial^{[\underline{k}]}(s)\|^{-1/|\underline{k}|}.$$

Then, there exists N such that, for $|k| \geq N$, we have $\delta \leq \|\partial^{[k]}(s)\|^{-1/|\underline{k}|}$ and consequently $\|\partial^{[\underline{k}]}(s)\|\delta^{|\underline{k}|} \leq 1$. Since $\eta < \delta$, it follows that

$$\|\partial^{[\underline{k}]}(s)\|\eta^{|\underline{k}|} \to 0$$

and s is η-convergent. □

Note however that if $V = \cup_i V_i$ is an admissible covering, then

$$R(\mathcal{E}, V) \neq \inf_i R(\mathcal{E}, V_i)$$

in general because this is not even true for the trivial crystal on the unit disc with respect to the standard coordinate.

In the good reduction case, we can define the notion of radius of convergence globally even when there are no étale coordinates.

Definition 4.4.9 *Let P be a smooth formal S-scheme. If $\mathcal{E}$ is a coherent $\mathcal{O}_{P_K}$-module with an integrable connection and $s \in \Gamma(P_K, \mathcal{E})$, the* radius of convergence *of s on V is*

$$R(s) = \sup\{\eta, s \text{ is } \eta\text{-}convergent\ on\ P_K\}.$$

And the radius of convergence *of $\mathcal{E}$ is*

$$R(\mathcal{E}, P_K) = \inf_{s \in \Gamma(P_K, \mathcal{E})} R(s).$$

We can now use Theorem 4.3.9 to show that, in the case of curves, the overconvergence condition may be expressed in the same terms as the "soluble" condition of Christol and Mebkhout ([28], [29], [30], [31]). We do it in a more general setting.

Proposition 4.4.10 *Let $\mathcal{Y}$ be an affine formal scheme over S with special fiber Y and X the complement of a hypersurface Z in Y. As usual, we write $V^\lambda :=]Y[_P \setminus]Z[_{P\lambda}$.*

Let V be a smooth strict neighborhood of $]X[_\mathcal{Y}$ in $\mathcal{Y}_K$. Assume that we have some coordinates $t_1, \ldots, t_d$ defined on $\mathcal{Y}$ and étale on X.

Then, an integrable connection on a coherent $\mathcal{O}_V$-module $\mathcal{E}$ is overconvergent if and only if it satisfies the following condition

$$\lim_{\lambda \overset{<}{\to} 1} R(\mathcal{E}, V^\lambda) = 1.$$

It is actually sufficient that

$$\lim_{\lambda \overset{<}{\to} 1} R(s_{|V^\lambda}) = 1$$

when s runs through a set of generators of $\mathcal{E}$ on V.

Proof In the case $Y = P_k$, Theorem 4.3.9 tells us that $\mathcal{E}$ is overconvergent if and only if it satisfies the following condition

$$\forall \eta < 1, \exists \lambda_0 < 1, \forall \lambda \geq \lambda_0, \forall s \in \Gamma(V^\lambda, \mathcal{E}), \quad \|\underline{\partial}^{[k]}(s)\|_\lambda \eta^{|\underline{k}|} \to 0$$

with $\| - \|_\lambda$ a Banach norm on V^λ. This means that given $\eta < 1$, $\mathcal{E}$ is η-convergent on V^λ when λ is sufficiently close to 1. This is clearly equivalent to the first condition.

Now, the trivial module with the trivial connection is overconvergent by definition, and it follows that, for fixed η, V_λ is of η-type for λ close to 1. Using the last assertion of Proposition 4.4.2, one sees that, in order to check η-convergence, it is sufficient to consider a set of generators. □

In practice, we might simply write $R(\mathcal{E}, \lambda)$ or $R(s, \lambda)$ instead of $R(\mathcal{E}, V^\lambda)$ or $R(s_{|V^\lambda})$.

Corollary 4.4.11 *Let $\mathcal{X}$ be a smooth formal scheme over S. An integrable connection ∇ on a coherent $\mathcal{O}_{\mathcal{X}_K}$-module $\mathcal{E}$ is convergent if and only if*

$$R(\mathcal{E}, \mathcal{X}_K) = 1.$$

Proof The question is local and therefore reduces to the particular case of Proposition 4.4.10 where $X = Y$. □

Of course, the condition of the corollary means that $\mathcal{E}$ is η-convergent for each $\eta < 1$.

We can come back to our examples. In particular, we assume that K has mixed characteristic p.

First of all, we consider the frame

$$(\mathbf{A}^1_k \subset \mathbf{P}^1_k \subset \widehat{\mathbf{P}}^1_{\mathcal{V}})$$

and $\mathcal{L}_\alpha$ on $\mathbf{D}(0, \lambda^{-1})$ with $\lambda < 1$. If θ denotes the generator of $\mathcal{L}_\alpha$ and we set $\|\theta\|_\lambda = 1$, we have

$$\|\partial^{[k]}(\theta)\|_\lambda^{-1/k} = \left\|\frac{(-\alpha)^k}{k!}\theta\right\|_\lambda^{-1/k} = \frac{|k!|^{1/k}}{|\alpha|}.$$

Using the usual estimates

$$|p|^{\frac{1}{p-1}} \leq |k!|^{1/k} \leq |p|^{\frac{1}{p-1} - \frac{\log_p k}{k}},$$

it follows that

$$\underline{\lim}_k \|\partial^{[k]}(\theta)\|_\lambda^{-1/k} = \frac{|p|^{\frac{1}{p-1}}}{|\alpha|}.$$

And therefore

$$R(\theta, \lambda) = \inf\{1, \underline{\lim}_k \|\partial^{[\underline{k}]}(\theta)\|^{-1/|\underline{k}|}\} = \inf(1, \frac{|p|^{\frac{1}{p-1}}}{|\alpha|}).$$

Thus, we see that $\mathcal{L}_\alpha$ is overconvergent if and only if $|\alpha| \leq |p|^{\frac{1}{p-1}}$. Actually, it is overconvergent and non trivial if and only if $|\alpha| = |p|^{\frac{1}{p-1}}$.

For the second example, we consider the frame

$$(\mathbf{A}^1_k \setminus 0 \subset \mathbf{P}^1_k \subset \widehat{\mathbf{P}}^1_{\mathcal{V}})$$

and $\mathcal{K}_\alpha$ on $\mathbf{A}(0, \lambda, \lambda^{-1})$. If θ denotes the generator and we set $\|\theta\|_\lambda = 1$, we have

$$\|\partial^{[k]}(\theta)\|_\lambda^{-1/k} = \left\|\binom{\beta}{k}\frac{1}{t^k}\theta\right\|_\lambda^{-1/k} = \frac{\lambda}{|\binom{\beta}{k}|^{1/k}}.$$

If $\beta \in \mathbf{Z}_p$, then $|\binom{\beta}{k}| \leq 1$ and it follows that

$$\|\partial^{[k]}(\theta)\|_\lambda^{-1/k} \geq \lambda.$$

Thus, we see that $R(\theta, \lambda) \geq \lambda$ and therefore, $\mathcal{K}_\beta$ is overconvergent.

Before going any further, note that replacing the generator θ by $t^n\theta$ for some $n \in \mathbf{Z}$ corresponds to turning β into $\beta + n$. In other words, we may always add or subtract an integer n to β without changing the nature of the connection.

Assume now that $\beta \in \mathcal{V} \setminus \mathbf{Z}_p$. We follow Berthelot in [10], 2.3.ii). First of all, we may assume that

$$\forall i \in \mathbf{Z}, \quad 0 < |\beta| \leq |\beta - i|.$$

More precisely, since $\mathbf{Z}_p$ is the closure of $\mathbf{Z}$ in $\mathcal{V}$, there exists $a \in \mathbf{N}$ such that $|p|^a < \inf_{i \in \mathbf{Z}} |\beta - i|$. It follows that $|\beta - i| = |\beta - j|$ whenever $i \equiv j$

mod p^a because then, $|i-j|<|p|^a$. In particular, $|\beta-i|$ takes a finite number of values and therefore, there exists $i_0\in\mathbf{Z}$ such that

$$\forall i\in\mathbf{Z},\quad 0<|\beta-i_0|\leq|\beta-i|.$$

We may then replace θ by $t^{i_0}\theta$ and get the asserted inequalities. Note that we may choose a such that $|p|^a<|\beta|\leq|p|^{a-1}$. Then, I claim that

$$|\beta-i|=\begin{cases}|\beta| & \text{if} \quad p^a|i\\ |i| & \text{if} \quad p^a\not|i\end{cases}.$$

If $p^a|i$, then $|i|\leq|p|^a<|\beta|$ and therefore, $|\beta-i|=|\beta|$. On the contrary, if $p^a\not|i$, then $|i|\geq|p|^{a-1}\geq|\beta|$. If the composite inequality is strict, then $|\beta-i|=|i|$ as asserted. Otherwise, we have $|i|=|\beta|$ and since $|\beta-i|\geq|\beta|$, then necessarily,

$$|\beta-i|=|\beta|=|i|.$$

Thus, if we write $k=k'p^a+r$ with $1\leq r\leq p^a$, we have

$$\left|\prod_{i=0}^{k-1}(\beta-i)\right|=\prod_{j=1}^{k'}\left|\frac{\beta}{p^aj}\right|\prod_{i=0}^{k-1}|i|=\left|\frac{\beta^{k'}(k-1)!}{p^{k'a}k'!}\right|$$

and therefore

$$\left|\binom{\beta}{k}\right|^{-1/k}=\left|\frac{\beta^{k'}}{kp^{k'a}k'!}\right|^{-1/k}.$$

Using our usual estimate $|k'!|\sim|p|^{k'/(p-1)}$ and $k'/k\sim 1/p^a$, we obtain

$$R(\theta,\lambda)=\left|\frac{p^{a+\frac{1}{p-1}}}{\beta}\right|^{1/p^a}\lambda<|p|^{\frac{1}{p^a(p-1)}}\lambda$$

and it follows that

$$\lim_\lambda R(\theta,\lambda)<|p|^{\frac{1}{p^a(p-1)}}<1$$

which shows that the connection is *not* overconvergent.

Finally, if $\beta\in K\setminus\mathcal{V}$, then $|\beta|>1\geq i$ for all $i\in\mathbf{Z}$ and therefore

$$\left|\binom{\beta}{k}^{-1/k}\right|=\left|\frac{\beta^k}{k!}\right|^{-1/k}\to\left|\frac{p^{1/(p-1)}}{\beta}\right|$$

and since $|\beta|>1>|p|^{1/(p-1)}$, we see that here again, our module is not overconvergent.

In other words, $\mathcal{K}_\beta$ is overconvergent and non trivial if and only if $\beta\in\mathbf{Z}_p\setminus\mathbf{Z}$. Note however that β might be a Liouville number!

As before, we may pull back Dwork and Kummer crystals and mix them. More precisely, it follows from Proposition 4.3.5 that if we are given a morphism of frames

$$\begin{array}{ccccc} X & \hookrightarrow & Y & \hookrightarrow & P \\ \downarrow & & \downarrow & & \downarrow u \\ \mathbf{A}^1_k & \hookrightarrow & \mathbf{P}^1_k & \hookrightarrow & \widehat{\mathbf{P}}^1_{\mathcal{V}} \end{array}$$

and $|\alpha| \leq |p|^{1/p-1}$, then the stratification of $u_K^* \mathcal{L}_\alpha$ is overconvergent on some strict neighborhood of $]X[_P$ in $]Y[_P$. For the same reason, if $\beta \in \mathbf{Z}_p$ and we are given a morphism of frames

$$\begin{array}{ccccc} X & \hookrightarrow & Y & \hookrightarrow & P \\ \downarrow & & \downarrow & & \downarrow v \\ \mathbf{A}^1_k \setminus 0 & \hookrightarrow & \mathbf{P}^1_k & \hookrightarrow & \widehat{\mathbf{P}}^1_{\mathcal{V}} \end{array}$$

then the stratification of $v_K^* \mathcal{K}_\beta$ is overconvergent on some strict neighborhood of $]X[_P$ in $]Y[_P$. Any tensor product of such modules will define a module with an overconvergent connection. They give rise to general exponential sums.

Another example on the frame

$$(\mathbf{A}^1_k \setminus 0 \subset \mathbf{P}^1_k \subset \widehat{\mathbf{P}}^1_{\mathcal{V}})$$

is given as follows : $F \in \mathcal{V}[X]$ is a split separable polynomial of degree $d > 0$, and we consider, as we already did, the curve U defined by

$$y^r = F(x) \quad \text{with} \quad y \in \mathbf{A}(0, \lambda, \lambda^{-1})$$

when $p \not| r$ and its projection p onto

$$V_\lambda := \{x \in \mathbf{D}(0, \lambda^{-r/d}), \quad |F(x)| > \lambda^r\}.$$

Then, $p_* \mathcal{O}_V$ is overconvergent. This is a general fact because p is finite and étale but it also follows from

$$p_* \mathcal{O}_V = \mathcal{O}_V \oplus \mathcal{K}_{\frac{1}{r}, F}$$

where F comes from the morphism of frames

$$\begin{array}{ccccc} \mathbf{A}^1_k \setminus \{\overline{c}_1, \cdots, \overline{c}_d\} & \hookrightarrow & \mathbf{P}^1_k & \hookrightarrow & \widehat{\mathbf{P}}^1_{\mathcal{V}} \\ \downarrow \overline{F} & & \downarrow \overline{F} & & \downarrow F \\ \mathbf{A}^1_k \setminus 0 & \hookrightarrow & \mathbf{P}^1_k & \hookrightarrow & \widehat{\mathbf{P}}^1_{\mathcal{V}} \end{array}$$

It would also be interesting to prove the overconvergence of the Legendre differential equation or more general hypergeometric differential equations as well as the Bessel differential equation.

Our next step consists is the study of the Monsky–Washnitzer situation. We could develop a theory for weakly formal schemes and dagger spaces (see [67] and [47]) but will stick to the affine situation. We first need to recall the following definition of a *weakly complete algebra* ([72], [69] and [70] (see also [83])). First, we define

$$\mathcal{V}[\underline{T}]^\dagger := \{\sum_{\underline{k}} a_{\underline{k}} \underline{T}^{\underline{k}} \in \mathcal{V}\{\underline{T}\}, \exists \rho > 1, |a_{\underline{k}}|\rho^{|\underline{k}|} \to 0\}.$$

A weakly complete algebra is a quotient $A := \mathcal{V}[\underline{T}]^\dagger/I$ where I is an ideal of finite type. Almost everything below should be independent of the presentation of A as a quotient of $\mathcal{V}[\underline{T}]^\dagger$. We will however always assume that such a presentation is fixed. We may consider the completion $\widehat{A} = \mathcal{V}\{\underline{T}\}/I\mathcal{V}\{\underline{T}\}$ of A as well as its generic fiber $A_K = K \otimes_{\mathcal{V}} A$ and we have $A := \widehat{A} \cap A_K$. Actually, we will denote by $K[\underline{T}]^\dagger$ the generic fiber of $\mathcal{V}[\underline{T}]^\dagger$.

As usual, we will write for $\rho \geq 1$,

$$K\{\underline{T}/\rho\} := \{\sum_{\underline{k}} a_{\underline{k}} \underline{T}^{\underline{k}} \in K[[\underline{T}]], |a_{\underline{k}}|\rho^{|\underline{k}|} \to 0\}.$$

If we set

$$A_\rho := K\{\underline{T}/\rho\}/IK\{\underline{T}/\rho\},$$

then, we have

$$A_K = \varinjlim_{\rho \overset{>}{\to} 1} A_\rho.$$

Later on, we will need the following lemma:

Lemma 4.4.12 *Let A be a weakly complete $\mathcal{V}$-algebra. If we endow A_ρ with the quotient norm $\| - \|_\rho$ and $f \in A_K$, we have*

$$\inf_{\rho \overset{>}{\to} 1} \|f\|_\rho = \|f\|_1.$$

Proof If $F \in K[\underline{T}]^\dagger$, there exists ρ_0 such that $F \in K\{\underline{T}/\rho_0\}$. If we write $F =: \sum_{\underline{k}} a_{\underline{k}} \underline{T}^{\underline{k}}$, we have $\|F\|_{\rho_0} = \sup_{\underline{k}} |a_{\underline{k}}|\rho_0^{|\underline{k}|}$. There exists a multiindex $\underline{k}_0$ such that $|a_{\underline{k}}|\rho_0^{|\underline{k}|} \leq |a_{\underline{k}_0}|\rho_0^{|\underline{k}_0|}$ whenever $|\underline{k}| \geq |\underline{k}_0|$. It follows that if $\rho \leq \rho_0$, we also have $|a_{\underline{k}}|\rho^{|\underline{k}|} \leq |a_{\underline{k}_0}|\rho^{|\underline{k}_0|}$ whenever $|\underline{k}| \geq |\underline{k}_0|$ because then, $\rho^{|k|-|k_0|} \leq \rho_0^{|k|-|k_0|}$.

We may therefore write

$$\|F\|_\rho = \sup_{|\underline{k}|\leq|\underline{k}_0|} |a_{\underline{k}}|\rho^{|\underline{k}|}$$

and thus,

$$\lim_{\rho>1} \|F\|_\rho = \sup_{|\underline{k}|\leq|\underline{k}_0|} \lim_{\rho>1} |a_{\underline{k}}|\rho^{|\underline{k}|} = \sup_{|\underline{k}|\leq|\underline{k}_0|} |a_{\underline{k}}| = \|F\|_1.$$

The general case follows from the previous one because

$$\|f\|_\rho = \inf_{\overline{F}=f} \|F\|_\rho.$$

□

As a first application, we can prove the following result that we will only use later on.

Proposition 4.4.13 *If $\varphi : A \to A'$ is a morphism of weakly complete $\mathcal{V}$-algebra and $\rho > 1$, there exists $\rho' > 1$ such that φ_K sends A_ρ inside $A'_{\rho'}$.*

Proof We write A as a quotient of $\mathcal{V}[\underline{T}]^\dagger$. Then, there exists ρ' such that, for $i = 1, \ldots, N$, $\varphi(T_i) \in A_{\rho'}$. Also, since $\|\varphi(T_i)\|_1 \leq 1$, we may assume thanks to Lemma 4.4.12 that $\|\varphi(T_i\|_{\rho'} \leq \rho$ and we are done. □

For any ring A, we denote by Coh(A) the category of coherent (left) modules on A. When A is a weakly complete $\mathcal{V}$-algebra with a given presentation and corresponding A_ρ's, we have an equivalence of categories

$$\mathrm{Coh}(A_K) \simeq \varinjlim \mathrm{Coh}(A_\rho).$$

Assume now that R is a commutative ring, A is a commutative R-algebra, and that there exists a universal module of differential $\Omega^1_{A/R}$ for derivations into *coherent* modules. Then, we say that $t_1, \ldots, t_m \in A$ are *étale coordinates*, if $dt_1, \ldots, dt_m$ form a basis of $\Omega^1_{A/R}$. In general, a connection on a coherent module always means with respect to *this* $\Omega^1_{A/R}$ and we will denote by MIC(A/R), or simply MIC(A) if R is understood from the context, the category of coherent modules with an integrable connection on A/R. It is important to mention that if A is a formally smooth weakly complete $\mathcal{V}$-algebra, we have

$$\mathrm{MIC}(A_K) \simeq \varinjlim \mathrm{MIC}(A_\rho).$$

Definition 4.4.14 *Let A be a formally smooth weakly complete $\mathcal{V}$-algebra with étale coordinates $t_1, \ldots, t_m$ on A_K. A coherent A_K-module M with an integrable connection is η-*convergent *if there exists $\rho_0 > 1$ and an A_{ρ_0}-module* *with an integrable connection M_0 such that*

$$M = A_K \otimes_{A_{\rho_0}} M_0$$

and, for $1 < \rho \leq \rho_0$, the module

$$M_\rho := A_\rho \otimes_{A_{\rho_0}} M_0$$

is η-convergent with respect to $t_1, \ldots, t_m$.

Proposition 4.4.15 *Let A be a formally smooth weakly complete $\mathcal{V}$-algebra with étale coordinates $t_1, \ldots, t_m$, $\mathcal{M}$ a coherent A-module with an integrable connection and $\eta < 1$. Then $K \otimes_{\mathcal{V}} \mathcal{M}$ is η-convergent if* $\operatorname{Char} k = 0$ *or $\eta < |p|^{\frac{1}{p-1}}$ and* $\operatorname{Char} k = p > 0$.

Proof For ρ close enough to 1, there exists an essentially unique A_ρ-module with an integrable connection M_ρ such that

$$K \otimes_{\mathcal{V}} \mathcal{M} = A_K \otimes_{A_\rho} M_\rho.$$

We fix some $\eta < |p|^{\frac{1}{p-1}}$ (or $\eta < 1$ if $p = 0$) and we want to show that M_ρ is η-convergent when ρ is close enough to 1.

We endow A_ρ with a quotient norm and M_ρ with the norm inherited from a free presentation of $\mathcal{M}$. We want to show that when ρ is close enough to 1, then for all $s \in M_\rho$, we have

$$\|\underline{\partial}^{[\underline{k}]}(s)\|_\rho \eta^{|\underline{k}|} \to 0.$$

Since we always have

$$|\frac{1}{k!}| \leq |p|^{-\frac{k}{p-1}},$$

it is sufficient to show that

$$\|\underline{\partial}^{\underline{k}}(s)\|_\rho \left(\frac{\eta}{|p|^{\frac{1}{p-1}}}\right)^{|\underline{k}|} \to 0.$$

We can choose some $\frac{\eta}{|p|^{\frac{1}{p-1}}} < \epsilon < 1$ and it is sufficient to show that the sequence $\|\underline{\partial}^{\underline{k}}(s)\|_\rho \epsilon^{|\underline{k}|}$ is bounded. We can even give a precise bound.

Claim 4.4.16

$$\forall \epsilon < 1, \exists \rho_0 > 1, \forall 1 < \rho \leq \rho_0, \forall \underline{k} \in \mathbf{N}^m, \forall s \in M_\rho, \quad \|\underline{\partial}^{\underline{k}}(s)\|_\rho \epsilon^{|\underline{k}|} \leq \|s\|_\rho.$$

Since we use the quotient norm, if $f \in \widehat{A}_K$, there exists $\alpha \in K$ such that $\|f\|_1 = |\alpha|$. It follows that $\|\alpha^{-1} f\|_1 = 1$ and in particular $\alpha^{-1} f \in \widehat{A}$. Since our coordinates are defined on $\widehat{A}$, for all $\underline{k}$, we have $\partial^{[\underline{k}]}(\alpha^{-1} f) \in \widehat{A}$ which means that $\|\partial^{[\underline{k}]}(\alpha^{-1} f)\|_1 \leq 1$ and therefore

$$\|\partial^{[\underline{k}]}(f)\|_1 \leq |\alpha| = \|f\|_1.$$

Now, if $f \in A_K$, since $\| - \|_\rho \to \| - \|_1$ when $\rho \to 1$, it follows that if we fix $\epsilon < 1$, there exists $\rho_0 > 1$ such that

$$\|\partial^{[\underline{k}]}(f)\|_{\rho^\epsilon} \leq \|f\|_\rho$$

when $1 < \rho \leq \rho_0$. A priori, ρ_0 depends on k and f, but it is always possible to choose a ρ_0 in such a way that

$$\forall i = 1, \dots, m, \forall j = 1, \dots, N, \quad \|\partial_i(\overline{T}_j)\|_{\rho^\epsilon} \leq \|\overline{T}_j\|_\rho.$$

Now, if $f = \overline{F}$ with $F = \sum_{\underline{k}} a_{\underline{k}} \underline{T}^{\underline{k}}$, we have

$$\partial_i(f) = \sum_j \sum_{\underline{k}} a_{\underline{k}} \overline{\underline{T}}^{\underline{k}-1_j} \partial_i(\overline{T}_j)$$

and therefore

$$\|\partial_i(f)\|_{\rho^\epsilon} \leq \sup_j \sup_{\underline{k}} |a_{\underline{k}}| \|\overline{\underline{T}}\|_\rho^{|\underline{k}|-1} \|\partial_i(\overline{T}_j)\|_{\rho^\epsilon} = \sup_{\underline{k}} |a_{\underline{k}}| \|\overline{\underline{T}}\|_\rho^{|\underline{k}|} \leq \|F\|_\rho.$$

This being true for any F with $\overline{F} = f$, the claim is proved when ($\underline{k} = 1_i$ and) $\mathcal{M} = A$.

Now, we endowed M_ρ with the norm coming from a free presentation of $\mathcal{M}$. It means, that we chose generators $\{s_j\}$ of $\mathcal{M}$ and let

$$\|s\| := \inf\{\sup_j \|f_j\|, s = \sum_j f_j s_j\}.$$

Since the connection is defined on $\mathcal{M}$, we may assume that $\|\partial_i(s_j)\|_{\rho^\epsilon} \leq 1$ for all i. If $s = \sum f_j s_j \in M_\rho$, the Leibnitz rule gives

$$\partial_i(s) = \sum_j f_j \partial_i(s_j) + \partial_i(f_j) s_j.$$

Since

$$\|f_j \partial_i(s_j)\|_{\rho^\epsilon} \leq \|f_j\|_\rho \|\partial_i(s_j)\|_{\rho^\epsilon} \leq \|f_j\|_\rho$$

and

$$\|\partial_i(f_j) s_j\|_{\rho^\epsilon} \leq \|\partial_i(f_j)\|_\rho \|s_j\|_{\rho^\epsilon} \leq \|f_j\|_\rho$$

we have

$$\|\partial_i(s)\|_{\rho^\epsilon} \leq \sup_j \|(f_j)\|_\rho$$

and therefore

$$\|\partial_i(s)\|_\rho \epsilon \leq \|s\|_\rho.$$

This being true for all $i = 1, \ldots, m$, it follows that for each $\underline{k} \in \mathbf{N}^m$, we have

$$\|\partial^k(s)\|_\rho \epsilon^k \leq \|s\|_\rho$$

□

We may also define the radius of convergence on weakly complete algebras.

Definition 4.4.17 *Let A be a formally smooth weakly complete $\mathcal{V}$-algebra with étale coordinates $t_1, \ldots, t_m$ on A_K and M be a coherent A_K-module with an integrable connection. Let $\rho_0 > 1$ and M_0 a coherent A_{ρ_0}-module with an integrable connection such that*

$$M = A_K \otimes_{A_{\rho_0}} M_0$$

and for each $\rho \leq \rho_0$, let

$$M_\rho = A_\rho \otimes_{A_{\rho_0}} M_0.$$

If $s \in M$, the ρ-radius of convergence *of s is*

$$R(s, \rho) = R(s \in M_\rho).$$

The ρ-radius of convergence *of M is*

$$R(M, \rho) = R(M_\rho).$$

Finally, M is said overconvergent *with respect to $t_1, \ldots, t_m$ if*

$$\lim_{\rho \overset{>}{\to} 1} R(M, \rho) = 1.$$

It might be better to define

$$R(s, \rho) = \inf(\rho^{-1}, R(s \in M_\rho))$$

and

$$R(M, \rho) = \inf(\rho^{-1}, R(M_\rho)).$$

This would not change the notion of overconvergence.

Now we have the following corollary to Proposition 4.4.15:

Corollary 4.4.18 *Let A be a formally smooth weakly complete $\mathcal{V}$-algebra with étale coordinates $t_1, \ldots, t_m$, $\mathcal{M}$ a coherent A-module with an integrable* *connection and* $\eta < 1$. *If* $\operatorname{Char} k = p > 0$, *we have*

$$\lim_{\rho \overset{>}{\to} 1} R(K \otimes_{\mathcal{V}} \mathcal{M}, \rho) \geq |p|^{\frac{1}{p-1}}.$$

Proof This is clear. □

Of course, we may consider our usual examples. We have Dwork module $K[t]^{\dagger}\theta$ with

$$\nabla(\theta) = -\alpha\theta \otimes dt.$$

which is non trivial and overconvergent if and only if $|\alpha| = |p|^{\frac{1}{p-1}}$. We also have Kummer module $K[t, t^{-1}]^{\dagger}\theta$ with

$$\nabla(\theta) = -\beta\theta \otimes \frac{dt}{t}.$$

which is non trivial and overconvergent if and only if $|\beta| \in \mathbf{Z}_p \setminus \mathbf{Z}$.

We now turn to the Robba ring situation. It is possible, and even necessary in order to prove many results in rigid cohomology, to introduce Robba rings of higher dimension on weakly complete algebras. However, we will stick here to the classical case. We should also notice that, in the literature, the valuation is generally assumed to be discrete or at least, one requires that K is maximally complete. Recall that the *Robba ring* on K (see for example [33] or [81]) is defined as

$$\mathcal{R} := \left\{ \sum_{-\infty}^{+\infty} a_k t^k, a_k \in K, \begin{cases} \exists \lambda < 1, |a_k|\lambda^k \to 0 & \text{for} \quad k \to -\infty \\ \forall \epsilon < 1, |a_k|\epsilon^k \to 0 & \text{for} \quad k \to +\infty \end{cases} \right\}.$$

Again, it might be convenient to write $\mathcal{R} = \cup_{\lambda \overset{<}{\to} 1} R_{\lambda}$ where

$$\mathcal{R}^{\lambda} = \left\{ \sum_{-\infty}^{+\infty} a_k t^k, a_k \in K, \begin{cases} |a_k|\lambda^k \to 0 & \text{for} \quad k \to -\infty \\ \forall \epsilon < 1, |a_k|\epsilon^k \to 0 & \text{for} \quad k \to +\infty \end{cases} \right\}$$

is the ring of rigid analytic functions on the annulus $\mathbf{A}(0, \lambda^+, 1^-)$. Thus, we have

$$\mathcal{R}^{\lambda} = \cap_{\lambda \leq \epsilon \overset{<}{\to} 1} K\{t/\epsilon, \lambda/t\}$$

and therefore,

$$\mathcal{R} = \cup_{\lambda \overset{<}{\to} 1} \cap_{\lambda \leq \epsilon \overset{<}{\to} 1} K\{t/\epsilon, \lambda/t\}.$$

It is not difficult to see that, with our above conventions (namely with respect to derivations into coherent modules), we have

$$\Omega^1_{\mathcal{R}} := \Omega^1_{\mathcal{R}/K} = \mathcal{R}dt/t$$

and that $d : \mathcal{R} \to \mathcal{R}dt/t$ is the obvious map. We will consider the category $\mathrm{Coh}(\mathcal{R})$ of coherent modules (necessarily free when the valuation is discrete because then, $\mathcal{R}$ is a Bézout ring) as well as the category $\mathrm{MIC}(\mathcal{R}) = \mathrm{MIC}(\mathcal{R}/K)$ of coherent modules with a connection (necessary integrable) on $\mathcal{R}$. Of course, we have

$$\mathrm{Coh}(\mathcal{R}) = \varinjlim_{\lambda \overset{<}{\to} 1} \mathrm{Coh}(\mathcal{R}_\lambda). \quad \text{and} \quad \mathrm{MIC}(\mathcal{R}) = \varinjlim_{\lambda \overset{<}{\to} 1} \mathrm{MIC}(\mathcal{R}_\lambda)$$

and also

$$\mathrm{Coh}(\mathcal{R}_\lambda) = \varprojlim_{\lambda \le \epsilon \overset{<}{\to} 1} \mathrm{Coh}(\mathcal{K}\{t/\epsilon, \lambda/t\}) \quad \text{and} \quad \mathrm{MIC}(\mathcal{R}) = \varprojlim_{\lambda \le \epsilon \overset{<}{\to} 1} \mathrm{MIC}(\mathcal{K}\{t/\epsilon, \lambda/t\})$$

so that

$$\mathrm{Coh}(\mathcal{R}) = \varinjlim_{\lambda \overset{<}{\to} 1} \varprojlim_{\lambda \le \epsilon \overset{<}{\to} 1} \mathrm{Coh}(K\{t/\epsilon, \lambda/t\})$$

and

$$\mathrm{MIC}(\mathcal{R}) = \varinjlim_{\lambda \overset{<}{\to} 1} \varprojlim_{\lambda \le \epsilon \overset{<}{\to} 1} \mathrm{MIC}(K\{t/\epsilon, \lambda/t\}).$$

Definition 4.4.19 *Let $\mathcal{R}$ be the Robba ring of K and $\eta < 1$. A coherent $\mathcal{R}$-module M with an integrable connection is* η-convergent *if there exists $\lambda_0 < 1$ and an $\mathcal{R}^{\lambda_0}$-module with an integrable connection M_0 such that*

$$M = \mathcal{R}_K \otimes_{\mathcal{R}^{\lambda_0}} M_0$$

and, for $\lambda_0 \le \lambda \le \epsilon < 1$, the module

$$M_{\lambda,\epsilon} := K\{t/\epsilon, \lambda/t\} \otimes_{\mathcal{R}^{\lambda_0}} M_0$$

is η-convergent (with respect to t).

Of course, this means that the corresponding module on $\mathbf{A}(0, \lambda^+, 1^-)$ is η-convergent. Also, it is clearly sufficient to consider the case $\lambda = \epsilon$ in our definition and we will do it in the future.

We may also define the radius of convergence on Robba rings.

Definition 4.4.20 *Let $\mathcal{R}$ be the Robba ring of K and M a coherent $\mathcal{R}$-module with an integrable connection. Let $\lambda_0 < 1$ and M_0 be a coherent $\mathcal{R}_{\lambda_0}$-module with an integrable connection such that*

$$M = \mathcal{R}_K \otimes_{\mathcal{R}^{\lambda_0}} M_0.$$

For each $\lambda_0 \leq \lambda < 1$, *let*

$$M_\lambda := K\{t/\lambda, \lambda/t\} \otimes_{\mathcal{R}^{\lambda_0}} M_0.$$

If $s \in M$, *the* λ-radius of convergence *of* s *is*

$$R(s, \lambda) = R(s \in M_\lambda).$$

The λ-radius of convergence *of* M *is*

$$R(M, \lambda) = R(M_\lambda).$$

Finally, M *is said* overconvergent *or* solvable *if*

$$\lim_{\lambda \overset{>}{\to} 1} R(M, \lambda) = 1.$$

Here again, it might be better to define

$$R(s, \lambda) = \inf(\lambda, R(s \in M_\lambda))$$

and

$$R(M, \lambda) = \inf(\lambda, R(M_\lambda)).$$

This would not change the notion of overconvergence.

Again, we may consider our usual examples such as Dwork module $\mathcal{R}\theta$ with

$$\nabla(\theta) = -\alpha\theta \otimes dt.$$

which is non trivial and overconvergent if and only if $|\alpha| = |p|^{\frac{1}{p-1}}$. We also have Kummer module $\mathcal{R}\theta$ with

$$\nabla(\theta) = \beta\theta \otimes \frac{dt}{t}.$$

which is non trivial and overconvergent if and only if $\beta \in \mathbf{Z}_p \setminus \mathbf{Z}$.

The theory of overconvergent modules on Robba rings is very rich with important contributions by Dwork, Robba, Christol and Mebkhout culminating in [66].

5

Overconvergent sheaves

As long as we are interested in differential equations, the setting of the previous chapter is essentially sufficient. However, in order to have a solid theory with a general formalism, it is necessary to insert the notion of strict neighborhood into the constructions of differential calculus. We will make this more precise later.

We fix a complete ultrametric field K with $\mathcal{V}$, k and π as usual.

5.1 Overconvergent sections

We start with a very naive definition:

Definition 5.1.1 *Let V be a rigid analytic variety over K and T an admissible open subset of V. A sheaf of sets $\mathcal{E}$ on V is* overconvergent along T *if $\mathcal{E}_{|T} = 0$ (and not $\emptyset$).*

Of course, 0 denotes the constant sheaf $V' \mapsto 0$ which is the initial object of the category of sheaves. And we have very obvious properties:

Proposition 5.1.2 *Let V be a rigid analytic variety, $T \subset V$ an admissible open subset and $u : V' \to V$ a morphism of rigid analytic varieties.*

(i) *If $\mathcal{E}$ is a an overconvergent sheaf on V along T, then $u^{-1}\mathcal{E}$ is overconvergent along any admissible open subset $T' \subset u^{-1}(T)$.*
(ii) *If $\mathcal{F}$ is an overconvergent sheaf along some admissible open subset T' of V' containing $u^{-1}(T)$, then $u_*\mathcal{F}$ is overconvergent along T.*

Proof In both cases, we may assume that $T' = u^{-1}(T)$ and call $u' : T' \to T$ the induced map. In the first case, we have

$$(u^{-1}\mathcal{E})_{|T'} = u'^{-1}\mathcal{E}_{|T} = u'^{-1}(0) = 0$$

because u'^{-1} is exact and 0 is an initial object. And in the second one, we see that

$$(u_*\mathcal{F})_{|T} = u'_*\mathcal{F}_{|T'} = u_*(0) = 0$$

because u_* is left exact. □

Now, we want to show that given any sheaf of sets $\mathcal{E}$ on V, there exists an overconvergent sheaf $j^\dagger_{V\setminus T}\mathcal{E}$ which is universal for morphisms $\mathcal{E} \to \mathcal{F}$ with $\mathcal{F}$ overconvergent along T. The reader who wishes may take Proposition 5.1.12, (i)) below as the definition of $j^\dagger$ and give alternate direct proofs for all the statements. We feel however that introducing the closed subtopos makes the relation between overconvergent sections and sections in the neighborhood of a compact subset more than just an analogy. It should also be noticed that all the results quoted from [1], Exposé IV, are very easy to prove in the setting of rigid analytic geometry.

Recall that the rigid topos of a rigid analytic variety V is simply the category $\widetilde{V}$ of all sheaves of sets on V. If T is an admissible open subset of V, the *closed complement of* $\widetilde{T}$ *in* $\widetilde{V}$ is defined in from [1], Exposé IV, 9.3, as the full subcategory $\widetilde{V} \setminus \widetilde{T}$ of sheaves of sets $\mathcal{E}$ on V such that $\mathcal{E}_{|T} = 0$, that is, overconvergent sheaves along T. The inclusion functor $i_* : \widetilde{V} \setminus \widetilde{T} \hookrightarrow \widetilde{T}$ is fully faithful and has a left adjoint i^{-1} which is exact. By definition, this left adjoint is characterized by a natural isomorphism

$$\mathrm{Hom}_{\widetilde{V}\setminus\widetilde{T}}(i^{-1}\mathcal{E}, \mathcal{F}) \simeq \mathrm{Hom}_{\widetilde{V}}(\mathcal{E}, i_*\mathcal{F}).$$

Proposition 5.1.3 *Let V be a rigid analytic variety, T an admissible open subset of V and $\mathcal{E}$ a sheaf on V. Then, there exists a sheaf $j^\dagger_{V\setminus T}\mathcal{E}$ characterized by the following universal property:*

Any morphism $\mathcal{E} \to \mathcal{F}$ with $\mathcal{F}$ overconvergent along T factors uniquely through the canonical map $\mathcal{E} \to j^\dagger_{V\setminus T}\mathcal{E}$.

Proof If $i : \widetilde{V} \setminus \widetilde{T} \hookrightarrow \widetilde{T}$ denotes as above the embedding of toposes, we set

$$j^\dagger_{V\setminus T}\mathcal{E} := i_* i^{-1}\mathcal{E}$$

The universal property is then a formal consequence of adjointness between i^{-1} and i_*. □

Definition 5.1.4 *If V is a rigid analytic variety and T an admissible open subset of V, then $j^\dagger_{V\setminus T}\mathcal{E}$ is the sheaf of* overconvergent sections of $\mathcal{E}$ on V along T.

The topos setting is very convenient to deal with the functoriality of the construction. More precisely, it easily follows from [1], Exposé IV, 9.3 that if V is a rigid analytic variety, $T \subset V$ an admissible open subset, $u : V' \to V$ a morphism of rigid analytic varieties and T' an admissible open subset of V containing $u^{-1}(T)$, we have a commutative diagram

$$\begin{array}{ccc} \widetilde{V}' \setminus \widetilde{T}' & \xrightarrow{u'} & \widetilde{V} \setminus \widetilde{T} \\ \Big\downarrow{\scriptstyle i'} & & \Big\downarrow{\scriptstyle i} \\ \widetilde{V}' & \xrightarrow{u} & \widetilde{V} \end{array}$$

In particular, if $\mathcal{F} \in \widetilde{V} \setminus \widetilde{T}$, there exists a canonical map

$$u^{-1}i_*\mathcal{F} \to i'_*u'^{-1}\mathcal{F}.$$

This map is bijective when $T' = u^{-1}(T)$ and, in general, the induced map

$$i'^{-1}u^{-1}i_*\mathcal{F} \to i'^{-1}i'_*u'^{-1}\mathcal{F}$$

is always bijective.

Proposition 5.1.5 *Let V be a rigid analytic variety, $T \subset V$ an admissible open subset, $u : V' \to V$ a morphism of rigid analytic varieties and T' an admissible open subset of V containing $u^{-1}(T)$. If $\mathcal{E}$ is a sheaf on V, then,*

(i) *There is a canonical map*

$$u^{-1}j^\dagger_{V\setminus T}\mathcal{E} \to j^\dagger_{V'\setminus T'}u^{-1}\mathcal{E}.$$

(ii) *When $T' = u^{-1}(T)$, this map is an isomorphism*

$$u^{-1}j^\dagger_{V\setminus T}\mathcal{E} \simeq j^\dagger_{V'\setminus T'}u^{-1}\mathcal{E}.$$

(iii) *We have*

$$j^\dagger_{V'\setminus T'}u^{-1}j^\dagger_{V\setminus T}\mathcal{E} \simeq j^\dagger_{V'\setminus T'}u^{-1}\mathcal{E}.$$

Proof With the notations above, there is a canonical map

$$u^{-1}j^\dagger_{V\setminus T}\mathcal{E} = u^{-1}i_*i^{-1}\mathcal{E} \to i'_*u'^{-1}i^{-1}\mathcal{E} = i'_*i'^{-1}u^{-1}\mathcal{E} = j^\dagger_{V'\setminus T'}u^{-1}\mathcal{E}.$$

which is bijective when $T' = u^{-1}(T)$. Moreover, we always have

$$j^\dagger_{V'\setminus T'}u^{-1}j^\dagger_{V\setminus T}\mathcal{E} = i'_*i'^{-1}u^{-1}i_*i^{-1}\mathcal{E} = i'_*i'^{-1}i'_*u'^{-1}i^{-1}\mathcal{E}$$
$$i'_*i'^{-1}u^{-1}\mathcal{E} = j^\dagger_{V'\setminus T'}u^{-1}\mathcal{E}.$$

□

Corollary 5.1.6 *Let V be a rigid analytic variety and T, T' two admissible open subsets of V. If a sheaf $\mathcal{E}$ on V is overconvergent along T', then $j^\dagger_{V\setminus T}\mathcal{E}$ is also overconvergent along T'.*

Proof We have

$$(j^\dagger_{V\setminus T}\mathcal{E})_{|T'} = j^\dagger_{T'\setminus(T\cap T')}\mathcal{E}_{|T'} = j^\dagger_{T'\setminus(T\cap T')}(0) = 0.$$

□

Proposition 5.1.7 *Let V be a rigid analytic variety, T an admissible open subset of V and $T = T_1 \cup T_2$ an admissible open covering. Then, we have*

$$j^\dagger_{V\setminus T_1} \circ j^\dagger_{V\setminus T_2} = j^\dagger_{V\setminus T_2} \circ j^\dagger_{V\setminus T_1} = j^\dagger_{V\setminus T}.$$

Proof Of course, since the assertion is symmetric in T_1 and T_2, it is sufficient to show that if $\mathcal{E}$ is a sheaf on V, then $j^\dagger_{V\setminus T_1}j^\dagger_{V\setminus T_2}\mathcal{E}$ has the same universal property as $j^\dagger_{V\setminus T}\mathcal{E}$. First of all, $(j^\dagger_{V\setminus T_1}j^\dagger_{V\setminus T_2}\mathcal{E})_{|T_1} = 0$ because $j^\dagger_{V\setminus T_1}j^\dagger_{V\setminus T_2}\mathcal{E}$ is overconvergent along T_1. Since $j^\dagger_{V\setminus T_2}\mathcal{E}$ is overconvergent along T_2, it follows from Corollary 5.1.6 that $(j^\dagger_{V\setminus T_1}j^\dagger_{V\setminus T_2}\mathcal{E})_{|T_2}$ is also overconvergent along T_2 which means that $(j^\dagger_{V\setminus T_1}j^\dagger_{V\setminus T_2}\mathcal{E})_{|T_2} = 0$ also. Since the open covering is admissible, we see that $(j^\dagger_{V\setminus T_1}j^\dagger_{V\setminus T_2}\mathcal{E})_{|T} = 0$ which means that $j^\dagger_{V\setminus T_1}j^\dagger_{V\setminus T_2}\mathcal{E}$ is overconvergent along T. The second part consists in showing that it satisfies the universal property. If $\mathcal{F}$ is overconvergent along T, it is also overconvergent along $T_2 \subset T$ and any map $\mathcal{E} \to \mathcal{F}$ will factor uniquely through $j^\dagger_{V\setminus T_2}\mathcal{E}$. For the same reason, $\mathcal{F}$ is also overconvergent along T_1 and our map $j^\dagger_{V\setminus T_2}\mathcal{E} \to \mathcal{F}$ will therefore factor uniquely through $j^\dagger_{V\setminus T_1}j^\dagger_{V\setminus T_2}\mathcal{E}$. □

Corollary 5.1.8 *Let V be a rigid analytic variety, T an admissible open subset of V and T' an admissible open subset of T. If $\mathcal{E}$ is a sheaf on V, there is a canonical isomorphism*

$$j^\dagger_{V\setminus T'}j^\dagger_{V\setminus T}\mathcal{E} \simeq j^\dagger_{V\setminus T'}\mathcal{E}.$$

Proof This is just a very simple case of Proposition 5.1.7 with $T = T_1$ and $T' = T_2$.
□

Proposition 5.1.9 *Let V be a rigid analytic variety and T an admissible open subset of V. If $\mathcal{E}$ and $\mathcal{F}$ are two sheaves on V with $\mathcal{F}$ overconvergent along T, then*

$$\mathcal{H}om(j_X^\dagger\mathcal{E}, \mathcal{F}) = \mathcal{H}om(\mathcal{E}, \mathcal{F}).$$

and this is an overconvergent sheaf.

Proof First of all,

$$\mathcal{H}om(\mathcal{E}, \mathcal{F})_{|T} = \mathcal{H}om(\mathcal{E}_{|T}, \mathcal{F}_{|T}) = \mathcal{H}om(\mathcal{E}_{|T}, 0) = 0$$

which shows that we do have an overconvergent sheaf. Now, in order to prove the equality, it is sufficient to consider global sections on some admissible open subset V'. If $j_{VV'} : V' \hookrightarrow V$ denotes the inclusion map and $T' = T \cap V'$, it follows from the second part of Proposition 5.1.5 that the left-hand side is

$$\mathrm{Hom}(j_{VV'}^{-1} j_{V\setminus T}^\dagger \mathcal{E}, j_{VV'}^{-1}\mathcal{F}) = \mathrm{Hom}(j_{V'\setminus T'}^\dagger j_{VV'}^{-1}\mathcal{E}, j_{VV'}^{-1}\mathcal{F})$$

and, since $j_{VV'}^{-1}\mathcal{F}$ is overconvergent, that the right-hand side gives

$$\mathrm{Hom}(j_{VV'}^{-1}\mathcal{E}, j_{VV'}^{-1}\mathcal{F}) = \mathrm{Hom}(j_{V'\setminus T'}^\dagger j_{VV'}^{-1}\mathcal{E}, j_{VV'}^{-1}\mathcal{F}).$$

□

We will mainly apply the notion of overconvergent sheaf and overconvergent sections in the context of frames. More precisely, let $(X \subset Y \subset P)$ be a frame, V an admissible open subset of $]Y[_P$ and $T := V \cap (]Y[_P \setminus]X[_P)$. If Z denotes a closed complement for X in Y, we have $T = V \cap]Z[_P$ which is therefore open in V. We may thus consider sheaves $\mathcal{E}$ on V that are *overconvergent* (along T). We may also make use of the functor $j_{V\cap]X[_P}^\dagger$ (or $j_X^\dagger$ for short) of *overconvergent sections on* V (along T). We will essentially use this notion when V is a strict neighborhood of $]X[_P$ in $]Y[_P$, or even $V =]Y[_P$, but it is necessary to work it out for more general admissible open subsets of $]Y[_P$.

Note that in this geometric situation, $V \setminus T$ is also an admissible open subset of V.

Proposition 5.1.10 *Let $(X \subset Y \subset P)$ be a frame, V an admissible open subset of $]Y[_P$ and $\mathcal{E}$ a sheaf on V. Then, the restriction of $j_X^\dagger\mathcal{E}$ to $V \cap]X[_P$ is identical to the restriction of $\mathcal{E}$ and its restriction to its complement $V \cap (]Y[_P \setminus]X[_P)$ is zero.*

Proof Of course, $j_X^\dagger\mathcal{E}$ is overconvergent and this exactly means that its restriction to $V \cap (]Y[_P \setminus]X[_P)$ is zero. Moreover, it follows from Proposition 5.1.5

that

$$(j_X^\dagger \mathcal{E})_{|V\cap]X[_P} = j^\dagger_{V\cap]X[_P} \mathcal{E}_{|V\cap]X[_P} = \mathcal{E}_{|V\cap]X[_P}.$$

□

Proposition 5.1.11 *Let $(X \subset Y \subset P)$ be a frame, V an admissible open subset of $]Y[_P$ and $\mathcal{E}$ a sheaf of sets on V. Assume that $X = X_1 \cap X_2$ where X_1 and X_2 are two open subsets of Y. Then, with obvious notations, we have*

$$j^\dagger_{X_1} j^\dagger_{X_2} \mathcal{E} = j^\dagger_{X_2} j^\dagger_{X_1} \mathcal{E} = j^\dagger_X \mathcal{E}.$$

Proof If Z_1 and Z_2 are closed complements of X_1 and X_2 respectively in Y, then $Z = Z_1 \cup Z_2$ is a closed complement for X in Y. Since $Z = Z_1 \cup Z_2$ is a closed covering, it follows from Proposition 2.2.15 that the covering $]Z[_P =]Z_1[_P \cup]Z_2[_P$ is an open admissible covering. Our assertion then follows from Proposition 5.1.7. □

We now explain the relation between overconvergence and strict neighborhoods.

Proposition 5.1.12 *Let $(X \subset Y \subset P)$ be a frame, V an admissible open subset of $]Y[_P$ and $\mathcal{E}$ a sheaf of sets on V. Then*

(i) *We have*

$$j^\dagger_X \mathcal{E} = \varinjlim j_{VV'*} j^{-1}_{VV'} \mathcal{E}$$

where V' runs through all the strict neighborhoods of $]X[_P$ in $]Y[_P$ and

$$j_{VV'} : V \cap V' \hookrightarrow V$$

denotes the inclusion map.

(ii) *If W is a quasi-compact admissible open subset of V, then*

$$\Gamma(W, j^\dagger_X \mathcal{E}) = \varinjlim_{V' \subset V} \Gamma(W \cap V', \mathcal{E})$$

where V' runs through the strict neighborhoods of $]X[_P$ in $]Y[_P$.

(iii) *Actually, if P is quasi-compact, we have*

$$\Gamma(W, j^\dagger_X \mathcal{E}) = \varinjlim_{\lambda < 1} \Gamma(W \cap V^\lambda, \mathcal{E})$$

where as usual, $V^\lambda :=]Y[_P \setminus]Z[_\lambda$ with Z a closed complement for X in Y.

(iv) *If moreover, V itself is quasi-compact, then*

$$j^\dagger_X \mathcal{E} := \varinjlim j_{\lambda *} j_\lambda^{-1} \mathcal{E}$$

where $j_\lambda : V \cap V^\lambda \hookrightarrow V$ denotes the inclusion map.

Proof The first question is local on V and we may therefore assume that the frame is quasi-compact. Shrinking P if necessary, we may also add this assumption to the second question. Thus, from now on, we assume P quasi-compact.

If W is a quasi-compact admissible open subset of V, it is a quasi-compact quasi-separated rigid analytic variety and therefore, the functor $\Gamma(W, -)$ commutes to filtered direct limits. It follows that

$$\Gamma(W, \varinjlim j_{VV'*}j_{VV'}^{-1}\mathcal{E}) = \varinjlim_{V'\subset V} \Gamma(W \cap V', \mathcal{E})$$

when V' runs through all strict neighborhoods of $]X[_P$ in $]Y[_P$. Moreover, since W is quasi-compact, the subsets $W \cap V^\lambda$ are cofinal in the set of all $W \cap V'$ thanks to Proposition 3.3.2. This shows that the second assertion follows from the first. For the same reason, the last assertion also follows from the first.

Now, the next point consists in showing that $\varinjlim j_{VV'*}j_{VV'}^{-1}\mathcal{E}$ is overconvergent on V. If $W \subset T := V \cap (]Y[_P \backslash]X[_P)$ is an affinoid open subset, it is quasi-compact and it follows from Corollary 3.2.5 that there exists $\lambda < 1$ with $W \cap V^\lambda = \emptyset$. Thus, we have

$$\Gamma(W, \varinjlim j_{VV'*}j_{VV'}^{1}\mathcal{E}) = \varinjlim_{\lambda<1} \Gamma(W \cap V^\lambda, \mathcal{E}) = 0.$$

Now, we want to show that any morphism $\varphi : \mathcal{E} \to \mathcal{F}$ with $\mathcal{F}$ overconvergent on V factors uniquely through the canonical morphism

$$\mathcal{E} \to \varinjlim j_{VV'*}j_{VV'}^{-1}\mathcal{E}$$

in order to give some morphism ψ. Let W be an affinoid open subset of V and

$$s \in \Gamma(W, \varinjlim j_{VV'*}j_{VV'}^{-1}\mathcal{E}) = \varinjlim_{V'\subset V} \Gamma(W \cap V', \mathcal{E}).$$

By definition of direct limits, there exists a strict neighborhood V' of $]X[_P$ in $]Y[_P$ such that $s \in \Gamma(W \cap V', \mathcal{E})$. Necessarily, we must have $\psi(s)_{|V'} = \varphi(s)$ and of course, we must also have $\psi(s)_{|T} = 0$ because $\mathcal{F}_{|T} = 0$. Since

$$W = (W \cap V') \cup (T \cap V')$$

is an admissible covering, this proves uniqueness. Conversely, we may always define $\psi(s)$ in this way because, necessarily, $\varphi(s)_{|V'\cap T} = 0$. □

Proposition 5.1.13 *Let $(X \subset Y \subset P)$ be a frame, V an admissible open subset of $]Y[_P$, V' a strict neighborhood of $]X[_P$ in $]Y[_P$ and $j_{VV'} : V \cap V' \hookrightarrow V$ the inclusion map.*

(i) *If $\mathcal{E}'$ is a sheaf on $V \cap V'$, we have*

$$j_X^\dagger j_{VV'*}\mathcal{E}' \simeq j_{VV'*} j_X^\dagger \mathcal{E}'.$$

(ii) *If $\mathcal{E}$ is a sheaf on V, we have*

$$(j_{VV'*} j_{VV'}^{-1}) j_X^\dagger \mathcal{E} \simeq j_X^\dagger (j_{VV'*} j_{VV'}^{-1})\mathcal{E} \simeq j_X^\dagger \mathcal{E}.$$

(iii) *The functors $j_{VV'}^{-1}$ and $j_{VV'*}$ induce an equivalence between the categories of overconvergent sheaves on V and $V \cap V'$.*

Proof We first want to show that if $\mathcal{E}'$ is a sheaf on $V \cap V'$, we have

$$j_X^\dagger j_{VV'*}\mathcal{E}' \simeq j_{VV'*} j_X^\dagger \mathcal{E}'.$$

Since both sheaves are 0 on $T := V \cap (]Y[_P \backslash]X[_P)$, it is sufficient to show that they coincide on $V \cap V'$. In other words, we have to show that

$$j_{VV'}^{-1} j_X^\dagger j_{VV'*}\mathcal{E}' \simeq j_{VV'}^{-1} j_{VV'*} j_X^\dagger \mathcal{E}'.$$

Since $j_{VV'}^{-1} j_{VV'*} = Id_{V'}$, this follows from Proposition 5.1.5:

$$j_{VV'}^{-1} j_X^\dagger j_{VV'*}\mathcal{E}' \simeq j_X^\dagger j_{VV'}^{-1} j_{VV'*}\mathcal{E}'.$$

Using this first result and Proposition 5.1.5 again, we see that if $\mathcal{E}$ is a sheaf on V, we have

$$j_{VV'*} j_{VV'}^{-1} j_X^\dagger \mathcal{E} \simeq j_{VV'*} j_X^\dagger j_{VV'}^{-1} \mathcal{E} \simeq j_X^\dagger j_{VV'*} j_{VV'}^{-1} \mathcal{E}.$$

Thus, in order to prove the second assertion, it remains to show that

$$(j_{VV'*} j_{VV'}^{-1}) j_X^\dagger \mathcal{E} \simeq j_X^\dagger \mathcal{E}.$$

Since this is trivially true both on T and $V \cap V'$, we are done.

We turn now to the last assertion. If $\mathcal{E}$ is an overconvergent sheaf on V, we know from Proposition 5.1.5 that $j_{VV'}^{-1}\mathcal{E}$ is overconvergent on $V \cap V'$. Moreover, we know from the second part that

$$j_{VV'*} j_{VV'}^{-1} \mathcal{E} = \mathcal{E}.$$

Conversely, if $\mathcal{E}'$ is an overconvergent sheaf on $V \cap V'$, then

$$j_{VV'*}\mathcal{E}' = j_{VV'*} j_X^\dagger \mathcal{E}' = j_X^\dagger j_{VV'*}\mathcal{E}'$$

is overconvergent on V. And of course, we have

$$j_{VV'}^{-1} j_{VV'*}\mathcal{E}' = \mathcal{E}'.$$

□

We will state functoriality in a general form and give a more usual description just after.

Proposition 5.1.14 *Let $(X \subset Y \subset P)$ (resp. $(X' \subset Y' \subset P')$) be a frame and V (resp. V') an admissible open subset of $]Y[_P$ (resp. $]Y'[_{P'}$). Let $f : X' \to X$ and $u : V' \to V$ be a pair of compatible morphisms (see below). Then, there is a canonical morphism*

$$u^{-1} j_X^\dagger \mathcal{E} \to j_{X'}^\dagger u^{-1} \mathcal{E}$$

which is an isomorphism when

$$V' \cap]X'[_{P'} = u^{-1}(V \cap]X[_P).$$

In general, we always have

$$j_{X'}^\dagger u^{-1} j_X^\dagger \mathcal{E} = j_{X'}^\dagger u^{-1} \mathcal{E}.$$

Proof Recall from Definition 3.4.7 that compatibility means that

$$\forall x \in V' \cap]X'[_{P'}, \quad f(\overline{x}) = \overline{u(x)}$$

and this implies that

$$u(V' \cap]X'[_{P'}) \subset V \cap]X[_P.$$

Thus, our assertion follows immediately form Proposition 5.1.5 with

$$T := V \cap (]Y[_P \backslash]X[_P) \quad \text{and} \quad T' := V' \cap (]Y'[_{P'} \backslash]X'[_{P'}).$$

The condition $u^{-1}(T) \subset T'$ is clearly satisfied with equality when

$$V' \cap]X'[_{P'} = u^{-1}(V \cap]X[_P).$$

□

Corollary 5.1.15 *Let*

$$\begin{array}{ccccc} X' & \hookrightarrow & Y' & \hookrightarrow & P' \\ \downarrow & & \downarrow g & & \downarrow u \\ X & \hookrightarrow & Y & \hookrightarrow & P \end{array}$$

be a morphism of frames. Let V be an admissible open subset of $]Y[_P$ and V' an admissible open subset of $]Y'[_{P'} \cap u_K^{-1}(V)$.

If we still call $u_K : V' \to V$ the induced map and $\mathcal{E}$ is any sheaf on V, there is a canonical morphism

$$u_K^{-1} j_X^\dagger \mathcal{E} \to j_{X'}^\dagger u_K^{-1} \mathcal{E}$$

which is an isomorphism when the morphism of frames is cartesian.

In general, there is a canonical isomorphism

$$j^\dagger_{X'} u_K^{-1} j^\dagger_X \mathcal{E} \simeq j^\dagger_{X'} u_K^{-1} \mathcal{E}.$$

Proof Of course, in this case, the compatibility condition is automatic. Moreover, when the morphism is cartesian, we have

$$]Y'[_{P'} \cap u_K^{-1}(]X[_P) =]X'[_{P'}$$

and it follows that the other condition too is satisfied. □

Corollary 5.1.16 *Let $(X \subset Y \subset P)$ be a frame and $(X' \subset Y' \subset P')$ a cartesian open subframe. Let V be an admissible open subset of $]Y[_P$ and V' an admissible open subset of $]Y'[_{P'} \cap V$. If $\mathcal{E}$ is any sheaf on V, then $(j^\dagger_X \mathcal{E})_{|V'} \simeq j^\dagger_{X'} \mathcal{E}$.*

Proof This is just the cartesian case of the proposition when u is an open immersion. □

Note that this applies in particular when the morphism is strict, which means that $Y' = Y \cap P'_k$.

Proposition 5.1.17 *Let $(X \subset Y \subset P)$ (resp. $(X' \subset Y' \subset P')$) be a frame and V (resp. V') an admissible open subset of $]Y[_P$ (resp. $]Y'[_{P'}$). Let $f : X' \to X$ and $u : V' \to V$ be a pair of compatible morphisms. If $\mathcal{F}$ is an overconvergent sheaf on V', then $u_*\mathcal{F}$ is also overconvergent.*

Proof Follows from Proposition 5.1.2. □

Later on, we will also need the following lemma.

Lemma 5.1.18 *Let $(X \subset Y \subset P)$ be a frame and V an admissible open subset of $]Y[_P$. Let $\sigma : K \hookrightarrow K'$ be an isometric embedding and $\varpi : \widetilde{V}^\sigma \to \widetilde{V}$ the projection.*

(i) *If $\mathcal{F}$ is an overconvergent sheaf on V^σ, then $\varpi_*\mathcal{F}$ is overconvergent on V.*
(ii) *If $\mathcal{E}$ is a sheaf on V, there is a canonical isomorphism*

$$\varpi^{-1} j^\dagger_X \mathcal{E} \simeq j^\dagger_{X^\sigma} \varpi^{-1} \mathcal{E}.$$

In particular, if $\mathcal{E}$ is overconvergent, so is $\varpi^{-1}\mathcal{E}$.

Proof Recall that if $\mathcal{F}$ is a sheaf on V^σ and $W \subset V$ is an admissible open subset, we have by definition,

$$\Gamma(W, \varpi_*\mathcal{F}) = \Gamma(W^\sigma, \mathcal{F}).$$

Moreover, if $T := V \cap (]Y[_P \setminus]X[_P)$, then $T^\sigma = V^\sigma \cap (]Y[^\sigma_P \setminus]X[^\sigma_P)$. The first assertion immediately follows.

Now, by functoriality, if we also denote by $\varpi : \widetilde{T}^\sigma \to \widetilde{T}$ the projection and if $\mathcal{E}$ is a sheaf on V, we have

$$(\varpi^{-1} j_X^\dagger \mathcal{E})_{|T^\sigma} = \varpi^{-1}(j_X^\dagger \mathcal{E})_{|T}$$

and it follows that $\varpi^{-1} j_X^\dagger \mathcal{E}$ is overconvergent. Now, if $\varphi : \varpi^{-1}\mathcal{E} \to \mathcal{F}$ is any morphism with $\mathcal{F}$ overconvergent, the corresponding map $\mathcal{E} \to \varpi_* \mathcal{F}$ factors uniquely through a morphism $j_X^\dagger \mathcal{E} \to \varpi_* \mathcal{F}$ that corresponds to a morphism $\varpi^{-1} j_X^\dagger \mathcal{E} \to \mathcal{F}$. This shows that $\varpi^{-1} j_X^\dagger \mathcal{E}$ has the same universal property as $j_{X^\sigma}^\dagger \varpi^{-1}\mathcal{E}$ and they are therefore equal. □

We finish this section with two examples. We first show how the smooth affine case can be understood via weakly complete algebras (Monsky–Washnitzer setting) and how the one-dimensional case reduces to the study of Robba rings.

First of all, we need to recall the notion of *weak completion* $A^\dagger$ of a $\mathcal{V}$-algebra of finite presentation A: if $A := \mathcal{V}[\underline{T}]/I$ is a finite presentation of A, we set $A^\dagger := \mathcal{V}[\underline{T}]^\dagger / I\mathcal{V}[\underline{T}]^\dagger$. Note that the weak completion $A^\dagger$ of A only depends on A and not on the choice of the presentation. The link between the Monsky–Washnitzer setting and this weak completion is given by the next proposition.

Lemma 5.1.19 *Let $X = \operatorname{Spec} A$ be an affine $\mathcal{V}$-scheme, $X \hookrightarrow \mathbf{A}^N_\mathcal{V}$ be a closed immersion and*

$$V_\rho := \mathbf{B}^N(0, \rho^+) \cap X_K^{\mathrm{rig}}.$$

If I is the ideal of X in $\mathbf{A}^N_\mathcal{V}$, then $V_\rho = \operatorname{Spm} A_\rho$ with

$$A_\rho := K\{\underline{T}/\rho\}/IK\{\underline{T}/\rho\}$$

and we have

$$A_K^\dagger = \varinjlim_{\rho \overset{>}{\to} 1} A_\rho.$$

Proof This is simply the geometric description of A_ρ. □

Proposition 5.1.20 *Let $X = \operatorname{Spec} A$ be an affine $\mathcal{V}$-scheme, $X \hookrightarrow A^N_\mathcal{V}$ a closed immersion and $Y \subset \mathbf{P}^N_\mathcal{V}$ the projective closure of X. For $\rho > 1$, we let*

$$V_\rho := \mathbf{B}^N(0, \rho^+) \cap X_K^{\mathrm{rig}} \quad \text{and} \quad A_\rho := \Gamma(V_\rho, \mathcal{O}_V).$$

If we consider the frame $(X_k \subset Y_k \subset \widehat{Y})$, then we have

$$\Gamma(Y_K^{\mathrm{rig}}, j_X^\dagger \mathcal{O}_{Y_K^{\mathrm{rig}}}) = A_K^\dagger.$$

Proof We saw in Proposition 3.3.9 that the admissible open subsets V_ρ form a cofinal system of strict neighborhoods of $\widehat{X}_K$ in Y_K^{rig}. Since Y_K^{rig} is quasi-compact, we get from Proposition 5.1.12,

$$\Gamma(Y_K^{\mathrm{rig}}, j_X^\dagger \mathcal{O}_{Y_K^{\mathrm{rig}}}) = \varinjlim_\rho \Gamma(Y_K^{\mathrm{rig}}, j_{Y_K^{\mathrm{rig}} V_\rho *} \mathcal{O}_{V_\rho})$$

and we have for each ρ,

$$\Gamma(Y_K^{\mathrm{rig}}, j_{Y_K^{\mathrm{rig}} V_\rho *} \mathcal{O}_{V_\rho}) = \Gamma(V_\rho, \mathcal{O}_{V_\rho}) = A_\rho.$$

It is therefore sufficient to recall from Proposition 5.1.19 that

$$A_K^\dagger = \cup_{\rho \xrightarrow{>} 1} A_\rho.$$

□

We now turn to the Robba ring case. In order to make things simpler, we only consider rational points. In general, one may always make a finite extension in order to achieve this.

Proposition 5.1.21 *Let $\mathcal{Y}$ be a flat formal $\mathcal{V}$-scheme whose special fiber Y is a connected curve and let X be a non empty open subset of Y. Let x be a smooth rational point of Y not in X and*

$$h_x :]x[_{\mathcal{Y}} \hookrightarrow \mathcal{Y}_K$$

the inclusion map. If $\mathcal{R}$ denotes the Robba ring of K, we have an isomorphism

$$\mathcal{R} \simeq \Gamma(\mathcal{Y}_K, j_X^\dagger h_{x*} \mathcal{O}_{]x[_{\mathcal{Y}}}).$$

Proof Since $\mathcal{Y}_K$ is quasi-compact, we have

$$\begin{aligned} \Gamma(\mathcal{Y}_K, j_X^\dagger h_{x*} \mathcal{O}_{]x[_{\mathcal{Y}}}) &= \varinjlim \Gamma(\mathcal{Y}_K, j_{\mathcal{Y}_K V *} j_{\mathcal{Y}_K V}^{-1} h_{x*} \mathcal{O}_{]x[_{\mathcal{Y}}}) \\ &= \varinjlim \Gamma(V \cap]x[_{\mathcal{Y}}, \mathcal{O}_{]x[_K}) \end{aligned}$$

when V runs through all strict neighborhoods of $]X[_{\mathcal{Y}}$ in $\mathcal{Y}_K$. Note now that we may replace $\mathcal{Y}$ by any open subset $\mathcal{Y}'$ containing x (and consequently X by $X' := X \cap \mathcal{Y}'$). More precisely, we have $]x[_{\mathcal{Y}'} =]x[_{\mathcal{Y}}$ and we saw in Proposition 3.4.6 that if V' is a strict neighborhood of $]X'[_{\mathcal{Y}'}$ in $]Y'[_{\mathcal{Y}'}$, there exists a strict neighborhood V of of $]X[_{\mathcal{Y}}$ in $\mathcal{Y}_K$ such that $V \cap \mathcal{Y}'_K \subset V'$. It follows that

$$V \cap]x[_{\mathcal{Y}} \subset V' \cap]x[_{\mathcal{Y}}.$$

Conversely of course, if V is a strict neighborhood of $]X[_{\mathcal{Y}}$ in $\mathcal{Y}_K$, it is also a strict neighborhood of $]X'[_{\mathcal{Y}'}$ in $\mathcal{Y}'_K$.

We may therefore assume that we have an étale coordinate $t : \mathcal{Y} \to \widehat{\mathbf{A}}^1_{\mathcal{V}}$ with

$$Y \setminus X = \{x\} = t^{-1}(0).$$

Thank to Proposition 3.3.11, we may also only consider strict neighborhoods of the form

$$V^{\lambda} := t^{-1}(\mathbf{A}(0, \lambda^{+}, 1^{+})).$$

And we know from Proposition 2.3.16 that t induces an isomorphism between $]x[_{\mathcal{Y}}$ and $\mathbf{D}(0, 1^{-})$. It follows that t induces an isomorphism

$$\Gamma(\mathcal{Y}_K, j_X^{\dagger} h_{x*}\mathcal{O}_{]x[_K}) = \varinjlim_{\lambda<1} \Gamma(\mathbf{A}(0, \lambda^{+}, 1^{-}), \mathcal{O}_{\mathbf{D}(0,1^{-})}) = \mathcal{R}.$$

□

Definition 5.1.22 *With the hypothesis and notations of the proposition, the ring*

$$\mathcal{R}(x) := \Gamma(\mathcal{Y}_K, j_X^{\dagger} h_{x*}\mathcal{O}_{]x[_{\mathcal{Y}}})$$

is called the Robba ring *at* x.

5.2 Overconvergence and abelian sheaves

Let V be a rigid analytic variety and T an admissible open subset of V. We consider again the embedding of toposes $i : \widetilde{V} \setminus \widetilde{T} \hookrightarrow \widetilde{T}$. Since i_* and i^{-1} are both left exact, they preserve algebraic structures and in particular, they induce a pair of adjoint functors on abelian sheaves. Moreover, we know from [1], Exposé IV, 14 that i_* has also a right adjoint $i^!$. We also know from [1], Exposé IV, 9.3, that the morphism $\mathcal{E} \to i_* i^{-1}\mathcal{E}$ is an epimorphism.

Proposition 5.2.1 *Let V be a rigid analytic variety and T an admissible open subset. On abelian sheaves, the functor $j_{V\setminus T}^{\dagger}$ is exact with a right adjoint. Moreover, if $\mathcal{E}$ is any abelian sheaf on V, the canonical map $\mathcal{E} \to j_{V\setminus T}^{\dagger}\mathcal{E}$ is an epimorphism.*

Proof Using the remark preceding the statement, this is an immediate consequence of the definitions. □

Note however that left exactness is not valid for sheaves of sets because $j_{V\setminus T}^{\dagger}$ applied to the empty sheaf does not give the empty sheaf unless $T = \emptyset$.

Proposition 5.2.2 *Let V be a rigid analytic variety and T an admissible open subset. Let V' be a strict neighborhood of $V \setminus T$ in V and*

$$j_{VV'} : V' \hookrightarrow V$$

the inclusion map. If $\mathcal{E}'$ is an overconvergent abelian sheaf on V', we have

$$Rj_{VV'*}\mathcal{E}' = j_{VV'*}\mathcal{E}'.$$

Proof Since derived functors commute to localization, this is trivially true on T where $j_{VV'*}$ becomes the 0 functor and also on V' where it becomes the identity. As usual, the conclusion comes from the fact that $V = V' \cup T$ is an admissible covering. □

If V is a rigid analytic variety and T is an admissible open subset of V, the inclusion map $h : T \hookrightarrow V$ may be seen as an open embedding of toposes $h : \widetilde{T} \hookrightarrow \widetilde{V}$ which is given by the pair of adjoint functors (h^{-1}, h_*). As explained in [1], Exposé IV, 11, the functor induced by h^{-1} on abelian sheaves has an exact left adjoint $h_!$ (this is also true for sheaves of sets but we get a different $h_!$). If, as before, $i : \widetilde{V} \setminus \widetilde{T} \hookrightarrow \widetilde{V}$ denotes the embedding of the closed complement, it is shown in [1], Exposé IV, 14, that there is an exact sequence

$$0 \to h_!h^{-1}\mathcal{E} \to \mathcal{E} \to i_*i^{-1}\mathcal{E} \to 0.$$

Definition 5.2.3 *If $h : T \hookrightarrow V$ is an open immersion of rigid analytic varieties and $\mathcal{E}$ an abelian sheaf on V, then*

$$\underline{\Gamma}^\dagger_T\mathcal{E} := h_!h^{-1}\mathcal{E}$$

is the sheaf of overconvergent sections on V with support in T.

Proposition 5.2.4 *Let V be a rigid analytic variety and T an admissible open subset. Then:*

(i) *$\underline{\Gamma}^\dagger_T$ is exact.*

(ii) *If $\mathcal{E}$ an abelian sheaf on V, there is a short exact sequence*

$$0 \to \underline{\Gamma}^\dagger_T\mathcal{E} \to \mathcal{E} \to j^\dagger_{V\setminus T}\mathcal{E} \to 0.$$

(iii) *If $T = T_1 \cap T_2$ is the intersection of two admissible open subsets, then*

$$\underline{\Gamma}^\dagger_{T_2} \circ \underline{\Gamma}^\dagger_{T_1} = \underline{\Gamma}^\dagger_{T_1} \circ \underline{\Gamma}^\dagger_{T_2} = \underline{\Gamma}^\dagger_T.$$

(iv) *If T' is an admissible open subset of V disjoint from T and $\mathcal{E}$ an abelian sheaf on V, then*

$$\Gamma^\dagger_T\mathcal{E} \simeq \Gamma^\dagger_T j^\dagger_{V\setminus T'}\mathcal{E}.$$

Proof Using basic properties of open immersions, the first two assertions are immediate consequences of the definitions. In particular, $\underline{\Gamma}^\dagger_T$ is the composite of two exact functors.

We now prove the third assertion. Denote by

$$h : T \hookrightarrow V, \quad h_1 : T_1 \hookrightarrow V, \quad h_2 : T_2 \hookrightarrow V$$

the inclusion maps. If $h'_1 : T \hookrightarrow T_2$ and $h'_2 : T \hookrightarrow T_1$ denote the inclusion maps, we have $h_2^{-1}h_{1*} = h'_{1*}h_2'^{-1}$. By adjunction, we also have $h_2^{-1}h_{1!} = h'_{1!}h_2'^{-1}$. Composing by $h_{2!}$ on the left and h_1^{-1} on the right gives

$$\begin{aligned}\underline{\Gamma}^{\dagger}_{T_2} \circ \underline{\Gamma}^{\dagger}_{T_1} &= (h_{2!}h_2^{-1})(h_{1!}h_1^{-1}) = h_{2!}(h_2^{-1}h_{1!})h_1^{-1} \\ &= h_{2!}(h'_{1!}h_2'^{-1})h_1^{-1} = (h_{2!}h'_{1!})(h_2'^{-1}h_1^{-1}) = h_!h^{-1} = \underline{\Gamma}^{\dagger}_{T}.\end{aligned}$$

Finally, the last assertion is a consequence of the others. More precisely, since $\underline{\Gamma}^{\dagger}_{T}$ is exact thanks to the first assertion, we obtain from the second assertion a short exact sequence

$$0 \to \underline{\Gamma}^{\dagger}_{T}\underline{\Gamma}^{\dagger}_{T'}\mathcal{E} \to \underline{\Gamma}^{\dagger}_{T}\mathcal{E} \to \underline{\Gamma}^{\dagger}_{T}j^{\dagger}_{V\setminus T'}\mathcal{E} \to 0.$$

And we may now use the third assertion which gives

$$\underline{\Gamma}^{\dagger}_{T}\underline{\Gamma}^{\dagger}_{T'}\mathcal{E} = \underline{\Gamma}^{\dagger}_{\emptyset}\mathcal{E} = 0.$$

□

Proposition 5.2.5 *Let V be a rigid analytic variety, $T \subset V$ an admissible open subset, $u : V' \to V$ a morphism of rigid analytic varieties and T' an admissible open subset of V containing $u^{-1}(T)$. If $\mathcal{E}$ is a sheaf on V, there is a canonical map*

$$u^{-1}\underline{\Gamma}^{\dagger}_{T}\mathcal{E} \to \underline{\Gamma}^{\dagger}_{T'}u^{-1}\mathcal{E}.$$

which is an isomorphism when $T' = u^{-1}(T)$.

Proof We have an exact sequence

$$0 \to \underline{\Gamma}^{\dagger}_{T}u^{-1}\mathcal{E} \to u^{-1}\mathcal{E} \to j^{\dagger}_{V\setminus T}u^{-1}\mathcal{E} \to 0$$

and since u^{-1} is exact, we also have another exact sequence

$$0 \to u^{-1}\underline{\Gamma}^{\dagger}_{T}\mathcal{E} \to u^{-1}\mathcal{E} \to u^{-1}j^{\dagger}_{V\setminus T}\mathcal{E} \to 0.$$

Our assertion therefore follows from Proposition 5.1.5.

□

Again, we will mainly apply this to frames. More precisely, when $(X \subset Y \subset P)$ is a frame, V is an admissible open subset of $]Y[_P$ and $T := V \setminus (V\cap]X[_P)$, we will consider the functor $\underline{\Gamma}^{\dagger}_{T}$ of overconvergent sections on V with support in T. If Z denotes a closed complement of X in Y, so that $T = V\cap]Z[_P$, we may also write $\underline{\Gamma}^\dagger_Z$ for short.

For future reference, note the following:

Proposition 5.2.6 *Let $(X \subset Y \subset P)$ be a frame and V an admissible open subset of $]Y[_P$. Let X_1 and X_2 be two open subsets of Y with closed complements Z_1 and Z_2 respectively. Then,*

$$\underline{\Gamma}^\dagger_{Z_1} \circ j^\dagger_{X_2} = j^\dagger_{X_2} \circ \underline{\Gamma}^\dagger_{Z_1}.$$

Proof We have an exact sequence

$$0 \to \underline{\Gamma}^\dagger_{Z_1}\mathcal{E} \to \mathcal{E} \to j^\dagger_{X_1}\mathcal{E} \to 0$$

and we may apply $j^\dagger_{X_2}$ which is exact in order to get another exact sequence

$$0 \to j^\dagger_{X_2}\underline{\Gamma}^\dagger_{Z_1}\mathcal{E} \to j^\dagger_{X_2}\mathcal{E} \to j^\dagger_{X_2}j^\dagger_{X_1}\mathcal{E} \to 0.$$

And we saw in Proposition 5.1.11 that

$$j^\dagger_{X_1} \circ j^\dagger_{X_2} = j^\dagger_{X_2} \circ j^\dagger_{X_1}.$$

Thus we have an exact sequence

$$0 \to j^\dagger_{X_2}\underline{\Gamma}^\dagger_{Z_1}\mathcal{E} \to j^\dagger_{X_2}\mathcal{E} \to j^\dagger_{X_1}j^\dagger_{X_2}\mathcal{E} \to 0$$

and it follows that $j^\dagger_{X_2}\underline{\Gamma}^\dagger_{Z_1}\mathcal{E}$ is identical to $\underline{\Gamma}^\dagger_{Z_1}j^\dagger_{X_2}\mathcal{E}$ as asserted. □

Lemma 5.2.7 *Let $(X \subset Y \subset P)$ be a frame, V an admissible open subset of $]Y[_P$ and $X = \cup_{i=1}^n X_i$ a finite open covering. If $\mathcal{E}$ is an abelian sheaf on V, the sequence*

$$0 \to j^\dagger_X\mathcal{E} \to \oplus_{i=1}^n j^\dagger_{X_i}\mathcal{E} \to \cdots \to j^\dagger_{X_{1\dots n}}\mathcal{E} \to 0$$

is exact.

Proof This is proved by induction on n, the case $n = 1$ being trivial. Using the second statement of Proposition 5.2.4, one sees that, if $T := V\setminus]X[_P$, the assertion is equivalent to the exactness of

$$0 \to \Gamma^\dagger_T\mathcal{E} \to \mathcal{E} \to \oplus_{i=1}^n j^\dagger_{X_i}\mathcal{E} \to \cdots \to j^\dagger_{X_{1\dots n}}\mathcal{E} \to 0$$

or, if we define

$$K_n := [\mathcal{E} \to \oplus_{i=1}^n j^\dagger_{X_i}\mathcal{E} \to \cdots \to j^\dagger_{X_{1\dots n}}\mathcal{E}]$$

to the fact that the natural morphism of complexes $\Gamma^\dagger_T\mathcal{E} \to K_n$ is a quasi-isomorphism, or even, an isomorphism in the derived category.

It follows from Proposition 5.1.7, that the cone of the morphism

$$K_{n-1} \to j^\dagger_{X_n} K_{n-1}$$

is exactly $K_n[-1]$. In other words, we have an exact triangle

$$K_n \to K_{n-1} \to j^\dagger_{X_n} K_{n-1} \to \cdots$$

On the other hand, if $X' = \cup_{i=1}^{n-1} X_i$ and $T' := V\backslash]X'[_P$, and if we also let $T_n := V\backslash]X_n[_P$, we have an exact sequence

$$0 \to \underline{\Gamma}^\dagger_{T_n} \Gamma^\dagger_{T'} \mathcal{E} \to \Gamma^\dagger_{T'} \mathcal{E} \to j^\dagger_{X_n} \Gamma^\dagger_{T'} \mathcal{E} \to 0.$$

It follows from Proposition 5.2.4 that

$$\underline{\Gamma}^\dagger_{T_n} \Gamma^\dagger_{T'} \mathcal{E} = \underline{\Gamma}^\dagger_{T} \mathcal{E}$$

and we get an exact triangle

$$\underline{\Gamma}^\dagger_{T} \mathcal{E} \to \Gamma^\dagger_{T'} \mathcal{E} \to j^\dagger_{Xn} \Gamma^\dagger_{T'} \mathcal{E} \to \cdots$$

By construction, we have a morphism of exact triangles

$$\begin{array}{ccccccc}
\underline{\Gamma}^\dagger_{T} \mathcal{E} & \longrightarrow & \Gamma^\dagger_{T'} \mathcal{E} & \longrightarrow & j^\dagger_{X_n} \Gamma^\dagger_{T'} \mathcal{E} & \longrightarrow & \cdots \\
\downarrow & & \downarrow & & \downarrow & & \downarrow \\
K_n & \longrightarrow & K_{n-1} & \longrightarrow & j^\dagger_{X_n} K_{n-1} & \longrightarrow & \cdots
\end{array}$$

By induction, the second arrow is an isomorphism, and since $j^\dagger_{X_n}$ is exact, the third also is. It follows that the first one is also an isomorphism and we are finished. □

Proposition 5.2.8 *Let $(X \subset Y \subset P)$ be a frame, V an admissible open subset of $]Y[_P$ and $\mathcal{E}$ an abelian sheaf on V. Assume that we are given an open or finite mixed covering*

$$\begin{array}{ccccc}
X_i & \hookrightarrow & Y_i & \hookrightarrow & P_i \\
\downarrow & & \downarrow & & \downarrow \\
X & \hookrightarrow & Y & \hookrightarrow & P
\end{array}$$

Then, there is a long exact sequence

$$0 \to j^\dagger_X \mathcal{E} \to \oplus h_{i*} j^\dagger_{X_i} \mathcal{E}_{|V_i} \to \cdots$$

with $V_i := V \cap]Y_i[_{P_i}$ and $h_i : V_i \hookrightarrow V$.

Proof Using Corollary 5.1.16, one sees that this assertion is local on V. We may therefore assume that $Y_i = Y$ for each i, that $P_i = P$ for each i and that the covering $X = \cup_i X_i$ is finite open (and also closed in the second case). We are now exactly in the situation of the lemma. □

Lemma 5.2.9 *Let*

$$\begin{array}{ccccc} X' & \hookrightarrow & Y' & \hookrightarrow & P' \\ \downarrow & & \downarrow & & \downarrow \\ X & \hookrightarrow & Y & \hookrightarrow & P \end{array}$$

be a mixed cartesian immersion of frames and $U := X \setminus X'$. Let V be a strict neighborhood of $]X[_P$ in $]Y[_P$ and $T' := V \cap]Y'[_{P'}$. If $\mathcal{E}$ is an abelian sheaf on V, then $\underline{\Gamma}^\dagger_{T'} j^\dagger_X \mathcal{E}$ only depends on X' and not on Y' or P'. Actually, there is a short exact sequence

$$0 \to \underline{\Gamma}^\dagger_{T'} j^\dagger_X \mathcal{E} \to j^\dagger_X \mathcal{E} \to j^\dagger_U \mathcal{E} \to 0.$$

Proof We may clearly assume that $P' = P$. If we write $U' := Y \setminus Y'$, we have a short exact sequence

$$0 \to \underline{\Gamma}^\dagger_{T'} j^\dagger_X \mathcal{E} \to j^\dagger_X \mathcal{E} \to j^\dagger_{U'} j^\dagger_X \mathcal{E} \to 0.$$

Since $U' \cap X = U$, it follows from Proposition 5.1.7 that $j^\dagger_{U'} j^\dagger_X \mathcal{E} = j^\dagger_U \mathcal{E}$ □

Definition 5.2.10 *With the notations of the lemma, we will write*

$$\underline{\Gamma}^\dagger_{X'} j^\dagger_X \mathcal{E} := \underline{\Gamma}^\dagger_{T'} j^\dagger_X \mathcal{E}.$$

The above lemma becomes a particular case of the next proposition.

Proposition 5.2.11 *Let $(X \subset Y \subset P)$ be a frame and V a strict neighborhood of $]X[_P$ in $]Y[_P$. Let*

$$X'' \hookrightarrow X' \hookrightarrow X$$

be a sequence of closed immersions, $U := X \setminus X''$ and $U' := X' \setminus X''$. If $\mathcal{E}$ is an abelian sheaf on V, there is an exact sequence

$$0 \to \underline{\Gamma}^\dagger_{X''} j^\dagger_X \mathcal{E} \to \underline{\Gamma}^\dagger_{X'} j^\dagger_X \mathcal{E} \to \underline{\Gamma}^\dagger_{U'} j^\dagger_U \mathcal{E} \to 0.$$

Proof It is sufficient to apply the snake lemma to the diagram

$$\begin{array}{ccccccccc} & & & & j_X^\dagger \mathcal{E} & = & j_X^\dagger \mathcal{E} & & \\ & & & & \downarrow & & \downarrow & & \\ 0 & \longrightarrow & \underline{\Gamma}_{U'}^\dagger j_U^\dagger \mathcal{E} & \longrightarrow & j_U^\dagger \mathcal{E} & \longrightarrow & j_{X \setminus X''}^\dagger \mathcal{E} & \longrightarrow & 0 \end{array}$$

□

Overconvergent sections with support in a closed subset only depend on an open neighborhood of the closed subset as the following shows:

Proposition 5.2.12 *Let $(X \subset Y \subset P)$ be a frame and V a strict neighborhood of $]X[_P$ in $]Y[_P$. Let X' be a closed subset of X and U an open neighborhood of X' in X. If $\mathcal{E}$ is an abelian sheaf on V, we have*

$$\Gamma_{X'}^\dagger j_X^\dagger \mathcal{E} = \Gamma_{X'}^\dagger j_U^\dagger \mathcal{E}.$$

Proof If we denote by U' the complement of X' in U, we have an open covering $X = U \cup U'$ with $U \cap U' = U \setminus X'$. Thus, it follows from Lemma 5.2.7 that the square

$$\begin{array}{ccc} j_X^\dagger \mathcal{E} & \longrightarrow & j_{X \setminus X'}^\dagger \mathcal{E} \\ \downarrow & & \downarrow \\ j_U^\dagger \mathcal{E} & \longrightarrow & j_{U \setminus X'}^\dagger \mathcal{E} \end{array}$$

is cartesian. We obtain an isomorphism

$$\Gamma_{X'}^\dagger j_X^\dagger \mathcal{E} = \Gamma_{X'}^\dagger j_U^\dagger \mathcal{E}$$

on the kernels. □

We will now study a notion that will be useful later on when we define rigid cohomology with compact support.

As before, if V is a rigid analytic variety and T is an admissible open subset of V, we can consider the open embedding of toposes $h : \widetilde{T} \hookrightarrow \widetilde{V}$ as well as the embedding $i : \widetilde{V} \setminus \widetilde{T} \hookrightarrow \widetilde{V}$ of the closed complement. The functor *sections with support in $V \setminus T$* is defined on abelian sheaves on V by

$$\underline{\Gamma}_{V \setminus T} \mathcal{E} := i_* i^! \mathcal{E}.$$

It is shown in [1], Exposé IV, 14, that there is a left exact sequence†

$$0 \to \underline{\Gamma}_{V\setminus T}\mathcal{E} \to \mathcal{E} \to h_* h^{-1}\mathcal{E}.$$

We will also consider the functor *global sections with support in* $V \setminus T$:

$$\Gamma_{V\setminus T}(V, \mathcal{E}) := \Gamma(V, \underline{\Gamma}_{V\setminus T}\mathcal{E}).$$

Later on, we will need the following functoriality property that we prove now:

Proposition 5.2.13 *Let V be a rigid analytic variety, T an open subset of V, $u : V' \to V$ a morphism and $T' := u^{-1}(T)$. If $\mathcal{E}'$ is an abelian sheaf on V', we have*

$$\Gamma_{V'\setminus T'}(V', \mathcal{E}') = \Gamma_{V\setminus T}(V, u_*\mathcal{E}').$$

Moreover, if u is finite and $\mathcal{E}'$ is a coherent $\mathcal{O}_{V'}$-module, then

$$R\Gamma_{V'\setminus T'}(V', \mathcal{E}') = R\Gamma_{V\setminus T}(V, u_*\mathcal{E}').$$

Proof We have a commutative diagram with left exact rows

$$\begin{array}{ccccccc}
0 \longrightarrow & \Gamma_{V'\setminus T'}(V', \mathcal{E}') & \longrightarrow & \Gamma(V', \mathcal{E}') & \longrightarrow & \Gamma(T', \mathcal{E}') . \\
 & & & \| & & \| \\
0 \longrightarrow & \Gamma_{V\setminus T}(V, u_*\mathcal{E}') & \longrightarrow & \Gamma(V, u_*\mathcal{E}') & \longrightarrow & \Gamma(T, u_*\mathcal{E}')
\end{array}$$

and the first assertion follows. In particular, we have

$$R\Gamma_{V'\setminus T'}(V', \mathcal{E}') = R\Gamma_{V\setminus T}(V, Ru_*\mathcal{E}'),$$

but with our additional hypothesis, $R^i u_*\mathcal{E}' = 0$ for $i > 0$. □

As usual, we shall apply these considerations only in the context of strict neighborhoods.

Proposition 5.2.14 *Let $(X \subset Y \subset P)$ be a frame, V be a strict neighborhood of $]X[_P$ in $]Y[_P$ and $\mathcal{E}$ an abelian sheaf on V. Then*

$$\Gamma_{]X[_P}(V, \mathcal{E}) = \ker(\Gamma(V, \mathcal{E}) \to \Gamma(V\setminus]X[_P, \mathcal{E}))$$

and if $h : V\setminus]X[_P \subset V$ denotes the inclusion map,

$$\underline{\Gamma}_{]X[_P}\mathcal{E} = \ker(\mathcal{E} \to h_* h^{-1}\mathcal{E}).$$

† There is a misprint: it is written $i_! i^{-1}$ instead of $i_* i^!$.

Proof By definition, we have a left exact sequence

$$0 \to \underline{\Gamma}_{]X[_P}\mathcal{E} \to \mathcal{E} \to h_*h^{-1}\mathcal{E}$$

that give us the second equality. From this sequence and the left exactness of $\Gamma(V, -)$, we derive another left exact sequence

$$0 \to \Gamma_{]X[_P}(V, \mathcal{E}) \to \Gamma(V, \mathcal{E}) \to \Gamma(V, h_*h^{-1}\mathcal{E})$$

and the first assertion results from the equality

$$\Gamma(V, h_*h^{-1}\mathcal{E}) = \Gamma(V\backslash]X[_P, \mathcal{E})).$$

□

Proposition 5.2.15 *Let $(X \subset Y \subset P)$ be a frame, V be a strict neighborhood of $]X[_P$ in $]Y[_P$ and $\mathcal{E}$ an abelian sheaf on V. Then, $\underline{\Gamma}_{]X[_P}\mathcal{E}$ is overconvergent. Moreover, if $j : V' \hookrightarrow V$ is the inclusion of a smaller strict neighborhood, we have*

$$(\underline{\Gamma}_{]X[_P}\mathcal{E})_{|V'} = \underline{\Gamma}_{]X[_P}\mathcal{E}_{|V'}$$

and

$$j_*\underline{\Gamma}_{]X[_P}\mathcal{E}_{|V'} = \underline{\Gamma}_{]X[_P}\mathcal{E}.$$

Proof If $h : T := V\backslash]X[_P \hookrightarrow V$ denotes the inclusion map, we clearly have

$$\mathcal{E}_{|T} \simeq (h_*h^{-1}\mathcal{E})_{|T}$$

and it follows that $(\underline{\Gamma}_{]X[_P}\mathcal{E})_{|T} = 0$ which exactly means that $\underline{\Gamma}_{]X[_P}\mathcal{E}$ is overconvergent.

Now, if we denote by $h' : T' := V'\backslash]X[_P \hookrightarrow V'$ the inclusion map, we have

$$(h_*h^{-1}\mathcal{E})_{|V'} = h'_*h'^{-1}\mathcal{E}_{|V'}$$

and it follows that

$$(\underline{\Gamma}_{]X[_P}\mathcal{E})_{|V'} = \underline{\Gamma}_{]X[_P}\mathcal{E}_{|V'}.$$

Finally, since $\underline{\Gamma}_{]X[_P}\mathcal{E}$ is overconvergent, we have

$$j_*\underline{\Gamma}_{]X[_P}\mathcal{E}_{|V'} = j_*(\underline{\Gamma}_{]X[_P}\mathcal{E})_{|V'} = \underline{\Gamma}_{]X[_P}\mathcal{E}.$$

□

Proposition 5.2.16 *Let $(X \subset Y \subset P)$ be a frame and $\mathcal{E}$ an abelian sheaf on some strict neighborhood V of $]X[_P$ in $]Y[_P$. Then*

(i) *If X' is an open subset of X, there is a canonical map*

$$\underline{\Gamma}_{]X'[_P}\mathcal{E} \to \underline{\Gamma}_{]X[_P}\mathcal{E}.$$

(ii) *If $X = X_1 \cup X_2$ is an open covering and $X_{12} := X_1 \cap X_2$, there is a left exact sequence*

$$0 \to \underline{\Gamma}_{]X_{12}[_P}\mathcal{E} \to \underline{\Gamma}_{]X_1[_P}\mathcal{E} \oplus \underline{\Gamma}_{]X_2[_P}\mathcal{E} \to \underline{\Gamma}_{]X[_P}\mathcal{E}.$$

Actually, there is an exact triangle

$$R\underline{\Gamma}_{]X_{12}[_P}\mathcal{E} \to R\underline{\Gamma}_{]X_1[_P}\mathcal{E} \oplus R\underline{\Gamma}_{]X_2[_P}\mathcal{E} \to R\underline{\Gamma}_{]X[_P}\mathcal{E} \to .$$

Proof There is a commutative diagram

$$\begin{array}{ccccc}]X'[_{P'} & \hookrightarrow & V & \xleftarrow{h'} & V\setminus]X'[_P \\ \downarrow & & \| & & \uparrow \\ ]X[_P & \hookrightarrow & V & \xleftarrow{h} & V\setminus]X[_P \end{array}$$

from which we derive the following commutative diagram

$$\begin{array}{ccccccc} 0 & \longrightarrow & \underline{\Gamma}_{]X[_P}\mathcal{E} & \longrightarrow & \mathcal{E} & \longrightarrow & h_*h^{-1}\mathcal{E} \\ & & \uparrow & & \| & & \uparrow \\ 0 & \longrightarrow & \underline{\Gamma}_{]X'[_{P'}}\mathcal{E} & \longrightarrow & \mathcal{E} & \longrightarrow & h'_*h'^{-1}\mathcal{E}. \end{array}$$

We now turn to the second assertion. We have

$$V\setminus]X_{12}[_P = (V\setminus]X_1[_P) \cup (V\setminus]X_2[_P)$$

and

$$(V\setminus]X_1[_P) \cap (V\setminus]X_2[_P) = V\setminus]X[_P.$$

Thus, if we denote by h, h_1, h_2, h_{12} the open immersions into V of

$$V\setminus]X[_P, \quad V\setminus]X_1[_P, \quad V\setminus]X_2[_P, \quad V\setminus]X_{12}[_P,$$

respectively, we obtain a commutative diagram with exact columns

$$\begin{array}{ccccccc}
R\underline{\Gamma}_{]X_{12}[}\mathcal{E} & \longrightarrow & R\underline{\Gamma}_{]X_1[}\mathcal{E} \oplus R\underline{\Gamma}_{]X_2[}\mathcal{E} & \longrightarrow & R\underline{\Gamma}_{]X[}\mathcal{E} & \longrightarrow & \\
\downarrow & & \downarrow & & \downarrow & & \\
\mathcal{E} & \longrightarrow & \mathcal{E} \oplus \mathcal{E} & \longrightarrow & \mathcal{E} & \longrightarrow & \\
\downarrow & & \downarrow & & \downarrow & & \\
h_{12*}h_{12}^{-1}\mathcal{E} & \longrightarrow & h_{1*}h_1^{-1}\mathcal{E} \oplus h_{2*}h_2^{-1}\mathcal{E} & \longrightarrow & h_*h^{-1}\mathcal{E} & \longrightarrow & \\
\downarrow & & \downarrow & & \downarrow & &
\end{array}$$

Since the bottom rows are exact, so is the top one. □

Proposition 5.2.17 *Let*

$$\begin{array}{ccccc}
X' & \hookrightarrow & Y' & \hookrightarrow & P' \\
\downarrow{\scriptstyle f} & & \downarrow{\scriptstyle g} & & \downarrow{\scriptstyle u} \\
X & \hookrightarrow & Y & \hookrightarrow & P
\end{array}$$

be a cartesian morphism of frames. Let V be a strict neighborhood of $]X[_P$ in $]Y[_P$ and V' a strict neighborhood of $]X'[_{P'}$ in $u_K^{-1}(V) \cap]Y'[_{P'}$. If $\mathcal{E}$ is an abelian sheaf and if $u_K : V' \to V$ denotes the induced map, there is a canonical base change map

$$u_K^{-1}\underline{\Gamma}_{]X[_P}\mathcal{E} \to \underline{\Gamma}_{]X'[_{P'}}u_K^{-1}\mathcal{E}.$$

If our morphism is an open or a mixed immersion, then we obtain an isomorphism.

Proof Since the morphism is cartesian, there is a commutative diagram (with cartesian squares)

$$\begin{array}{ccccc}
]X'[_{P'} & \hookrightarrow & V' & \overset{h'}{\hookleftarrow} & V' \setminus]X'[_{P'} \\
\downarrow & & \downarrow & & \downarrow \\
]X[_P & \hookrightarrow & V & \overset{h}{\hookleftarrow} & V \setminus]X[_P
\end{array}$$

and it follows that there is a canonical adjunction map

$$u_K^{-1}h_*h^{-1}\mathcal{E} \to h'_*h'^{-1}u_K^{-1}\mathcal{E}$$

from which we derive the following commutative diagram

$$\begin{array}{ccccccc} 0 & \longrightarrow & u_K^{-1}\underline{\Gamma}_{]X[_P}\mathcal{E} & \longrightarrow & u_K^{-1}\mathcal{E} & \longrightarrow & u_K^{-1}h_*h^{-1}\mathcal{E} \\ & & \downarrow & & \| & & \downarrow \\ 0 & \longrightarrow & \underline{\Gamma}_{]X'[_{P'}}u_K^{-1}\mathcal{E} & \longrightarrow & u_K^{-1}\mathcal{E} & \longrightarrow & h'_*h'^{-1}u_K^{-1}\mathcal{E} \end{array}$$

Finally, when u_K is an open immersion, then the right arrow is bijective and it follows that we get an isomorphism on the left. □

Proposition 5.2.18 *Let $(X \subset Y \subset P)$ be a frame and $\sigma : K \hookrightarrow K'$ an isometric embedding. Let V be a strict neighborhood of $]X[_P$ in $]Y[_P$ and $\mathcal{E}$ an abelian sheaf on V. If $\varpi : \widetilde{V}^\sigma \to \widetilde{V}$ denotes the canonical morphism of toposes, there is a canonical base change map*

$$\varpi^{-1}\underline{\Gamma}_{]X[_P}\mathcal{E} \to \underline{\Gamma}_{]X^\sigma[_{P^\sigma}}\varpi^{-1}\mathcal{E}.$$

Proof As usual, if $h : V\backslash]X[_P \hookrightarrow V$ denotes the inclusion map, we may consider the adjunction map

$$\varpi^{-1}h_*h^{-1}\mathcal{E} \to h_*^\sigma(h^\sigma)^{-1}\varpi^{-1}\mathcal{E}$$

and easily derive our canonical morphism. □

Proposition 5.2.19 *Let*

$$\begin{array}{ccccc} X' & \hookrightarrow & Y' & \hookrightarrow & P' \\ \downarrow & & \downarrow & & \downarrow \\ X & \hookrightarrow & Y & \hookrightarrow & P \end{array}$$

be a mixed cartesian immersion of frames and $U := X \setminus X'$. Let V be a strict neighborhood of $]X[_P$ in $]Y[_P$ and $\mathcal{E}$ an abelian sheaf on V. We will denote by

$$h : V\backslash]X[_P \hookrightarrow V \quad \text{and} \quad j : V\backslash]U[_P \hookrightarrow V$$

the inclusion maps. Then,

(i) *If Y'' is a closed complement for U in Y, then X' is open in Y'' and $V\backslash]U[_P$ is a strict neighborhood of $]X'[_P$ in $]Y''[_P$.*
(ii) *We have*

$$j_*\underline{\Gamma}_{]X'[_P}j^{-1}\mathcal{E} = \ker(j_*j^{-1}\mathcal{E} \mapsto h_*h^{-1}\mathcal{E}).$$

(iii) *There is a natural left exact sequence*

$$0 \to \underline{\Gamma}_{]U[_P}\mathcal{E} \to \underline{\Gamma}_{]X[_P}\mathcal{E} \to j_*\underline{\Gamma}_{]X'[_{P'}}j^{-1}\mathcal{E}$$

which is exact on the right also when $\mathcal{E}$ is injective.

(iv) *We have an exact triangle*

$$R\underline{\Gamma}_{]U[_P}\mathcal{E} \to R\underline{\Gamma}_{]X[_P}\mathcal{E} \to Rj_*R\underline{\Gamma}_{]X'[_{P'}}j^{-1}\mathcal{E} \to .$$

Proof As usual, we may assume that $P' = P$. It is convenient to choose a closed complement Z for X in Y. Then, Z is also a closed complement for X' in Y'' and in particular, X' is open in Y''. Now, by definition, the covering $]Y[_P = V \cup]Z[_P$ is admissible and it follows that the covering

$$]Y''[_P = (V \cap]Y''[_P) \cup]Z[_P$$

is also admissible. Since we have

$$V \setminus]U[_P = V \cap]Y''[_P,$$

this is a strict neighborhood of $]X'[_P$ in $]Y''[_P$.

We now turn to the second assertion. If we denote by

$$h' : V \setminus]X[_P \hookrightarrow V \setminus]U[_P$$

the inclusion map, we have a left exact sequence

$$0 \to \underline{\Gamma}_{]X'[_P}j^{-1}\mathcal{E} \to j^{-1}\mathcal{E} \to h'_*h'^{-1}j^{-1}\mathcal{E}$$

and since $h = j \circ h'$ and j_* is left exact, we get a left exact sequence

$$0 \to j_*\underline{\Gamma}_{]X'[_P}j^{-1}\mathcal{E} \to j_*j^{-1}\mathcal{E} \mapsto h_*h^{-1}\mathcal{E}$$

as asserted.

If we apply the snake lemma to the diagram

$$\begin{array}{ccccccc} & & & & \mathcal{E} & = & \mathcal{E} \\ & & & & \downarrow & & \downarrow \\ 0 & \longrightarrow & j_*\underline{\Gamma}_{]X'[_P}j^{-1}\mathcal{E} & \longrightarrow & j_*j^{-1}\mathcal{E} & \longrightarrow & h_*h^{-1}\mathcal{E} \end{array}$$

we get the left exact sequence

$$0 \to \underline{\Gamma}_{]U[_P}\mathcal{E} \to \underline{\Gamma}_{]X[_P}\mathcal{E} \to j_*\underline{\Gamma}_{]X'[_P}j^{-1}\mathcal{E}.$$

When $\mathcal{E}$ is injective, the middle arrow in our diagram is surjective and it follows that our sequence is exact on the right also. Thus, we get an exact triangle

$$R\underline{\Gamma}_{]U[_P}\mathcal{E} \to R\underline{\Gamma}_{]X[_P}\mathcal{E} \to Rj_*R\underline{\Gamma}_{]X'[_P}j^{-1}\mathcal{E} \to .$$

□

We finish this section with a computation of cohomology with support that will be used later when we consider rigid cohomology with compact support in the Monsky–Washnitzer setting. We will write

$$K[\underline{t}]_c := \{\sum_{\underline{i}<0} a_{\underline{i}}\underline{t}^{\underline{i}}, \forall \eta < 1, |a_{\underline{i}}|\eta^{-|\underline{i}|} < \infty\}.$$

It might also be convenient to set for $\eta = 1$,

$$K[\underline{t}]_\eta := \{\sum_{\underline{i}<0} a_{\underline{i}}\underline{t}^{\underline{i}}, |a_{\underline{i}}|\eta^{-|\underline{i}|} < \infty\}$$

so that $K[\underline{t}]_c = \cap K[\underline{t}]_\eta$. Of course, one may also use the condition $|a_{\underline{i}}|\eta^{-|\underline{i}|} \to 0$ and get the same result.

Lemma 5.2.20 *We have*

(i) $H^d_{\mathbf{B}^d(0,1^+)}(\mathbf{A}^{d,\mathrm{rig}}_K, \mathcal{O}_{\mathbf{A}^{d,\mathrm{rig}}_K}) = K[\underline{t}]_c$.
(ii) $H^i_{\mathbf{B}^d(0,1^+)}(\mathbf{A}^{d,\mathrm{rig}}_K, \mathcal{O}_{\mathbf{A}^{d,\mathrm{rig}}_K}) = 0$ *when* $i < d$.
(iii) $H^i_{\mathbf{B}^d(0,1^+)}(\mathbf{B}^d(0, \lambda^+), \mathcal{E}) = 0$ *whenever* $\mathcal{E}$ *is coherent and* $i > d$.

Proof Of course, we may replace $\mathbf{A}^{d,\mathrm{rig}}_K$ by $\mathbf{B}^d(0, \lambda^+)$ in the first two assertions. The complement of $\mathbf{B}^d(0, 1^+)$ in $\mathbf{B}^d(0, \lambda^+)$ has an admissible open covering by d quasi-Stein subsets

$$\mathbf{B}(0, \lambda^+) \times \cdots \times \mathbf{B}(0, \lambda^+) \times \mathbf{A}(0, 1^-, \lambda^+) \times \mathbf{B}(0, \lambda^+) \times \cdots \times \mathbf{B}(0, \lambda^+).$$

The last assertion therefore follows from the fact that coherent sheaves have no cohomology on quasi-Stein spaces. The first two are obtained by a direct computation of cohomology with support with respect to this covering. The details are left to the reader who may also use induction on d and completed tensor products if he wishes. □

Proposition 5.2.21 *Let* X *be a smooth affine scheme of pure dimension* d *on* $\mathcal{V}$, V *a strict neighborhood of* $\widehat{X}_K$ *in* X^{rig}_K *and* $\mathcal{E}$ *a coherent module on* V. *Then,*

(i) *We have*

$$H^i_{\widehat{X}_K}(V, \mathcal{E}) = 0 \quad \text{for} \quad i > d$$

and

$$H^d_{\widehat{X}_K}(V, \mathcal{E}) \simeq \Gamma(V, \mathcal{E}) \otimes_{\Gamma(V, \mathcal{O}_V)} H^d_{\widehat{X}_K}(V, \mathcal{O}_V).$$

(ii) *If* $\mathcal{E}$ *is locally free, then*

$$H^i_{\widehat{X}_K}(V, \mathcal{E}) = 0 \quad \text{for} \quad i \neq d.$$

(iii) *If $X = X_1 \cup X_2$ is an open affine covering and $X_{12} := X_1 \cap X_2$, we have a right exact sequence*

$$H^d_{\widehat{X}_{12K}}(V \cap X^{\mathrm{rig}}_{12K}, \mathcal{E}) \to H^d_{\widehat{X}_{1K}}(V \cap X^{\mathrm{rig}}_{1K}, \mathcal{E}) \oplus H^d_{\widehat{X}_{2K}}(V \cap X^{\mathrm{rig}}_{2K}, \mathcal{E}) \\ \to H^d_{\widehat{X}_K}(V, \mathcal{E}) \to 0.$$

It is also left exact when $\mathcal{E}$ is locally free.

Proof Noether's normalization theorem provides a finite map

$$X_k \to \mathbf{A}^d_k$$

which, shrinking V if necessary, lifts it to a finite map

$$\varphi : V \to \mathbf{B}^N(0, \lambda^+).$$

The reader who wants to check this can use for example Proposition 8.1.14 below. It then follows from assertion (iii) of Lemma 5.2.20 and Proposition 5.2.13 that

$$H^i_{\widehat{X}_K}(V, \mathcal{E}) = H^i_{\mathbf{B}^N(0,1^+)}(\mathbf{B}^N(0, \lambda^+), \varphi_*\mathcal{E}) = 0$$

for $i > d$. In particular, $H^d_{\widehat{X}_K}(V, -)$ is right exact and the second part of the assertion follows from the fact that $\mathcal{E}$ has a free presentation because V is affinoid.

Any locally free sheaf of finite rank on an affinoid variety is a direct factor of a free sheaf of finite rank. It is therefore sufficient to prove the second assertion when $\mathcal{E} = \mathcal{O}_{X^{\mathrm{rig}}_K}$. In the case of the affine space, our assertion therefore follows from part (ii) of Lemma 5.2.20. In general, we may choose a closed embedding $i : X \hookrightarrow \mathbf{A}^N_{\mathcal{V}}$ and a locally free resolution $\mathcal{E}_\bullet$ of minimal length $r = N - d$ of $i_*\mathcal{O}_X$. There is a spectral sequence

$$E_1^{p,q} = H^q_{\mathbf{B}^N(0,1^+)}(\mathbf{A}^{N\mathrm{rig}}_K, \mathcal{E}^{\mathrm{rig}}_{r-p,K}) \Rightarrow H^{p+q} := H^{p+q-r}_{\widehat{X}_K}(X^{\mathrm{rig}}_K)$$

that degenerates in E_2 more precisely, we have

$$H^{N+q-r}_{\widehat{X}_K}(X^{\mathrm{rig}}_K) = E_2^{N,q}$$

which is 0 unless $0 \leq q \leq r$ and in particular $H^i_{\widehat{X}_K}(X^{\mathrm{rig}}_K) = 0$ for $i = N + q - r < N - r = d$.

Finally, the last assertion follows from Proposition 5.2.16.

□

If $i : W \hookrightarrow V$ is a closed immersion of smooth rigid analytic varieties of relative dimension d and e, respectively, there is a *fundamental isomorphism*

$$i_*\Omega^d_W[d] \simeq R\mathcal{H}om_{\mathcal{O}_V}(i_*\mathcal{O}_W, \Omega^e_V[e]).$$

Recall that, if $\mathcal{I}$ denotes the ideal of W in V, there is a natural isomorphism

$$i_*\Omega^d_W \simeq \mathcal{H}om_{\mathcal{O}_V}(i_* \wedge^r (\mathcal{I}/\mathcal{I}^2), \Omega^e_V).$$

When X is defined in Y by a regular sequence $f_1, \ldots, f_r$, the fundamental isomorphism comes from the Koszul resolution

$$\mathcal{K}_\bullet(\underline{f}) \simeq i_*\mathcal{O}_W$$

with $\mathcal{K}_p(\underline{f}) = \bigwedge^p \mathcal{O}^r_V$ and the isomorphism

$$\mathcal{O}_W \simeq \wedge^r(\mathcal{I}/\mathcal{I}^2), \quad 1 \mapsto \overline{f}_1 \wedge \cdots \wedge \overline{f}_r.$$

If T is an admissible subset of V, the restriction map $\mathcal{O}_V \to i_*\mathcal{O}_W$ induces, thanks to the fundamental isomorphism, a morphism

$$i_*\Omega^d_W[d] \to \Omega^e_V[e]$$

in the derived category from which we get the *Gysin map*

$$H^d_{(V\backslash T)\cap W}(W, \Omega^d_W) \to H^e_{V\backslash T}(V, \Omega^e_V).$$

Proposition 5.2.22 *If $X \hookrightarrow Y$ is a closed immersion of smooth affine $\mathcal{V}$-schemes of respective dimension d and e, then the Gysin map is injective*

$$H^d_{\widehat{X}_K}(X^{\mathrm{rig}}_K, \Omega^d_{X^{\mathrm{rig}}_K}) \hookrightarrow H^e_{\widehat{Y}_K}(Y^{\mathrm{rig}}_K, \Omega^e_{Y^{\mathrm{rig}}_K}).$$

More precisely, if X is defined in Y by a regular sequence $f_1, \cdots, f_r$, we have

$$H^d_{\widehat{X}_K}(X^{\mathrm{rig}}_K, \Omega^d_{X^{\mathrm{rig}}_K}) \simeq \{\omega \in H^e_{\widehat{Y}_K}(Y^{\mathrm{rig}}_K, \Omega^e_{Y^{\mathrm{rig}}_K}), \forall i = 1, \ldots, r, f_i\omega = 0\}.$$

Proof The first assertion will result from the second one thanks to part (iii) of Proposition 5.2.21. Now, we may consider the spectral sequence associated to the Koszul resolution. More precisely, we let

$$\mathcal{K}^p := \mathcal{H}om_{\mathcal{O}_{X^{\mathrm{rig}}_K}}(\mathcal{K}_{r-p}(\underline{f}), \Omega^e_{Y^{\mathrm{rig}}_K}))$$

so that

$$\mathcal{K}^\bullet \simeq i_*\Omega^d_{X^{\mathrm{rig}}_K}[-r].$$

Then, the spectral sequence

$$E_1^{p,q} = H^q_{\widehat{Y}_K}(Y^{\mathrm{rig}}_K, \mathcal{K}^p) \Rightarrow H^{p+q} := H^{p+q-r}_{\widehat{X}_K}(X^{\mathrm{rig}}_K, \Omega^d_{X^{\mathrm{rig}}_K})$$

degenerates in E_2 thanks to assertion (ii) of Proposition 5.2.21 and thus

$$H^d_{\widehat{X}_K}(X^{\mathrm{rig}}_K, \Omega^d_{X^{\mathrm{rig}}_K}) = E^{0,e}_\infty = E^{0,e}_2 = \ker[E^{0,e}_1 \to E^{1,e}_2].$$

□

5.3 Dagger modules

In the next propositions, modules can be left or right modules as long as the statements have a meaning. Anyway, we do not intend to apply these results to non commutative rings and the reader who so wishes may therefore assume that the ring $\mathcal{A}$ is commutative.

Proposition 5.3.1 *Let $(X \subset Y \subset P)$ be a frame, V an admissible open subset of $]Y[_P$ and $\mathcal{A}$ a sheaf of rings on V. Then the categories of overconvergent $\mathcal{A}$-modules and $j_X^\dagger\mathcal{A}$-modules are equivalent.*

More precisely, the forgetful functor from $j_X^\dagger\mathcal{A}$-modules to $\mathcal{A}$-modules is fully faithful with exact left adjoint $j_X^\dagger$ and its image is the subcategory of overconvergent $\mathcal{A}$-modules.

Proof If $\mathcal{F}$ is a $j_X^\dagger\mathcal{A}$-module and $T := V \setminus (]X[_P \cap V)$, then, by restriction, $\mathcal{F}_{|T}$ is a module on the null ring and is therefore equal to 0. It follows that $\mathcal{F}$ is overconvergent.

Conversely, if $\mathcal{F}$ is an overconvergent $\mathcal{A}$-module, the bilinear map

$$\mathcal{A} \times \mathcal{F} \to \mathcal{F}$$

factors uniquely through

$$j_X^\dagger(\mathcal{A} \times \mathcal{F}) = j_X^\dagger\mathcal{A} \times j_X^\dagger\mathcal{F} = j_X^\dagger\mathcal{A} \times \mathcal{F}$$

and induces on $\mathcal{F}$ a structure of $j_X^\dagger\mathcal{A}$-module.

It follows that the categories of overconvergent $\mathcal{A}$-modules and $j_X^\dagger\mathcal{A}$-modules are equivalent. The other assertions easily follow. □

Proposition 5.3.2 *Let $(X \subset Y \subset P)$ be a frame, V an admissible open subset of $]Y[_P$ and $\mathcal{A}$ a sheaf of rings on V. If $\mathcal{E}$ and $\mathcal{F}$ are two $\mathcal{A}$-modules, we have*

$$\begin{aligned} j_X^\dagger(\mathcal{E} \otimes_{\mathcal{A}} \mathcal{F}) &\simeq j_X^\dagger\mathcal{E} \otimes_{\mathcal{A}} \mathcal{F} \simeq \mathcal{E} \otimes_{\mathcal{A}} j_X^\dagger\mathcal{F} \\ &\simeq j_X^\dagger\mathcal{E} \otimes_{\mathcal{A}} j_X^\dagger\mathcal{F} \simeq j_X^\dagger\mathcal{E} \otimes_{j_X^\dagger\mathcal{A}} j_X^\dagger\mathcal{F} \end{aligned}$$

(and in particular

$$j_X^\dagger\mathcal{A} \otimes_{\mathcal{A}} \mathcal{E} \simeq j_X^\dagger\mathcal{E}).$$

We also have

$$\mathcal{H}om_{j_X^\dagger\mathcal{A}}(j_X^\dagger\mathcal{E}, j_X^\dagger\mathcal{F}) = \mathcal{H}om_{\mathcal{A}}(j_X^\dagger\mathcal{E}, j_X^\dagger\mathcal{F}) = \mathcal{H}om_{\mathcal{A}}(\mathcal{E}, j_X^\dagger\mathcal{F})$$

Proof We do first the tensor product. Note that, since extension of scalars

$$\mathcal{E} \mapsto j_X^\dagger\mathcal{A} \otimes_{\mathcal{A}} \mathcal{E}$$

is also a right adjoint to the forgetful functor, we have

$$j_X^\dagger \mathcal{A} \otimes_{\mathcal{A}} \mathcal{E} = j_X^\dagger \mathcal{E}.$$

From this, we deduce the first isomorphism

$$j_X^\dagger(\mathcal{E} \otimes_{\mathcal{A}} \mathcal{F}) = j_X^\dagger \mathcal{A} \otimes_{\mathcal{A}} (\mathcal{E} \otimes_{\mathcal{A}} \mathcal{F}) = j_X^\dagger \mathcal{E} \otimes_{\mathcal{A}} \mathcal{F}$$

and we get the second by symmetry. The last one is a consequence of the surjectivity of $\mathcal{A} \to j_X^\dagger \mathcal{A}$. Finally, we have

$$j_X^\dagger \mathcal{E} \otimes_{\mathcal{A}} j_X^\dagger \mathcal{F} = j_X^\dagger(\mathcal{E} \otimes_{\mathcal{A}} j_X^\dagger \mathcal{F}) = \mathcal{E} \otimes_{\mathcal{A}} j_X^\dagger j_X^\dagger \mathcal{F} = \mathcal{E} \otimes_{\mathcal{A}} j_X^\dagger \mathcal{F}.$$

Concerning the assertions about internal Hom, we get the first equality from the surjectivity of $\mathcal{A} \to j_X^\dagger \mathcal{A}$ and the second one from Proposition 5.1.9. More precisely, if W is an open subset of V, we may see it as a sheaf of sets on V with

$$W(W') = 0 \quad \text{if} \quad W' \subset W \quad \text{and} \quad W(W') = \emptyset \quad \text{otherwise.}$$

Then, we have

$$\Gamma(W, \mathcal{H}om_{\mathcal{A}}(j_X^\dagger \mathcal{E}, j_X^\dagger \mathcal{F})) = \mathrm{Hom}_{\mathcal{A}}(j_X^\dagger \mathcal{E}, \mathcal{H}om(W, j_X^\dagger \mathcal{F}))$$

and thanks to Proposition 5.1.9,

$$= \mathrm{Hom}_{\mathcal{A}}(\mathcal{E}, \mathcal{H}om(W, j_X^\dagger \mathcal{F})) = \Gamma(W, \mathcal{H}om_{\mathcal{A}}(\mathcal{E}, j_X^\dagger \mathcal{F})).$$

□

Corollary 5.3.3 *Let $(X \subset Y \subset P)$ be a frame, V an admissible open subset of $]Y[_P$ and $\mathcal{A}$ a sheaf of rings on V. If $\mathcal{E}$ and $\mathcal{F}$ are two $\mathcal{A}$-modules with $\mathcal{F}$ flat in a strict neighborhood of $]X[_P$ in $]Y[_P$, then*

$$\underline{\Gamma}_X^\dagger \mathcal{E} \otimes_{\mathcal{A}} \mathcal{F} \simeq \underline{\Gamma}_X^\dagger(\mathcal{E} \otimes_{\mathcal{A}} \mathcal{F}).$$

Proof We may clearly assume $\mathcal{F}$ flat on V and tensor with $\mathcal{F}$ the exact sequence

$$0 \to \underline{\Gamma}^\dagger \mathcal{E} \to \mathcal{E} \to j^\dagger \mathcal{E} \to 0$$

in order to get

$$0 \to \underline{\Gamma}^\dagger \mathcal{E} \otimes_{\mathcal{A}} \mathcal{F} \to \mathcal{E} \otimes_{\mathcal{A}} \mathcal{F} \to j^\dagger \mathcal{E} \otimes_{\mathcal{A}} \mathcal{F} \to 0.$$

Finally, we use the fact that

$$j^\dagger \mathcal{E} \otimes_{\mathcal{A}} \mathcal{F} = j^\dagger(\mathcal{E} \otimes_{\mathcal{A}} \mathcal{F}).$$

□

Corollary 5.3.4 *Let $(X \subset Y \subset P)$ be a frame, V an admissible open subset of $]Y[_P$ and $\mathcal{A}$ a sheaf of rings on V. If $\mathcal{E}$ and $\mathcal{F}$ are two overconvergent $\mathcal{A}$-modules, then*

$$\mathcal{E} \otimes_{j_X^\dagger \mathcal{A}} \mathcal{F} = \mathcal{E} \otimes_{\mathcal{A}} \mathcal{F},$$
$$\mathcal{H}om_{j_X^\dagger \mathcal{A}}(\mathcal{E}, \mathcal{F}) = \mathcal{H}om_{\mathcal{A}}(\mathcal{E}, \mathcal{F})$$

and they are both overconvergent.

Proof This is an immediate consequence of Proposition 5.3.2 because here $\mathcal{E} = j_X^\dagger \mathcal{E}$ and $\mathcal{F} = j_X^\dagger \mathcal{F}$. □

Finally, it should be noticed that the functors $j^\dagger$ and $\underline{\Gamma}$ are not unrelated. This is the basis of Poincaré duality.

Proposition 5.3.5 *Let $(X \subset Y \subset P)$ be a frame, V an admissible open subset of $]Y[_P$ and $\mathcal{A}$ a sheaf of rings on V. If $\mathcal{E}$ and $\mathcal{F}$ are two $\mathcal{A}$-modules, there is a canonical isomorphism*

$$\mathcal{H}om_{\mathcal{A}}(j_X^\dagger \mathcal{E}, \mathcal{F}) \simeq \underline{\Gamma}_{]X[_P} \mathcal{H}om_{\mathcal{A}}(\mathcal{E}, \mathcal{F}).$$

Proof This is completely formal. If we denote as usual by

$$h : V \setminus]X[_P \hookrightarrow V$$

the inclusion map, we have

$$\mathcal{H}om_{\mathcal{A}}(h_! h^{-1} \mathcal{E}, \mathcal{F}) = h_* h^{-1} \mathcal{H}om_{\mathcal{A}}(\mathcal{E}, \mathcal{F}).$$

Since, by definition, $\underline{\Gamma}_X^\dagger \mathcal{E} := h_! h^{-1} \mathcal{E}$, the short exact sequence

$$0 \to \underline{\Gamma}_X^\dagger \mathcal{E} \to \mathcal{E} \to j_X^\dagger \mathcal{E} \to 0$$

gives a left exact sequence

$$0 \to \mathcal{H}om_{\mathcal{A}}(j_X^\dagger \mathcal{E}, \mathcal{F}) \to \mathcal{H}om_{\mathcal{A}}(\mathcal{E}, \mathcal{F}) \to h_* h^{-1} \mathcal{H}om_{\mathcal{A}}(\mathcal{E}, \mathcal{F}).$$

Thus, we get an identification of $\mathcal{H}om_{\mathcal{A}}(j_X^\dagger \mathcal{E}, \mathcal{F})$ with the kernel of the second map which is $\underline{\Gamma}_{]X[_P} \mathcal{H}om_{\mathcal{A}}(\mathcal{E}, \mathcal{F})$. □

Corollary 5.3.6 *If $\mathcal{A}$ is a commutative ring and $\mathcal{E}$, $\mathcal{F}$ and $\mathcal{G}$ three $\mathcal{A}$-modules, any $\mathcal{A}$-bilinear map $\mathcal{E} \times \mathcal{F} \to \mathcal{G}$ induces an $\mathcal{A}$-bilinear map*

$$j_X^\dagger \mathcal{E} \times \Gamma_{]X[_P} \mathcal{F} \to \Gamma_{]X[_P} \mathcal{G}.$$

Proof By adjunction, any morphism $\mathcal{E} \otimes_{\mathcal{A}} \mathcal{F} \to \mathcal{G}$ induces a morphism

$$\mathcal{F} \to \mathcal{H}om_{\mathcal{A}}(\mathcal{E}, \mathcal{G})$$

and by restriction, a morphism

$$\Gamma_{]X[_P}\mathcal{F} \to \Gamma_{]X[_P}\mathcal{H}om_{\mathcal{A}}(\mathcal{E}, \mathcal{G}).$$

The proposition tells us that this morphism may be seen as a morphism

$$\Gamma_{]X[_P}\mathcal{F} \to \mathcal{H}om_{\mathcal{A}}(j_X^\dagger\mathcal{E}, \mathcal{G})$$

which, by adjunction again, corresponds to a morphism

$$j_X^\dagger\mathcal{E} \otimes_{\mathcal{A}} \Gamma_{]X[_P}\mathcal{F} \to \mathcal{G}.$$

Since $j_X^\dagger\mathcal{E} \otimes_{\mathcal{A}} \Gamma_{]X[_P}\mathcal{F}$ is overconvergent, its restriction to $V\backslash]X[_P$ is zero, and our morphisms takes its values in $\Gamma_{]X[_P}\mathcal{G}$. □

We could write many more sorites about $j_X^\dagger\mathcal{A}$-modules for general rings $\mathcal{A}$, but this does not seem necessary for our purpose.

Proposition 5.3.7 *Let $(X \subset Y \subset P)$ be a frame, V an admissible open subset of $]Y[_P$ and V' a strict neighborhood of $]X[_P$ in $]Y[_P$. If $j_{VV'} : V \cap V' \hookrightarrow V$ denotes the inclusion map, then $j_{VV'}^{-1}$ and $j_{VV'*}$ induce an equivalence between $j_X^\dagger\mathcal{O}_V$-modules and $j_X^\dagger\mathcal{O}_{V\cap V'}$-modules.*

Moreover, if E is a $j_X^\dagger\mathcal{O}_V$-module, we have $Rj_{VV'*}j_{VV'}^{-1}E = E$.

Proof We know that $j_{VV'}^{-1}$ and $j_{VV'*}$ induce an equivalence between overconvergent sheaves on V and $V \cap V'$. Moreover, it follows from Propositions 5.1.5 and 5.1.13 that

$$j_{VV'}^{-1}j_X^\dagger\mathcal{O}_V = j_X^\dagger\mathcal{O}_{V\cap V'}$$

and

$$j_{VV'*}j_X^\dagger\mathcal{O}_{V\cap V'} = j_X^\dagger\mathcal{O}_V.$$

Finally, the last assertion immediately follows from Proposition 5.2.2. □

Note in particular that if V is a strict neighborhood of $]X[_P$ in $]Y[_P$, the category of $j_X^\dagger\mathcal{O}_V$-modules is canonically equivalent to the category of $j_X^\dagger\mathcal{O}_{]Y[_P}$-modules. In the future, when there is no risk of confusion, we will identify these two categories.

Proposition 5.3.8 *Let $(X \subset Y \subset P)$ (resp. $(X' \subset Y' \subset P')$) be a frame and V (resp. V') an admissible open subset of $]Y[_P$ (resp. $]Y'[_{P'}$). Let $f : X' \to X$ and $u : V' \to V$ be a pair of compatible morphisms. Then, there is a canonical morphism*

$$u^*j_X^\dagger\mathcal{E} \to j_{X'}^\dagger u^*\mathcal{E}$$

which is an isomorphism when

$$V' \cap]X'[_{P'} = u^{-1}(V \cap]X[_P).$$

Finally, in general, we always have

$$j^\dagger_{X'} u^* j^\dagger_X \mathcal{E} = j^\dagger_{X'} u^* \mathcal{E}.$$

Proof We saw in Proposition 5.1.14 that there is a canonical morphism

$$u^{-1} j^\dagger_X \mathcal{E} \to j^\dagger_{X'} u^{-1} \mathcal{E}$$

which is $u^{-1}\mathcal{O}_V$-linear by construction. We extend it by linearity to

$$\mathcal{O}_{V'} \otimes_{u^{-1}\mathcal{O}_V} u^{-1} j^\dagger_X \mathcal{E} \to \mathcal{O}_{V'} \otimes_{u^{-1}\mathcal{O}_V} j^\dagger_{X'} u^{-1} \mathcal{E}.$$

Of course, we have

$$u^* j^\dagger_X \mathcal{E} = \mathcal{O}_{V'} \otimes_{u^{-1}\mathcal{O}_V} u^{-1} j^\dagger_X \mathcal{E}$$

but also

$$\mathcal{O}_{V'} \otimes_{u^{-1}\mathcal{O}_V} j^\dagger_{X'} u^{-1} \mathcal{E} = j^\dagger_{X'} (\mathcal{O}_{V'} \otimes_{u^{-1}\mathcal{O}_V} u^{-1} \mathcal{E}) = j^\dagger_{X'} u^* \mathcal{E}$$

thanks to Proposition 5.3.2. The last assertion is proven exactly in the same way, using the last assertion of Proposition 5.1.14:

$$\begin{aligned} j^\dagger_{X'} u^* j^\dagger_X \mathcal{E} &= j^\dagger_{X'} (\mathcal{O}_{V'} \otimes_{u^{-1}\mathcal{O}_V} u^{-1} j^\dagger_X \mathcal{E}) = \mathcal{O}_{V'} \otimes_{u^{-1}\mathcal{O}_V} j^\dagger_{X'} u^{-1} j^\dagger_X \mathcal{E} \\ &= \mathcal{O}_{V'} \otimes_{u^{-1}\mathcal{O}_V} j^\dagger_{X'} u^{-1} \mathcal{E} = j^\dagger_{X'} (\mathcal{O}_{V'} \otimes_{u^{-1}\mathcal{O}_V} u^{-1} \mathcal{E}) = j^\dagger_{X'} u^* \mathcal{E}. \end{aligned}$$

□

Corollary 5.3.9 *Let*

$$\begin{array}{ccccc} X' & \hookrightarrow & Y' & \hookrightarrow & P' \\ \downarrow f & & \downarrow g & & \downarrow u \\ X & \hookrightarrow & Y & \hookrightarrow & P \end{array}$$

be a morphism of frames. Let V be an admissible open subset of $]Y[_P$ and V' an admissible open subset of $]Y'[_{P'} \cap u_K^{-1}(V)$. If $\mathcal{E}$ is an $\mathcal{O}_V$-module, and if we write $u_K : V' \to V$ for the induced morphism, there is a canonical morphism

$$u_K^* j^\dagger_X \mathcal{E} \to j^\dagger_{X'} u_K^* \mathcal{E}$$

which is an isomorphism when the morphism of frames is cartesian. In general, we always have

$$j^\dagger_{X'} u_K^* j^\dagger_X \mathcal{E} \simeq j^\dagger_{X'} u_K^* \mathcal{E}.$$

Proof This really is a particular case of Proposition 5.3.8. Alternatively, it can be derived as in the proposition from the corollary of Proposition 5.1.14. □

Corollary 5.3.10 *In the situation of the proposition, if X'' (resp. X''') is a closed subvariety of X (resp. X') with $f^{-1}(X'') \subset X'''$, there is a canonical morphism*

$$u^* j_X^\dagger \Gamma_{X''}^\dagger \mathcal{E} \to \Gamma_{X'''}^\dagger j_{X'}^\dagger u^* \mathcal{E}.$$

Proof If we denote by $U := X \setminus X''$ and $U' := X' \setminus X'''$, we have the following commutative diagram with exact rows:

$$\begin{array}{ccccccccc} & & u^* j_X^\dagger \Gamma_{X''}^\dagger \mathcal{E}\hat{\mathrm{E}} & \longrightarrow & u^* j_X^\dagger \mathcal{E}\hat{\mathrm{E}} & \longrightarrow & u^* j_U^\dagger \mathcal{E}\hat{\mathrm{E}} & \longrightarrow & 0 \\ & & \vdots & & \downarrow & & \downarrow & & \\ 0 & \longrightarrow & \Gamma_{X'''}^\dagger j_{X'}^\dagger u^* \mathcal{E} & \longrightarrow & j_{X'}^\dagger u^* \mathcal{E} & \longrightarrow & j_{U'}^\dagger u^* \mathcal{E} & \longrightarrow & 0 \end{array}$$

□

Definition 5.3.11 *Let*

$$\begin{array}{ccccc} X' & \hookrightarrow & Y' & \hookrightarrow & P' \\ \downarrow f & & \downarrow g & & \downarrow u \\ X & \hookrightarrow & Y & \hookrightarrow & P \end{array}$$

be a morphism of frames, V an admissible open subset of $]Y[_P$ and V' an admissible open subset of $]Y'[_{P'} \cap u_K^{-1}(V)$. The overconvergent pullback *of a $j_X^\dagger \mathcal{O}_V$-module E is*

$$u^\dagger E := j_{X'}^\dagger u_K^* E.$$

When u is an open immersion, we obtain $j_{X'}^\dagger E_{|V'}$. On the other hand, for a cartesian morphism, we get simply the usual pullback $u_K^* E$.

Proposition 5.3.12 *Let*

$$\begin{array}{ccccc} X' & \hookrightarrow & Y' & \hookrightarrow & P' \\ \downarrow f & & \downarrow g & & \downarrow u \\ X & \hookrightarrow & Y & \hookrightarrow & P \end{array}$$

be a morphism of frames, V be an admissible open subset of $]Y[_P$ and V' an admissible open subset of $]Y'[_{P'} \cap u_K^{-1}(V)$. Then, the pair

$$(u^\dagger, u_{K*}) : (V', j_{X'}^\dagger \mathcal{O}_{V'}) \to (V, j_X^\dagger \mathcal{O}_V)$$

defines a morphism of ringed spaces.

Proof We have to show that

$$u^\dagger E = j_{X'}^\dagger \mathcal{O}_{V'} \otimes_{u_K^{-1} j_X^\dagger \mathcal{O}_V} E.$$

It follows from Proposition 5.3.2

$$j_{X'}^\dagger \mathcal{O}_{V'} \otimes_{u_K^{-1} j_X^\dagger \mathcal{O}_V} E = j_{X'}^\dagger \mathcal{O}_{V'} \otimes_{j_{X'}^\dagger u_K^{-1} j_X^\dagger \mathcal{O}_V} E$$

and we saw in Corollary 5.1.15 that

$$j_{X'}^\dagger u_K^{-1} j_X^\dagger \mathcal{O}_V \simeq j_{X'}^\dagger u_K^{-1} \mathcal{O}_V.$$

Using Proposition 5.3.2 again, we get

$$j_{X'}^\dagger \mathcal{O}_{V'} \otimes_{u_K^{-1} j_X^\dagger \mathcal{O}_V} E = j_{X'}^\dagger \mathcal{O}_{V'} \otimes_{u_K^{-1} \mathcal{O}_V} E = j_{X'}^\dagger u_K^* E = u^\dagger E.$$

□

Proposition 5.3.13 *If*

$$\begin{array}{ccccc} X' & \hookrightarrow & Y' & \hookrightarrow & P' \\ \downarrow{\scriptstyle f} & & \downarrow{\scriptstyle g} & & \downarrow{\scriptstyle u} \\ X & \hookrightarrow & Y & \hookrightarrow & P \end{array}$$

is a flat morphism of frames, the functor $u^\dagger$ is exact.

Proof Of course, we are giving ourselves V and V' as usual. We saw in Corollary 3.3.6 that u_K induces a flat morphism $W' \to]Y[_P$ with W' a strict neighborhood of $]X'[_{P'}$ in $]Y'[_{P'}$. Moreover, we know from Proposition 5.1.13 that the inclusion map $W' \cap V' \hookrightarrow V'$ induces an equivalence between overconvergent sheaves on both sides. We may therefore assume that $u_K : V' \to V$ is flat. It follows that u_K^* is exact and we also know from Proposition 5.2.1 that $j_{X'}^\dagger$ is exact too. □

Proposition 5.3.14 *Let $(X \subset Y \subset P)$ be a frame, V an admissible open subset of $]Y[_P$ and $\sigma : K \hookrightarrow K'$ be an isometric embedding. If $\mathcal{E}$ is an $\mathcal{O}_V$-module, there is a canonical isomorphism*

$$(j_X^\dagger \mathcal{E})^\sigma \simeq j_{X^\sigma}^\dagger \mathcal{E}^\sigma.$$

Proof We already know from Lemma 5.1.18 that if $\varpi : \widetilde{V}^\sigma \to \widetilde{V}$ denotes the projection, there is a canonical isomorphism

$$\varpi^{-1} j_X^\dagger \mathcal{E} \simeq j_{X^\sigma}^\dagger \varpi^{-1} \mathcal{E}.$$

Thanks to Proposition 5.3.2, we get

$$(j_X^\dagger \mathcal{E})^\sigma = \mathcal{O}_{V_{K'}} \otimes_{\varpi^{-1}\mathcal{O}_V} \varpi^{-1} j_X^\dagger \mathcal{E} = \mathcal{O}_{V_{K'}} \otimes_{\varpi^{-1}\mathcal{O}_V} j_{X^\sigma}^\dagger \varpi^{-1} \mathcal{E}$$
$$= j_{X^\sigma}^\dagger (\mathcal{O}_{V_{K'}} \otimes_{\varpi^{-1}\mathcal{O}_V} \varpi^{-1}\mathcal{E}) = j_{X^\sigma}^\dagger \mathcal{E}^\sigma.$$

□

5.4 Coherent dagger modules

Proposition 5.4.1 *If $(X \subset Y \subset P)$ is a frame and V an admissible open subset of $]Y[_P$, the sheaf of rings $j_X^\dagger \mathcal{O}_V$ is coherent.*

Proof We have to show that if W is an admissible open subset of $]Y[_P$ and

$$\varphi : j_X^\dagger \mathcal{O}_W^N \to j_X^\dagger \mathcal{O}_W$$

a $j_X^\dagger \mathcal{O}_W$-linear morphism, then $\ker \varphi$ is a $j_X^\dagger \mathcal{O}_W$-module of finite type. The question is local on W and we may therefore assume that it is quasi-compact. The morphism φ is given by a sequence

$$f_1, \dots, f_N \in \Gamma(W, j_X^\dagger \mathcal{O}_W) = \varinjlim \Gamma(W \cap V', \mathcal{O}_V)$$

when V' runs through the strict neighborhoods of $]X[_P$ in $]Y[_P$. Thus, there exists V' such that

$$f_1, \dots, f_N \in \Gamma(W \cap V', \mathcal{O}_V).$$

This shows that

$$\varphi_{|W \cap V'} = j_X^\dagger \varphi'$$

for some morphism

$$\varphi' : \mathcal{O}_{W \cap V'}^N \to \mathcal{O}_{W \cap V'}.$$

Since $\mathcal{O}_{W \cap V'}$ is a coherent sheaf, $\ker \varphi'$ is an $\mathcal{O}_{W \cap V'}$-module of finite type. But, we saw in Proposition 5.3.1 that $j_X^\dagger$ is exact. Thus, we see first that $j_X^\dagger \ker \varphi'$ is a $j_X^\dagger \mathcal{O}_{W \cap V'}$-module of finite type but also that

$$j_X^\dagger \ker \varphi' = \ker \varphi_{|W \cap V'}.$$

It follows that $\ker \varphi_{|W \cap V'}$ is of finite type and the same is true of $\ker \varphi$ since it is 0 on $T := W \cap (]Y[_P \setminus]X[_P)$. □

Corollary 5.4.2 *Let $(X \subset Y \subset P)$ be a frame and V an admissible open subset of $]Y[_P$.*

(i) *If E, F, G are three coherent $j_X^\dagger \mathcal{O}_V$-modules, then $E \otimes_{j_X^\dagger \mathcal{O}_V} F$ and $\mathcal{H}om_{j_X^\dagger \mathcal{O}_V}(F, G)$ are coherent $j_X^\dagger \mathcal{O}_V$-modules and we have*

$$\mathrm{Hom}_{j_X^\dagger \mathcal{O}_V}(E \otimes_{j_X^\dagger \mathcal{O}_V} F, G) \simeq \mathrm{Hom}_{j_X^\dagger \mathcal{O}_V}(E, \mathcal{H}om_{j_X^\dagger \mathcal{O}_V}(F, G)).$$

(ii) *Let*

$$\begin{array}{ccccc} X' & \hookrightarrow & Y' & \hookrightarrow & P' \\ \downarrow f & & \downarrow g & & \downarrow u \\ X & \hookrightarrow & Y & \hookrightarrow & P \end{array}$$

be a morphism of frames and V' and admissible open subset of $]Y'[_{P'} \cap u_K^{-1}(V)$. If E is a coherent $j_X^\dagger \mathcal{O}_V$-module, then $u^\dagger E$ is a coherent $j_{X'}^\dagger \mathcal{O}_{V'}$-module.

(iii) *Let $\sigma : K \hookrightarrow K'$ be an isometric embedding. If E is a coherent $j_X^\dagger \mathcal{O}_V$-module, then E^σ is a coherent $j_{X^\sigma}^\dagger \mathcal{O}_{V^\sigma}$-module.*

Proof These are standard properties of coherent rings since by Proposition 5.3.12, in assertion (ii), we have a morphism of ringed spaces

$$(u^\dagger, u_{K*}) : (V', j_{X'}^\dagger \mathcal{O}_{V'}) \to (V, j_X^\dagger \mathcal{O}_V).$$

Also, by Lemma 5.3.14 and Proposition 5.3.2, in the last assertion, we have a morphism of ringed spaces

$$\varpi : (\widetilde{V}^\sigma, j_{X^\sigma}^\dagger \mathcal{O}_{V^\sigma}) \to (\widetilde{V}, j_X^\dagger \mathcal{O}_V).$$

□

Corollary 5.4.3 *If $\mathcal{E}$ is a coherent $\mathcal{O}_V$-module, then $j_X^\dagger \mathcal{E}$ is a coherent $j_X^\dagger \mathcal{O}_V$-module. Moreover, if $\mathcal{F}$ is any $j_X^\dagger \mathcal{O}_V$-module, then*

$$\mathcal{H}om_{j_X^\dagger \mathcal{O}_V}(j_X^\dagger \mathcal{E}, j_X^\dagger \mathcal{F}) = j_X^\dagger \mathcal{H}om_{\mathcal{O}_V}(\mathcal{E}, \mathcal{F}).$$

Proof Using the morphism of frames

$$\begin{array}{ccccc} X & & & & \\ \downarrow & \searrow & & & \\ & & Y & \hookrightarrow & P \\ & \mathrel{/\!/} & & & \\ Y & & & & \end{array}$$

the first assertion is a particular case of assertion (ii) of Corollary 5.4.2.

The second assertion is local and we may therefore assume that $\mathcal{E}$ has a finite presentation. Since $j_X^\dagger$ is exact, we are reduced to the case $\mathcal{E} = \mathcal{O}_V$ in which case it is trivial. □

We will now prove a converse to this corollary. If $\mathcal{A}$ is a sheaf of rings on some (Grothendieck) topological space, we will denote by Coh($\mathcal{A}$) the category of coherent $\mathcal{A}$-modules.

Theorem 5.4.4 *Let $(X \subset Y \subset P)$ be a quasi-compact frame and V an admissible open subset of $]Y[_P$. Then, the functors $j_X^\dagger$ induce an equivalence of categories*

$$\varinjlim_{V'} \mathrm{Coh}(\mathcal{O}_{V\cap V'}) \simeq \mathrm{Coh}(j_X^\dagger \mathcal{O}_{V'})$$

when V' runs through the strict neighborhoods of $]X[_P$ in $]Y[_P$.

Before giving the proof, we will explain this statement in detail in the next corollary.

Corollary 5.4.5 *We have the following*

(i) *If E is a coherent $j_X^\dagger \mathcal{O}_V$-module, there exists a strict neighborhood V' of $]X[_P$ in $]Y[_P$ and a coherent $\mathcal{O}_{V\cap V'}$-module $\mathcal{E}$ such that $E = j_X^\dagger \mathcal{E}$.*

(ii) *If $\mathcal{E}$ and $\mathcal{F}$ are coherent $\mathcal{O}_V$-modules and*

$$\varphi : j_X^\dagger \mathcal{E} \to j_X^\dagger \mathcal{F}$$

is any morphism, there exists a strict neighborhood V' of $]X[_P$ in $]Y[_P$ and a morphism

$$\psi : \mathcal{E}_{|V\cap V'} \to \mathcal{F}_{|V\cap V'}$$

such that $\varphi = j_X^\dagger \psi$.

(iii) *If $\mathcal{E}$ and $\mathcal{F}$ are coherent $\mathcal{O}_V$-modules and*

$$\psi, \psi' : \mathcal{E} \to \mathcal{F}$$

satisfy $j_X^\dagger \psi' = j_X^\dagger \psi$, then there exists a strict neighborhood V' of $]X[_P$ in $]Y[_P$ such that $\psi'_{|V\cap V'} = \psi_{|V\cap V'}$.

Proof Actually, this corollary is equivalent to the theorem. More precisely, assertion (i) tells you that the functor is essentially surjective, assertion (ii) that it is full and assertion (iii) that it is faithful. □

Note also that full faithfulness means that

$$\mathrm{Hom}(j_X^\dagger\mathcal{E}, j_X^\dagger\mathcal{F}) = \varinjlim_{V'} \mathrm{Hom}(\mathcal{E}_{|V\cap V'}, \mathcal{F}_{|V\cap V'}).$$

Proof (of the theorem) We will actually prove the corollary which is equivalent to the theorem.

Note first that assertion (i) is trivially true when E is free.

Also, assertions (ii) and (iii) are true when $\mathcal{E}$ is free on V quasi-compact: in this case, morphisms are given by finite sets of sections and we know from Proposition 5.1.12 that

$$\Gamma(V, j_X^\dagger\mathcal{F}) = \varinjlim_{V'\subset V} \Gamma(V\cap V', \mathcal{F}).$$

We will now prove all the assertions when V is quasi-compact. We do the (global) finite presentation case first. Thus, we start with assertion (i) when there exists a presentation

$$j_X^\dagger\mathcal{O}_V^N \to j_X^\dagger\mathcal{O}_V^M \to E \to 0.$$

Assertion (ii) in the free case gives us the existence of a strict neighborhood V' of $]X[_P$ in $]Y[_P$ such that the morphism of free $j_X^\dagger\mathcal{O}_V$-modules is of the form $j_X^\dagger\tau$ with

$$\tau : \mathcal{O}_{V\cap V'}^N \to \mathcal{O}_{V\cap V'}^M.$$

Since $j_X^\dagger$ is (right) exact, it is sufficient to choose $\mathcal{E} = \operatorname{coker}\tau$.

Now we prove assertion (ii) when there is a presentation

$$\mathcal{O}_V^N \xrightarrow{\tau} \mathcal{O}_V^M \xrightarrow{\pi} \mathcal{E} \to 0.$$

From the free case, we get

$$\varphi\circ j_X^\dagger\pi = j_X^\dagger\widetilde{\psi}$$

for some

$$\widetilde{\psi} : \mathcal{O}_{V\cap V'}^M \to \mathcal{F}$$

where V' is a strict neighborhood of $]X[_P$ in $]Y[_P$. And we have

$$j_X^\dagger(\widetilde{\psi}\circ\tau) = \varphi\circ j_X^\dagger(\pi\circ\tau) = \varphi\circ 0 = 0.$$

Moreover, thanks to Proposition 3.3.2, we may assume that $V\cap V'$ also is quasi-compact. Since we know assertion (iii) in the free case, we may therefore replace V' by a smaller strict neighborhood and get $\widetilde{\psi}\circ\tau = 0$. It follows that

$\widetilde{\psi}$ factors uniquely through $\mathcal{E}$ to give $\psi : \mathcal{E} \to \mathcal{F}$. Since $j_X^\dagger$ is (right) exact, $j_X^\dagger \pi$ is surjective and the equality

$$j_X^\dagger \psi \circ j_X^\dagger \pi = j_X^\dagger(\psi \circ \pi) = j_X^\dagger \widetilde{\psi} = \varphi \circ j_X^\dagger \pi$$

implies that $j_X^\dagger \psi = \varphi$. This is assertion (ii).

Still assuming V quasi-compact, we now prove assertion (iii) when $\mathcal{E}$ has a finite presentation as above. This assertion states that, the morphism ψ we just built is unique in the sense that any other morphism ψ' will coincide with ψ if we shrink V' a little more. But this is clear because $\widetilde{\psi}$ has this property and π is surjective.

When V is quasi-compact but there is not necessarily a global finite presentation, we may always find a finite covering by quasi-compact admissible open subsets V_i with a finite presentation on each. Assertion (iii) easily follows because, for each i we may find a strict neighborhood V_i' and then take the intersection. For assertion (ii), we may find for each i a strict neighborhood V_i' and a morphism ϕ_i on $V_i \cap V_i'$. Using assertion (iii), we may find for each i, j a strict neighborhood $V_{i,j}'$ such that ϕ_i and ϕ_j coincide on $V_i \cap V_j \cap V_{ij}'$. Again, we choose V' as the intersection of all those strict neighborhoods and glue. Finally, for assertion (i), we can find $\mathcal{E}_i$ on each V_i and, using assertion (ii), strict neighborhoods V_{ij}' such that $\mathcal{E}_i$ and $\mathcal{E}_j$ coincide on $V_i \cap V_j \cap V_{ij}'$. And again, we can glue.

We can now handle the general case. Of course, we may assume that V is connected. As a connected admissible open subset of a quasi-compact rigid analytic variety, V is countable at infinity and we can write V as an admissible countable increasing union of quasi-compact admissible open subsets W_n. For each n, there exists $\eta_n < 1$ such that $W_n \subset [Y]_{\eta_n}$.

We consider assertion (i). We will show by induction that there exists a sequence $\lambda_n \overset{<}{\to} 1$ and for each n a coherent module $\mathcal{E}_n$ on $W_n \cap V^{\lambda_n}$ such that

$$j_X^\dagger \mathcal{E}_n = E_{|W_n \cap V^{\lambda_n}} \quad \text{and} \quad \mathcal{E}_{n|W_{n-1} \cap V^{\lambda_{n-1}}} = \mathcal{E}_{n-1}.$$

Since W_n is quasi-compact, we know assertion (i) for W_n. Thus, we can find λ_n and $\mathcal{E}_n$ such that the first condition is satisfied. Moreover, assertion (iii) on W_n tells us that we may also assume that the second condition is satisfied. By construction, the $\mathcal{E}_n$ glue in order to give a sheaf $\mathcal{E}$ on

$$\cup_n (W_n \cap V^{\lambda_n}) = V \cap V_{\underline{\eta}}^{\underline{\lambda}}.$$

For assertion (ii), we proceed in the same way. Using assertions (ii) and (iii) for the W_n's, we can build a sequence $\lambda_n \overset{<}{\to} 1$ and for each n a morphism

$$\psi_n : \mathcal{E}_{|W_n \cap V^{\lambda_n}} \to \mathcal{F}_{|W_n \cap V^{\lambda_n}}$$

such that

$$j_X^\dagger \psi_n = \varphi_{|W_n \cap V^{\lambda_n}} \quad \text{and} \quad \psi_{n|W_{n-1} \cap V^{\lambda_{n-1}}} = \psi_{n-1}.$$

They glue in order to give $\psi : \mathcal{E}_{|V \cap V_{\underline{\eta}}^{\underline{\lambda}}} \to \mathcal{F}_{|V \cap V_{\underline{\eta}}^{\underline{\lambda}}}$.

Finally, assertion (iii) in the general case is easily proven. From assertion (iii) for the W_n's, we can find a sequence $\lambda_n \overset{<}{\to} 1$ such that for each n,

$$\psi'_{|W_n \cap V^{\lambda_n}} = \psi_{|W_n \cap V^{\lambda_n}}$$

and it follows that

$$\psi'_{|V \cap V_{\underline{\eta}}^{\underline{\lambda}}} = \psi_{|V \cap V_{\underline{\eta}}^{\underline{\lambda}}}.$$

□

Lemma 5.4.6 *Let $(X \subset Y \subset P)$ be a frame, V and admissible open subset of $]Y[_P$ and*

$$j : V \cap]X[_P \hookrightarrow V$$

the inclusion map. If $\mathcal{E}$ is an $\mathcal{O}_V$-module, the canonical morphism

$$j_X^\dagger \mathcal{E} \to j_* j^{-1} \mathcal{E}$$

is injective.

Proof It is sufficient to show that if W is a quasi-compact rigid analytic open subset of V, the canonical map

$$\varinjlim \Gamma(W \cap V', \mathcal{E}) \to \Gamma(W \cap]X[_P, \mathcal{E})$$

is injective. Thus, we choose some strict neighborhood V' of $]X[_P$ in $]Y[_P$ and $s \in \varinjlim \Gamma(W \cap V', \mathcal{E})$ whose image in $\Gamma(W \cap]X[_P, \mathcal{E})$ is zero. The support of the coherent submodule $\mathcal{O}_{W \cap V'} s$ of $\mathcal{E}$ is a closed analytic subset T of $W \cap V'$ and $T \cap]X[_P = \emptyset$. Using Proposition 3.4.11, we may assume that $T = \emptyset$ and therefore $s = 0$. □

Proposition 5.4.7 *If $(X \subset Y \subset P)$ is a frame and V and admissible open subset of $]Y[_P$, the restriction functor*

$$\mathrm{Coh}(j_X^\dagger \mathcal{O}_V) \to \mathrm{Coh}(\mathcal{O}_{V \cap]X[_P})$$

is faithful.

Proof We have to show that, if E and F are two $j_X^\dagger \mathcal{O}_V$-modules, the canonical map

$$\mathrm{Hom}_{j_X^\dagger \mathcal{O}_V}(E, F) \to \mathrm{Hom}_{\mathcal{O}_{V \cap]X[_P}}(E_{|V \cap]X[_P}, F_{|V \cap]X[_P})$$

is injective. If we let

$$H := \mathcal{H}om_{j_X^\dagger \mathcal{O}_V}(j_X^\dagger \mathcal{E}, j_X^\dagger \mathcal{F})$$

and denote by

$$j : V \cap]X[_P \hookrightarrow V$$

the inclusion map, we can rewrite our map on the form

$$\Gamma(V, H) \to \Gamma(V, j_* j^{-1} H)$$

and we know that it is injective from Lemma 5.4.6 (using left exactness of $\Gamma(V, -)$). □

As an example, we consider as usual the Monsky–Washnitzer situation.

Proposition 5.4.8 *Let $X = \operatorname{Spec} A$ be an affine $\mathcal{V}$-scheme. Let $X \hookrightarrow \mathbf{A}^N_{\mathcal{V}}$ be a closed immersion and Y be the projective closure of X in $\mathbf{P}^N_{\mathcal{V}}$. With respect to the frame $(X_k \subset Y_k \subset \widehat{Y})$, we have an equivalence*

$$\begin{array}{ccc} \mathrm{Coh}(j_X^\dagger \mathcal{O}_{X_K^{\mathrm{rig}}}) & \xrightarrow{\simeq} & \mathrm{Coh}(A_K^\dagger) \\ E & \longmapsto & M := \Gamma(V, E). \end{array}$$

Proof We saw in Proposition 3.3.9 that the admissible open subsets

$$V_\rho := \mathbf{B}^N(0, \rho^+) \cap X_K^{\mathrm{rig}} \subset Y_K^{\mathrm{rig}} = \widehat{Y}_K$$

form a cofinal family of (affinoid) strict neighborhoods of $]X_k[_P$ in P_K when $\rho > 1$. Applying Theorem 5.4.4, it follows that we have an equivalence

$$\varinjlim_\rho \mathrm{Coh}(\mathcal{O}_{V_\rho}) \simeq \mathrm{Coh}(j_X^\dagger \mathcal{O}_{X_K^{\mathrm{rig}}}).$$

But V_ρ is affinoid and we have $A_K^\dagger = \cup A_\rho$ with $A_\rho := \Gamma(V_\rho, \mathcal{O}_{V_\rho})$ and it follows that

$$\varinjlim_\rho \mathrm{Coh}(A_\rho) \simeq \mathrm{Coh}(A_K^\dagger).$$

And, of course, we have for each $\rho > 1$, $\mathrm{Coh}(\mathcal{O}_{V_\rho}) \simeq \mathrm{Coh}(A_\rho)$. □

Before going any further, we should also investigate a little bit the role of the Robba ring $\mathcal{R}$. It should be noticed that, when the valuation is discrete, or more generally, when K is maximally complete, then any torsion free coherent $\mathcal{R}$-module is free because $\mathcal{R}$ is a Bézout ring.

Proposition 5.4.9 *Let $\mathcal{Y}$ be a flat formal $\mathcal{V}$-scheme whose special fiber Y is a connected curve and X a non empty open subset of Y. Let E be a coherent $j_X^\dagger \mathcal{O}_{\mathcal{Y}_K}$-module and x a smooth rational point in $Y \setminus X$. If V is a strict neighborhood of $]X[_{\mathcal{Y}}$ in $\mathcal{Y}_K$,*

$$h_x :]x[_{\mathcal{Y}} \cap V \hookrightarrow V$$

the inclusion map and $\mathcal{E}$ a coherent $\mathcal{O}_V$-module with $E = j_X^\dagger \mathcal{E}$, then

$$E_x := \Gamma(\mathcal{Y}_K, j_X^\dagger h_{x*} h_x^{-1} \mathcal{E})$$

only depends on E and not on $\mathcal{E}$ and we have

$$H^q(\mathcal{Y}_K, j_X^\dagger h_{x*} h_x^{-1} \mathcal{E}) = 0 \quad \text{for} \quad q > 0.$$

Moreover, we get an exact functor

$$\begin{array}{ccc} \mathrm{Coh}(j_X^\dagger \mathcal{O}_{\mathcal{Y}_K}) & \longrightarrow & \mathrm{Coh}(\mathcal{R}(x)) \\ E & \longmapsto & E_x. \end{array}$$

Proof It follows from Theorem 5.4.4 that E_x only depends on E: if $E = j_X^\dagger \mathcal{E}'$ with $\mathcal{E}'$ coherent on some strict neighborhood V' of $]X[_y$ in $\mathcal{Y}_K$, there exists $V'' \subset V \cap V'$ such that $\mathcal{E}_{|V'} = \mathcal{E}'_{|V''}$. And we do get a functor $E \mapsto E_x$.

Now, by a standard argument, in order to show that our functor preserves coherence and that higher cohomology vanishes, it is sufficient to show that the functor

$$\mathcal{F} \mapsto \varinjlim_{\lambda} \Gamma(\mathbf{A}(0, \lambda^+, 1^-), \mathcal{F})$$

sends finitely presented modules on the annulus $\mathbf{A}(0, \lambda_0^+, 1^-)$ to coherent $\mathcal{R}$-modules and that

$$H^q(\mathbf{A}(0, \lambda^+, 1^-), \mathcal{F}) = 0 \quad \text{for} \quad q > 0$$

(more precisely, we use a lifting t of a local parameter at x to identify $]x[_P$ with $\mathbf{D}(0, 1^-)$). This is an exercise in rigid analytic geometry.

Finally, in order to show that our functor is exact, since we know that higher cohomology vanishes, it is sufficient to recall that h_x is quasi-Stein and therefore that the functor $\mathcal{E} \mapsto j_X^\dagger h_{x*} h_x^{-1} \mathcal{E}$ is exact on coherent sheaves. □

Definition 5.4.10 *We then say that E_x is the* Robba fiber *of E at x.*

The Robba fiber is functorial as the next proposition shows.

Proposition 5.4.11 *Let $\mathcal{Y}$ (resp. $\mathcal{Y}'$) be a flat formal $\mathcal{V}$-scheme whose special fiber Y (resp. Y') is a connected curve and X (resp. X') a non empty open* *subset of Y (resp. Y'). Let $u : \mathcal{Y}' \to \mathcal{Y}$ be a morphism a formal $\mathcal{V}$-schemes with special fiber $g : Y' \to Y$. Assume that $g(X') \subset X$ and let x' be a smooth rational point in $Y' \setminus X'$ such that $x := g(x') \notin X$. Then,*

(i) *The morphism u induces a morphism of rings $u^* : \mathcal{R}(x) \to \mathcal{R}(x')$.*
(ii) *If E is a coherent $j_X^\dagger \mathcal{O}_{\mathcal{Y}_K}$-module, then*

$$R(x') \otimes_{\mathcal{R}(x)} E_x \simeq (u^\dagger E)_{x'}.$$

Proof We can find a strict neighborhood V of $]X[_{\mathcal{Y}}$ in $\mathcal{Y}_K$ and a coherent $\mathcal{O}_V$-module $\mathcal{E}$ with $E = j_X^\dagger \mathcal{E}$. Then, $V' := u_K^{-1}(V)$ is a strict neighborhood of $]X'[_{\mathcal{Y}'}$ in $\mathcal{Y}'_K$ and, if we still denote by $u_K : V' \to V$ the induced map, then $u_K^* \mathcal{E}$ is a coherent $\mathcal{O}_{V'}$-module and we have $u^\dagger E = j_{X'}^\dagger u_K^* \mathcal{E}$.

If we denote by

$$h_x :]x[_{\mathcal{Y}} \cap V \hookrightarrow V \quad \text{and} \quad h_{x'} :]x'[_{\mathcal{Y}'} \cap V' \hookrightarrow V'$$

the inclusion maps, Corollary 5.3.9 gives us a canonical morphism

$$\begin{aligned} u_K^* j_X^\dagger h_{x*} h_x^{-1} \mathcal{E} &\to j_{X'}^\dagger u_K^* h_{x*} h_x^{-1} \mathcal{E} \\ &\to j_{X'}^\dagger h_{x'*} u_K^* h_x^{-1} \mathcal{E} = j_{X'}^\dagger h_{x'*} h_{x'}^{-1} u_K^* \mathcal{E} \end{aligned}$$

and by adjunction,

$$j_X^\dagger h_{x*} h_x^{-1} \mathcal{E} \to u_{K*} j_{X'}^\dagger h_{x'*} h_{x'}^{-1} u_K^* \mathcal{E}.$$

Taking global sections on $\mathcal{Y}_K$ give $u^* : E_x \to (u^\dagger E)_{x'}$. Applying this to the particular case $E := j_X^\dagger \mathcal{O}_{\mathcal{Y}_K}$, we obtain a natural map

$$u^* : \mathcal{R}(x) \to \mathcal{R}(x').$$

It remains to show that the canonical map

$$R(x') \otimes_{\mathcal{R}(x)} E_x \to (u^\dagger E)_{x'}$$

is bijective. We may choose a finite presentation for E. Now, the functor $j_{X'}^\dagger u_K^*$ is right exact as well as the Robba fiber functor. We may therefore assume that E is free and then, by additivity, that $E = j_X^\dagger \mathcal{O}_{\mathcal{Y}_K}$ in which case our assertion is trivial. □

In the next proposition, we show that the category $\mathrm{Coh}(j_X^\dagger \mathcal{O}_V)$ is local on $(X \subset Y \subset P)$.

Proposition 5.4.12 *Let*

$$\begin{array}{ccccc} X_i & \hookrightarrow & Y_i & \hookrightarrow & P_i \\ \downarrow & & \downarrow & & \downarrow \\ X & \hookrightarrow & Y & \hookrightarrow & P \end{array}$$

be an open covering or a finite mixed covering of frames. Let V be an admissible open subset of $]Y[_P$ and, for each $i \in I$, let $V_i := V \cap]Y_i[_{P_i}$ and so on, with usual multi-index notation.

Then, there is an equivalence of categories

$$\mathrm{Coh}(j_X^\dagger \mathcal{O}_V) \simeq \varprojlim \left[\prod_i \mathrm{Coh}(j_{X_i}^\dagger \mathcal{O}_{V_i}) \rightrightarrows \prod_{i,j} \mathrm{Coh}(j_{X_{ij}}^\dagger \mathcal{O}_{V_{ij}}) \right.$$

$$\left. \ \equiv\!\!\!> \prod_{i,j,k} \mathrm{Coh}(j_{X_{ijk}}^\dagger \mathcal{O}_{V_{ijk}}) \right]$$

Proof We prove the corollary below which is simply a down to earth description of the proposition. □

Corollary 5.4.13 *In the situation of the proposition, we have the following*

(i) *For each $i \in I$, let E_i be a coherent $j_{X_i}^\dagger \mathcal{O}_{V_i}$-module and, for each $i, j \in I$, let*

$$\varphi_{ij} : j_{X_i}^\dagger E_{j|V_{ij}} \simeq j_{X_j}^\dagger E_{i|V_{ij}},$$

be an isomorphism such that for each $i, j, k \in I$, we have on V_{ijk}:

$$j_{X_k}^\dagger \varphi_{ij} \circ j_{X_i}^\dagger \varphi_{jk} = j_{X_j}^\dagger \varphi_{ik}.$$

Then, there exits a coherent $j_X^\dagger \mathcal{O}_V$-module E and, for each $i \in I$, an isomorphism $\varphi_i : j_{X_i}^\dagger E_{|V_i} \simeq E_i$ such that for each $i, j \in I$, we have on V_{ij},

$$j_{X_j}^\dagger \varphi_i = \varphi_{ij} \circ j_{X_i}^\dagger \varphi_j.$$

Moreover, E is unique up to a unique isomorphism.

(ii) *Let E and F be two coherent $j_X^\dagger \mathcal{O}_V$-modules and for each $i \in I$, let*

$$\{\varphi_i : j_{X_i}^\dagger E_{|V_i} \to j_{X_i}^\dagger F_{|V_i}\}$$

be a morphism such that for each $i, j \in I$, we have on V_{ij}:

$$j_{X_j}^\dagger \varphi_i = j_{X_i}^\dagger \varphi_j.$$

Then, there exists a unique morphism $\varphi : E \to F$ such that, for each $i \in I$, we have on V_i : $j_{X_i}^\dagger \varphi = \varphi_i$.

Proof Thanks to Proposition 2.2.15, the question is clearly local on $]Y[_P$. We may therefore assume that for each $i \in I$, we have $Y_i = Y$ and $P_i = P$ so that $V_i = V$. In particular, a mixed covering is also an open covering. We may also assume that P is quasi-compact and therefore that I is finite. We will write $X_{ij} = X_i \cap X_j$ and so on...

We look at the first part. We can find for each $i \in I$, a strict neighborhood V_i of $]X_i[_P$ in $]Y[_P$ and a family $\{\mathcal{E}_i\}_{i\in I}$ of coherent $\mathcal{O}_{V\cap V_i}$-modules such that for each $i \in I$, we have $j^\dagger_{X_i}\mathcal{E}_i = E_i$. Now, for each $i, j \in I$, we may consider

$$\varphi_{ij} : j^\dagger_{X_{ij}}\mathcal{E}_j \simeq j^\dagger_{X_{ij}}\mathcal{E}_i$$

and we can find a strict neighborhood V_{ij} of $]X_{ij}[_P$ in $]Y[_P$ and a morphism

$$\psi_{ij} : \mathcal{E}_{j|V\cap V_{ij}} \simeq \mathcal{E}_{i|V\cap V_{ij}}$$

with $\varphi_{ij} := j^\dagger_{X_{ij}}\psi_{ij}$. Now, for each $i, j, k \in I$, we have

$$j^\dagger_{X_{ijk}}(\psi_{ij} \circ \psi_{jk}) = j^\dagger_{X_{ijk}}\psi_{ik}$$

and it follows that there exists a strict neighborhood V_{ijk} of $]X_{ijk}[_P$ in $]Y[_P$ such that

$$\psi_{ij|V\cap V_{ijk}} \circ \psi_{jk|V\cap V_{ijk}} = \psi_{ik|V\cap V_{ijk}}.$$

Shrinking some of the V_{ij}'s if necessary, we may assume thanks to Proposition 3.4.9 that for all $i, j, k \in I$, we have

$$V_{ij} \cap V_{jk} \cap V_{ik} \subset V_{ijk}$$

and shrinking some of the V_i's if necessary, we may also assume that for all $i, j \in I$, we have

$$V_i \cap V_j \subset V_{ij}.$$

Thanks to Proposition 3.4.10, we may shrink V again and assume that $V \subset \cup_i(V \cap V_i)$ is an admissible covering. Then, we can glue the $\mathcal{E}_i$'s in order to get a coherent $\mathcal{O}_V$ module $\mathcal{E}$ and we take $E := j^\dagger_X\mathcal{E}$. It is clearly a solution to our problem and uniqueness will follow from the second part.

For the second part, we proceed in the same way. Shrinking V if necessary, we can find two coherent $\mathcal{O}_V$-modules $\mathcal{E}$ and $\mathcal{F}$ such that $E = j^\dagger_X\mathcal{E}$ and $F = j^\dagger_X\mathcal{F}$. We can also find for each i, a strict neighborhood V_i of $]X_i[_P$ in $]Y[_P$ and a morphism

$$\psi_i : \mathcal{E}_{|V\cap V_i} \to \mathcal{F}_{|V\cap V_i}$$

such that $\varphi_i = j^\dagger_{X_i}\psi$. Moreover, we can find for each $i, j \in I$, a strict neighborhood V_{ij} of $]X_{ij}[_P$ in $]Y[_P$ such that $\psi_{i|V\cap V_{ij}} = \psi_{j|V\cap V_{ij}}$. Using Propositions

3.4.9 and 3.4.10 again, we may assume that $V = \cup V_i$ is an admissible covering and that

$$V_{ij} = V_i \cap (V \cap V_j).$$

We can therefore glue the ψ_i's in order to get ψ on V and we take $\varphi := j_X^\dagger \psi$. Assume now that there is another candidate φ' that satisfies the condition $j_{X_i}^\dagger \varphi' = \varphi_i$ for each $i \in I$. We can shrink V if necessary and write $\varphi' = j_X^\dagger \psi'$ with

$$\psi' : \mathcal{E} \to \mathcal{F}.$$

Then, we have for each $i \in I$, $j_{X_i}^\dagger \psi_{|V_i} = j_{X_i}^\dagger \psi'_{|V_i}$ and if we shrink some of the V_i's, we get $\psi'_{|V_i} = \psi_{|V_{ij}}$. It follows that $\psi' = \psi$ on V and therefore $\varphi' = \varphi$. □

Proposition 5.4.14 *Let $(X \subset Y \subset P)$ be a frame and V an admissible open subset of $]Y[_P$. Any admissible covering of V has an admissible affinoid refinement which is acyclic for coherent $j_X^\dagger \mathcal{O}_V$-modules. Actually, if X is the complement of a hypersurface in Y, any affinoid covering of V is acyclic for coherent $j_X^\dagger \mathcal{O}_V$-modules.*

We first prove a technical lemma.

Lemma 5.4.15 *In the situation of the proposition, assume that P is affine and let*

$$Z = V(g_1, \ldots, g_r) \cap P_k \subset P$$

be a closed complement for X in Y. If $\lambda < 1$ and $i = 1, \ldots, r$, let

$$V_i^\lambda := \{x \in]Y[_P, |g_i(x)| \geq \lambda, |g_1(x)|, \ldots, |g_r(x)|\}.$$

Let W be an affinoid open subset of $V_i^{\lambda_0}$ for some $\lambda_0 < 1$. Then,
We have

$$H^q(W, E) = 0 \quad \text{for} \quad q > 0.$$

Proof The assertion follows from Corollary 3.2.3: since W is quasi-compact, there exists $\lambda_0 \leq \mu < 1$ and a coherent $\mathcal{O}_{W \cap V^\mu}$-module $\mathcal{E}$ such that $E_{|W} = j_X^\dagger \mathcal{E}$. Since W is quasi-compact and quasi-separated, we have

$$H^q(W, E) = \varinjlim_{\mu \leq \lambda} H^q(W \cap V^\lambda, \mathcal{E})$$

and it is therefore sufficient to recall from Corollary 3.2.3 that $W \cap V^\lambda$ is affinoid. □

Proof (of the proposition) The question is local on $]Y[_P$ and we may therefore assume that P is affine. We saw in Lemma 3.2.2 that if $\lambda < 1$, there is an admissible covering

$$V^\lambda :=]Y[_P \backslash]Z[_{P\lambda} = \cup_{i=1}^r V_i^\lambda.$$

The proposition therefore results from the last assertion of the lemma. □

Corollary 5.4.16 *Let $(X \subset Y \subset P)$ be an S-frame, V an admissible open subset of $]Y[_P$ and E a coherent $j_X^\dagger \mathcal{O}_V$-module. If $K \hookrightarrow K'$ is a* finite *field extension, there is a canonical isomorphism*

$$K' \otimes_K H^q(V, E) \simeq H^q(V_{K'}, E_{K'}).$$

Proof Since any covering of V has an affinoid refinement which is acyclic for E, it is sufficient to prove that

$$K' \otimes_K \Gamma(W, E) \simeq \Gamma(W_{K'}, E_{K'})$$

when W is an affinoid open subset of V. There exists a strict neighborhood V_0 of $]X[_P$ such that $E = j^\dagger \mathcal{E}$ with $\mathcal{E}$ coherent on $W \cap V_0$. Since W is quasi-compact, we have

$$\Gamma(W, E) = \varinjlim \Gamma(W \cap V', \mathcal{E})$$

when V' runs through the strict neighborhoods of $]X[_P$ inside V_0. But it also follows from Proposition 3.4.5 that

$$\Gamma(W_{K'}, E_{K'}) = \varinjlim \Gamma(W_{K'} \cap V'_{K'}, \mathcal{E}_{K'}).$$

Since direct limits and tensor product commute with each other, our assertion follows from the fact that

$$K' \otimes_K \Gamma(W \cap V', \mathcal{E}) \simeq \Gamma(W_{K'} \cap V'_{K'}, \mathcal{E}_{K'})$$

because $\mathcal{E}$ is coherent. □

We prove now a very useful result concerning section with support in a tube of a coherent sheaf.

Proposition 5.4.17 *Let $(X \subset Y \subset P)$ be a frame and V a strict neighborhood of $]X[_P$ in $]Y[_P$. If $\mathcal{E}$ is a coherent $\mathcal{O}_V$-module and*

$$h : V \backslash]X[_P \hookrightarrow V$$

denotes the inclusion map, we have

$$R\underline{\Gamma}_{]X[_P} \mathcal{E} \simeq [\mathcal{E} \to h_* h^{-1} \mathcal{E}].$$

Proof If $\mathcal{I}$ is an injective abelian sheaf, the sequence

$$0 \to \underline{\Gamma}_{]X[_P}\mathcal{I} \to \mathcal{I} \to h_*h^{-1}\mathcal{I} \to 0$$

is exact on the right also. It follows that, in general, there is a canonical exact triangle

$$R\underline{\Gamma}_{]X[_P}\mathcal{E} \to \mathcal{E} \to Rh_*h^{-1}\mathcal{E}.$$

But it we saw in Proposition 2.2.13 that h is a quasi-Stein morphism. Since $\mathcal{E}$ is coherent, we have

$$h_*h^{-1}\mathcal{E} = Rh_*h^{-1}\mathcal{E}.$$

□

We will finish this section with the study of cohomology with support in the Monsky–Washnitzer setting, but we first need to recall how the cohomology of coherent modules on rigid analytic varieties comes with a natural topology as explained for example in Section 1.6 of [84].

If $\mathcal{E}$ is a coherent module on an affinoid variety V, then $\Gamma(V, \mathcal{E})$ is a module of finite type on the affinoid algebra $\Gamma(V, \mathcal{E})$, and as such, has a canonical Banach topology. Moreover, any morphism of coherent modules $\mathcal{E} \to \mathcal{F}$ induces a continuous map $\Gamma(V, \mathcal{E}) \to \Gamma(V, \mathcal{F})$ and any morphism of affinoid varieties $\varphi : W \to V$ induces a continuous map $\Gamma(V, \mathcal{E}) \to \Gamma(W, \varphi^*\mathcal{E})$. Recall now that a rigid analytic variety V is of *countable type* if it has a countable admissible affinoid covering. If $\mathcal{E}$ is a coherent module (or even a complex with coherent terms) on a separated variety of countable type, then the given covering induces a quasi-Fréchet topology on $H^i(V, \mathcal{E})$. And if follows from the Banach open mapping theorem that this structure is actually independent of the covering. More generally, if the inclusion $h : T \hookrightarrow V$ of an admissible open subset of countable type of V is quasi-Stein and $\mathcal{E}$ is a coherent sheaf on V, then

$$H^i_{V\setminus T}(V, \mathcal{E}) = H^i(V, \mathcal{E} \to h_*h^*\mathcal{E})$$

has a natural quasi-Fréchet topology. Again, all this is functorial.

Proposition 5.4.18 *Let $X = \operatorname{Spec} A$ be a smooth affine $\mathcal{V}$-scheme of pure dimension d and M a coherent $A_K^\dagger$-module. Let $\mathcal{E}$ be a coherent module on some strict neighborhood V of $\hat{X}_K$ in X_K^{rig} such that $M = \Gamma(V, j^\dagger\mathcal{E})$. Then,*

$$M_c := H^d_{\hat{X}_K}(V, \mathcal{E})$$

only depends on M and has a natural structure of topological $A_K^\dagger$-module.

Proof It follows from Theorem 5.4.4 that M_c does not depend on the choice of V or $\mathcal{E}$. Also, clearly, the $\Gamma(V, \mathcal{O}_V)$-structures when V varies are compatible and gives a structure of topological $A_K^\dagger$-module. □

In the case $M = A_K^\dagger$ (resp. $M = \Omega^d_{A_K^\dagger}$), we will just write A_c (resp. $\omega_{A,c}$).

Proposition 5.4.19 *Let $X = \operatorname{Spec} A$ be a smooth affine $\mathcal{V}$-scheme of pure dimension and M a coherent $A_K^\dagger$-module. Then, we have*

(i) $M_c \simeq M \otimes_{A_K^\dagger} A_c$.
(ii) *If M is projective, then M_c is Fréchet.*
(iii) *If $X = X_1 \cup X_2$ is an open affine covering and $X_{12} := X_1 \cap X_2$, we have, with obvious notations, a right exact sequence*

$$(A_{12,K}^\dagger \otimes_{A_K^\dagger} M)_c \to (A_{1,K}^\dagger \otimes_{A_K^\dagger} M)_c \oplus (A_{2,K}^\dagger \otimes_{A_K^\dagger} M)_c \to M_c \to 0.$$

It is also left exact when M is projective.
(iv) *If $X \hookrightarrow Y$ is a closed immersion of smooth affine $\mathcal{V}$-schemes of pure dimension, then the Gysin map is injective*

$$\omega_{A,c} \hookrightarrow \omega_{B,c}.$$

More precisely, if X is defined in Y by a regular sequence $f_1, \cdots, f_r$, we have

$$\omega_{B,c} \simeq \{\omega \in \omega_{A,c}, \quad \forall i = 1, \ldots, r, \quad f_i\omega = 0\}.$$

Proof The first part follows from the first result of Proposition 5.2.21. In order to show that M_c is Fréchet (i.e. separated) when M is locally free, one reduces to the case of the structural sheaf on the affine space where it is checked directly. The last two assertions follow directly from Propositions 5.2.21 and 5.2.22. □

We will need *Serre duality* in the Monsky–Washnitzer setting. If $t_1, \ldots, t_N$ are indeterminates, we will write $\underline{dt} := dt_1 \wedge \cdots \wedge dt_N$ and consider the *trace map* (or *residue map*)

$$\begin{array}{ccc} K[\underline{t}]_c\underline{dt} & \xrightarrow{\int} & K \\ \sum_{\underline{i}<\underline{0}} a_{\underline{i}}\underline{t}^{\underline{i}}\underline{dt} & \longmapsto & a_{-\underline{1}}. \end{array}$$

Proposition 5.4.20 *Let $X = \operatorname{Spec} A$ be a smooth affine $\mathcal{V}$-scheme of pure dimension d. Then,*

(i) *If we choose a closed embedding $X \hookrightarrow \mathbf{A}^N_{\mathcal{V}}$, the composite map*

$$\int_A : \omega_{A,c} \hookrightarrow K[\underline{t}]\underline{dt} \xrightarrow{\int} K$$

only depends on A and not on the embedding.

(ii) *If* $Y = \operatorname{Spec} B$ *is a smooth closed subscheme of* X, *the diagram*

$$\begin{array}{ccc} \omega_{B,c} & \hookrightarrow & \omega_{A,c} \\ & {\scriptstyle\int_B}\searrow \quad \swarrow{\scriptstyle\int_A} & \\ & K & \end{array}$$

is commutative.

Proof Of course, the second assertion immediately follows from the first, since we may then embed Y in $\mathbf{A}^N_{\mathcal{V}}$ via X. Moreover, in order to prove the first assertion, using the diagonal embedding, it is sufficient to prove the second one in the case of the canonical inclusion $\mathbf{A}^M_{\mathcal{V}} \hookrightarrow \mathbf{A}^N_{\mathcal{V}}$. By induction, we may assume $M = N - 1$. Then, by construction, the Gysin map is simply given by

$$\omega \mapsto \omega \wedge \frac{dt_N}{t_N}$$

and we are done. □

Proposition 5.4.21 *Let* $X = \operatorname{Spec} A$ *be a smooth affine* $\mathcal{V}$*-scheme of pure dimension* d. *Let* M *be a projective* $A^\dagger_K$*-module of finite rank. Then, the pairing*

$$\begin{array}{ccc} M^\vee \times M \otimes_{A^\dagger_K} \omega_{A,c} & \longrightarrow & K \\ (\phi, m \otimes \omega) & \longmapsto & \int_A \varphi(m)\omega \end{array}$$

is topologically perfect.

Proof First of all, since $M^\vee \times M \to A^\dagger_K$ is a perfect duality, we may clearly assume $M = A^\dagger_K$.

Now, we assume first that $X = A^d_{\mathcal{V}}$. Then, for $\lambda\eta = 1$, there is a perfect duality (convergent/bounded) of Banach spaces

$$\begin{array}{ccc} K\{\underline{t}/\lambda\} \times K[\underline{t}]_\eta \underline{dt} & \longrightarrow & K \\ (f, g\underline{dt}) & \longmapsto & \int fg\underline{dt}. \end{array}$$

Taking the intersection on the left and union on the right gives what we want. Assume now that X is defined by a regular sequence $f_1, \ldots, f_r$ in $A^N_{\mathcal{V}}$. Then, we have a right exact sequence induced by the f_i's

$$(K[\underline{t}]^\dagger)^r \to K[\underline{t}]^\dagger \to A^\dagger_K \to 0$$

and a left exact sequence

$$0 \to A_c \to K[\underline{t}]_c \to (K[\underline{t}]_c)^r.$$

We get our perfect pairing by taking the quotient on the left-hand side and kernel on the right-hand side. The general case reduces to this one thanks to the next to the last assertion of Proposition 5.4.19. □

It should be possible to define $\omega_{A,c}$ as the topological dual of A_K whenever A is a formally smooth weakly complete $\mathcal{V}$-algebra and work out Serre duality in this more general setting, even when the objects do not arise from a geometric situation.

Note also that, although we will not need it, one can give an alternative definition of the trace map as the composite

$$\omega_c = H^d_{\widehat{X}_K}(X_K^{\mathrm{rig}}, \Omega^d_{X_K^{\mathrm{rig}}}) = H^d_{\widehat{X}_K}(Y_K^{\mathrm{rig}}, \omega_{Y_K}^{\mathrm{rig}}[-d])$$
$$\to H^d(Y_K^{\mathrm{rig}}, \omega_{Y_K}^{\mathrm{rig}}[-d]) = H^d(Y_K, \omega_{Y_K}[-d]) \xrightarrow{Tr} K$$

where ω_{Y_K} is the dualizing complex of [50].

We will also need *Serre duality* on Robba rings. We consider first the *trace map* (or *residue map*)

$$\mathcal{R}dt/t \xrightarrow{\int} K, \quad \sum_{i=-\infty}^{+\infty} a_i t^i dt/t \longmapsto a_0.$$

Proposition 5.4.22 *Let $\mathcal{R}$ be the Robba ring of K and M be a free $\mathcal{R}$-module of finite rank. Then, the pairing*

$$\begin{aligned} M^\vee \times M dt/t &\longrightarrow K \\ (\varphi, m dt/t) &\longmapsto \int \varphi(m) dt/t \end{aligned}$$

is topologically perfect.

Proof Of course, it is sufficient to consider the case $M = \mathcal{R}$. As topological vector space, we have $\mathcal{R} = K[t]_c \oplus K[t]^\dagger$ and the pairing

$$\begin{aligned} \mathcal{R} \times \mathcal{R} &\longrightarrow K \\ (f, g) &\longmapsto \int fg dt/t \end{aligned}$$

is induced by the topological duality between$K[t]^\dagger$ and $K[t]_c$. □

6
Overconvergent calculus

We fix a complete ultrametric field K with $\mathcal{V}$, k and π as usual, and we denote by S a formal $\mathcal{V}$-scheme.

Although it is not necessary for most results, the reader may assume that $\mathrm{Char} K = 0$.

6.1 Stratifications and overconvergence

Starting at Proposition 6.1.10 below, we always assume that $\mathrm{Char} K = 0$.

Definition 6.1.1 *Let $(X \subset Y \subset P)$ be an S-frame and V an admissible open subset of $]Y[_P$. A* stratification on a $j_X^\dagger\mathcal{O}_V$-module *is a stratification as $\mathcal{O}_V$-module. And a morphism of stratified $j_X^\dagger\mathcal{O}_V$-modules is simply a morphism of stratified $\mathcal{O}_V$-modules.*

In other words, the category of stratified $j_X^\dagger\mathcal{O}_V$-modules over S is the full subcategory of stratified $\mathcal{O}_V$-modules over S_K that are overconvergent. Note that a morphism in this category is automatically $j_X^\dagger\mathcal{O}_V$-linear. One could also define the category of crystals in $j_X^\dagger\mathcal{O}_V$-modules and we would get and equivalent category when V is smooth.

Proposition 6.1.2 *Let $(X \subset Y \subset P)$ be an S-frame and V an admissible open subset of $]Y[_P$. Then, the stratified $j_X^\dagger\mathcal{O}_V$-modules form an abelian category and the forgetful functor to $j_X^\dagger\mathcal{O}_V$-modules is faithful. Moreover, if E, F are two stratified $j_X^\dagger\mathcal{O}_V$-modules, then $E \otimes_{j_X^\dagger\mathcal{O}_V} F$ has a natural structure of stratified $j_X^\dagger\mathcal{O}_V$-modules. The same results holds for $\mathcal{H}om_{j_X^\dagger\mathcal{O}_V}(E, F)$ when E is coherent.*

Proof Since the analog of the first assertion is true both for stratified $\mathcal{O}_V$-modules and $j_X^\dagger\mathcal{O}_V$-modules, it is clearly also true in our situation.

Now, we know from Corollary 5.3.4 that

$$E \otimes_{j_X^\dagger \mathcal{O}_V} F = E \otimes_{\mathcal{O}_V} F$$

is therefore canonically endowed with a stratification. Moreover, we also know from Corollary 5.3.4 that it is overconvergent.

For the last assertion, standard arguments allow to reduce to the case $E = j_X^\dagger \mathcal{O}_V$ in which case the assertion becomes trivial. The details are left to the reader. □

Proposition 6.1.3 *Let $(X \subset Y \subset P)$ be an S-frame and V an admissible open subset of $]Y[_P$. Then, the inclusion of stratified $j_X^\dagger \mathcal{O}_V$-modules into stratified $\mathcal{O}_V$-modules has a left adjoint induced by $j_X^\dagger$.*

Proof For $i = 1, 2$, the morphisms of frames

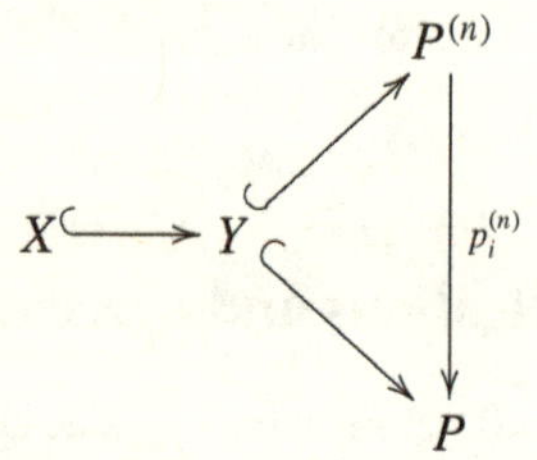

are obviously cartesians. It follows that if $\mathcal{E}$ is an $\mathcal{O}_V$-module, then

$$p_i^{(n)*} j_X^\dagger \mathcal{E} = j_X^\dagger p_i^{(n)*} \mathcal{E}.$$

Therefore, any stratification

$$\{\epsilon^{(n)} : p_2^{(n)*} \mathcal{E} \simeq p_1^{(n)*} \mathcal{E}\}_{n \in \mathbf{N}}$$

on $\mathcal{E}$ will induce a stratification

$$\{\epsilon^{(n)} : p_2^{(n)*} j_X^\dagger \mathcal{E} \simeq p_1^{(n)*} j_X^\dagger \mathcal{E}\}_{n \in \mathbf{N}}$$

on $j_X^\dagger \mathcal{E}$. This clearly defines an adjoint to the inclusion functor. □

Corollary 6.1.4 *Let $(X \subset Y \subset P)$ be an S frame and X' a closed subvariety of X. If V is a strict neighborhood of $]X[_P$ in $]Y[_P$ and E is a stratified $j^\dagger \mathcal{O}_V$-module, then $\underline{\Gamma}_{X'}^\dagger E$ is stable under the stratification of E.*

Proof If we write $U := X \setminus X'$, we know that

$$\underline{\Gamma}_{X'}^\dagger E = \ker(E \to j_U^\dagger E),$$

and the map on the left is clearly compatible to the stratifications. □

The next proposition shows that the category of stratified $j_X^\dagger \mathcal{O}_V$-modules over S is functorial in $(X \subset Y \subset P)/S$.

Proposition 6.1.5 *Let*

$$\begin{array}{ccccc} X' & \hookrightarrow & Y' & \hookrightarrow & P' \\ \downarrow{\scriptstyle f} & & \downarrow{\scriptstyle g} & & \downarrow{\scriptstyle u} \\ X & \hookrightarrow & Y & \hookrightarrow & P \end{array}$$

be a morphism from an S'-frame to an S-frame over some morphism $v : S' \to S$. If V is a strict neighborhood of $]X[_P$ in $]Y[_P$, V' is a strict neighborhood of $]X'[_{P'}$ in $]Y'[_{P'} \cap u_K^{-1}(V)$ and E is a stratified $j_X^\dagger \mathcal{O}_V$-module over S, then $u^\dagger E$ has a natural structure of stratified $j_{X'}^\dagger \mathcal{O}_{V'}$-module. And this is functorial in E.

Proof Since $u^\dagger E$ is overconvergent, we only have to show that it has a natural stratification. As usual, this can be split in two. When the morphism is cartesian, then $u^\dagger E = u_K^* E$ and there is nothing to do. If $u = \mathrm{id}_P$, then $u^\dagger E = j_{X'}^\dagger E$ and our assertion follows from Proposition 6.1.3. □

Proposition 6.1.6 *Let $(X \subset Y \subset P)$ be an S-frame, V an admissible open subset of $]Y[_P$ and $\sigma : K \hookrightarrow K'$ be an isometric embedding. If E is a stratified $j_X^\dagger \mathcal{O}_V$-module over S, then E^σ has a natural structure of stratified $j_{X^\sigma}^\dagger \mathcal{O}_{V^\sigma}$-module.*

Proof We know from Proposition 5.3.14 that E^σ is overconvergent and we also know that it has a natural stratification. □

In the next proposition, we denote by $\mathrm{Strat}(X \subset Y \subset P)$ the category of *coherent* stratified $j_X^\dagger \mathcal{O}_{]Y[_P}$-modules.

Proposition 6.1.7 *Let*

$$\begin{array}{ccccc} X_i & \hookrightarrow & Y_i & \hookrightarrow & P_i \\ \downarrow & & \downarrow & & \downarrow \\ X & \hookrightarrow & Y & \hookrightarrow & P \end{array}$$

be an open covering or a finite mixed covering of an S-frame. Then, there is an equivalence of categories

$$\mathrm{Strat}(X \subset Y \subset P) \simeq$$
$$\varprojlim \Big[\textstyle\prod_i \mathrm{Strat}(X_i \subset Y_i \subset P_i) \rightrightarrows \prod_{i,j} \mathrm{Strat}(X_{ij} \subset Y_{ij} \subset P_{ij})$$
$$\overrightarrow{\rightrightarrows} \textstyle\prod_{i,j,k} \mathrm{Strat}(X_{ijk} \subset Y_{ijk} \subset P_{ijk}) \Big].$$

Proof This is an immediate consequence of Proposition 5.4.12 because a stratification is simply a family of isomorphisms of coherent shaves on infinitesimal neighborhoods. □

Definition 6.1.8 *Let* $(X \subset Y \subset P)$ *be an S-frame and V an admissible open subset of* $]Y[_P$. *A* (integrable) connection on a $j_X^\dagger \mathcal{O}_V$-module *is a (integrable) connection as* $\mathcal{O}_V$*-module. And a morphism is simply a horizontal morphism in the usual sense.*

In other words, the category of $j_X^\dagger \mathcal{O}_V$-modules with a (integrable) connection over S_K is the full subcategory of $\mathcal{O}_V$-modules with a (integrable) connection over S_K that are overconvergent. Again, morphisms are automatically $j_X^\dagger \mathcal{O}_V$-linear.

Proposition 6.1.9 *Let* $(X \subset Y \subset P)$ *be a smooth S-frame and V an admissible open subset of* $]Y[_P$. *Then, the category of stratified* $j_X^\dagger \mathcal{O}_V$*-modules is equivalent to the category of left* $j_X^\dagger \mathcal{D}_{V/S_K}$*-modules. And any such module inherits an integrable connection.*

Proof Since the frame is smooth over S, it follows from Corollary 3.3.6 that there exists a smooth strict neighborhood of $]X[_P$ in $]Y[_P$ over S_K. Using Proposition 5.3.7, we may therefore assume that V is smooth. In particular, $\mathcal{D}_{V/S_K}$ is a sheaf of rings and our assertion has a meaning. We know that the category of stratified $j_X^\dagger \mathcal{O}_V$-modules is equivalent to the category of $\mathcal{D}_{V/S_K}$-modules that are overconvergent and, using Proposition 5.3.1 again, this is the same thing as the category of $j_X^\dagger \mathcal{D}_{V/S_K}$-modules. Of course, our module has a canonical integrable connection. □

It should be noticed that $j_X^\dagger \mathcal{D}_{V/S_K}$ is *not* the sheaf $\mathcal{D}^\dagger_{\mathcal{X}/S\mathbb{Q}}$ of [14]. We have only convergence conditions on the coefficients but not on the differential operators. Therefore, this is a far more naive object.

Until the end of the section, we assume that $\operatorname{Char} K = 0$.

Proposition 6.1.10 *Let* $(X \subset Y \subset P)$ *be a smooth S-frame and V an admissible open subset of* $]Y[_P$. *Then, the categories of stratified* $j_X^\dagger \mathcal{O}_V$*-modules, the category of left* $j_X^\dagger \mathcal{D}_{V/S_K}$*-modules and the category of* $j_X^\dagger \mathcal{O}_V$*-modules with an integrable connection are all equivalent.*

Proof We saw in Proposition 6.1.9 that the first two categories are equivalent. Using Proposition 5.3.1, the other equivalence is a direct consequence of the analog result for $\mathcal{O}_V$-modules. □

Corollary 6.1.11 *Let $(X \subset Y \subset P)$ be a smooth S-frame and V an admissible open subset of $]Y[_P$. Then,*

(i) *The category of $j_X^\dagger \mathcal{O}_V$-modules with an integrable connection is abelian.*
(ii) *If E and F are two $j_X^\dagger \mathcal{O}_V$-modules with an integrable connection, then $E \otimes_{j_X^\dagger \mathcal{O}_{]Y[_P}} F$ is also a $j_X^\dagger \mathcal{O}_V$-module with an integrable connection. And so is $\mathcal{H}om_{j_X^\dagger \mathcal{O}_{]Y[_P}}(E, F)$ when E is coherent.*
(iii) *If X' a closed subvariety of X and E is a $j^\dagger \mathcal{O}_V$-module with an integrable connection, then $\underline{\Gamma}_{X'}^\dagger E$ is stable under the connection of E.*
(iv) *Let*

$$\begin{array}{ccccc} X' & \hookrightarrow & Y' & \hookrightarrow & P' \\ \downarrow f & & \downarrow g & & \downarrow u \\ X & \hookrightarrow & Y & \hookrightarrow & P \end{array}$$

be a morphism from a smooth S'-frame over some morphism of $\mathcal{V}$-formal schemes $S' \to S$. If E a $j_X^\dagger \mathcal{O}_V$-module with an integrable connection and V' an admissible open subset of $]Y'[_{P'} \cap u_K^{-1}(V)$, then $u^\dagger E$ is a $j_{X'}^\dagger \mathcal{O}_{V'}$-modules with an integrable connection.
(v) *If $\sigma : K \hookrightarrow K'$ is an isometric embedding and E is a $j_X^\dagger \mathcal{O}_V$-module with an integrable connection over S, then E^σ is a $j_{X^\sigma}^\dagger \mathcal{O}_{V^\sigma}$-module with an integrable connection over S^σ.*

Proof We saw all these properties for stratified $j_X^\dagger \mathcal{O}_V$-modules. □

If $(X \subset Y \subset P)$ is a smooth S-frame, we will denote by $\mathrm{MIC}(X \subset Y \subset P)$ the category of coherent $j_X^\dagger \mathcal{O}_{]Y[_P}$-modules with an integrable connection.

Proposition 6.1.12 *Let $(X \subset Y \subset P)$ be a smooth S-frame. Then, the category $\mathrm{MIC}(X \subset Y \subset P)$ is an abelian subcategory of the category of $j_X^\dagger \mathcal{O}_{]Y[_P}$-modules with connection which is stable under $\otimes_{j_X^\dagger \mathcal{O}_{]Y[_P}}$ and $\mathcal{H}om_{j_X^\dagger \mathcal{O}_{]Y[_P}}$. Moreover, the forgetful functor*

$$\mathrm{MIC}(X \subset Y \subset P) \to \mathrm{Coh}(j_X^\dagger \mathcal{O}_{]Y[_P})$$

is faithful and exact.

Proof We know that coherent $j_X^\dagger \mathcal{O}_{]Y[_P}$-modules form an abelian subcategory stable under tensor product and internal Hom of the category of all $j_X^\dagger \mathcal{O}_{]Y[_P}$-modules. And we also know that the forgetful functor from $j_X^\dagger \mathcal{O}_{]Y[_P}$-modules with integrable connections to $j_X^\dagger \mathcal{O}_{]Y[_P}$-modules is exact. □

Proposition 6.1.13 *If*

$$\begin{array}{ccccc} X' & \hookrightarrow & Y' & \hookrightarrow & P' \\ \downarrow{\scriptstyle f} & & \downarrow{\scriptstyle g} & & \downarrow{\scriptstyle u} \\ X & \hookrightarrow & Y & \hookrightarrow & P \end{array}$$

is a morphism of smooth S-frames and $E \in \mathrm{MIC}(X \subset Y \subset P)$, then $u^\dagger E \in \mathrm{MIC}(X' \subset Y' \subset P')$.

Proof We know from Proposition 6.1.5 that $u^\dagger E$ is a $j_X^\dagger \mathcal{O}_{]Y[_P}$-module with an integrable connection and we saw in Corollary 5.4.2 that it is coherent. □

Proposition 6.1.14 *Let $(X \subset Y \subset P)$ be a smooth S-frame and $\sigma : K \hookrightarrow K'$ an isometric embedding. If $E \in \mathrm{MIC}(X \subset Y \subset P)$, then $E^\sigma \in \mathrm{MIC}(X^\sigma \subset Y^\sigma \subset P^\sigma)$.*

Proof We know from Proposition 6.1.6 that E^σ is a $j_{X^\sigma}^\dagger \mathcal{O}_{]Y^\sigma[_{P^\sigma}}$-module with an integrable connection and we have also seen in Corollary 5.4.2 that it is coherent. □

Proposition 6.1.15 *Let $(X \subset Y \subset P)$ be a smooth quasi-compact S-frame. Then, the functors $j_X^\dagger$ induce an equivalence of categories*

$$\varinjlim_V \mathrm{MIC}(V/S_K) \simeq \mathrm{MIC}(X \subset Y \subset P/S)$$

when V runs through the smooth strict neighborhoods of $]X[_P$ in $]Y[_P$. Moreover, the restriction functor

$$\mathrm{MIC}(X \subset Y \subset P/S) \to \mathrm{MIC}(]X[_P/S_K)$$

is faithful and exact.

Proof Let E be a coherent module with an integrable connection ∇. We know from Theorem 5.4.4 that we can write $E = j_X^\dagger \mathcal{E}$ with $\mathcal{E}$ a coherent $\mathcal{O}_V$-module on a strict neighborhood V of $]X[_P$ in $]Y[_P$. Shrinking V if necessary, we may also assume that the first-level Taylor morphism θ^1, which is linear, is defined over V. It follows that ∇ also is defined over V. This shows that the functor is essentially surjective. Full faithfulness is also a consequence of Theorem 5.4.4 since the functor that forgets the connection is faithful. For the same reason, the last assertion follows from Corollary 5.4.7. □

Proposition 6.1.16 *Let*

$$\begin{array}{ccccc} X_i & \hookrightarrow & Y_i & \hookrightarrow & P_i \\ \downarrow & & \downarrow & & \downarrow \\ X & \hookrightarrow & Y & \hookrightarrow & P \end{array}$$

be an open covering or a finite mixed covering of a smooth S-frame. Then, there is an equivalence of categories

$$\mathrm{MIC}(X \subset Y \subset P) \simeq$$
$$\varprojlim \Big[\prod_i \mathrm{MIC}(X_i \subset Y_i \subset P_i) \rightrightarrows \prod_{\{i,j\}} \mathrm{MIC}(X_{ij} \subset Y_{ij} \subset P_{ij})$$
$$\rightrightarrows\!\!\!\!\rightarrow \prod_{\{i,j,k\}} \mathrm{MIC}(X_{ijk} \subset Y_{ijk} \subset P_{ijk})\Big].$$

Proof Follows from the analogous result on stratifications. □

As an example, we consider again the Monsky–Washnitzer situation. Recall that we assume $\mathrm{Char} K = 0$ and that, in this situation, if A is a formally smooth weakly complete $\mathcal{V}$-algebra, then the category $\mathrm{MIC}(A_K)$ is equivalent to the category of stratified coherent A_K-modules (defined as usual).

Proposition 6.1.17 *Let $X = \mathrm{Spec} A$ be a smooth affine $\mathcal{V}$-scheme and Y the closure of X in some projective space for a given presentation of A. Then, we have an equivalence of categories*

$$\begin{array}{ccc} \mathrm{MIC}(X_k \subset Y_k \subset \widehat{Y}/\mathcal{V}) & \xrightarrow{\simeq} & \mathrm{MIC}(A_K^\dagger) \\ E & \longmapsto & \Gamma(Y_K^{\mathrm{rig}}, E). \end{array}$$

Proof Follows from Proposition 5.4.8 because, on both side, an integrable connection is equivalent to a stratification. □

Proposition 6.1.18 *Let $X = \mathrm{Spec} A$ be a smooth affine $\mathcal{V}$-scheme of pure dimension d and M a coherent $A_K^\dagger$-module with an integrable connection. Let $\mathcal{E}$ be a coherent module with an integrable connection on some strict neighborhood V of $\hat{X}_K$ in X_K^{rig} such that $M = \Gamma(V, j^\dagger\mathcal{E})$. Then, the map*

$$M_c = H^d_{\hat{X}_K}(V, \mathcal{E}) \to H^d_{\hat{X}_K}(V, \mathcal{E} \otimes_{\mathcal{O}_V} \Omega^1_V) \simeq M_c \otimes_{A_K^\dagger} \Omega^1_{A_K^\dagger}$$

induced by the connection of $\mathcal{E}$ is a continuous integrable connection that only depends on M. In particular, A_c has a natural connection and the isomorphism

$$M_c \simeq A_c \otimes_{A_K^\dagger} M$$

is compatible with the connections.

Proof Again, this follows from the linearity of the condition once translated into stratification language. □

One can also construct an object that describes the behavior at infinity of an overconvergent module with an integrable connection on a curve.

Proposition 6.1.19 *Let $\mathcal{Y}$ be a flat formal $\mathcal{V}$-scheme whose special fiber Y is a connected curve, X a non empty open subset of Y and x a smooth rational point in $Y \setminus X$. Then, we have*

$$(j_X^\dagger \Omega^1_{\mathcal{Y}_K})_x \simeq \Omega^1_{\mathcal{R}(x)}.$$

Moreover, if E is a coherent $j_X^\dagger \mathcal{O}_{\mathcal{Y}_K}$-module with a connection, then E_x has a canonical connection and we get a functor

$$\begin{array}{ccc} \mathrm{MIC}(X \subset Y \subset \mathcal{Y}) & \longrightarrow & \mathrm{MIC}(\mathcal{R}(x)). \\ E & \longmapsto & E_x \end{array}$$

Proof As usual, this reduces to an analog assertion on some semi-open annulus. Alternatively, once translated into the language of stratification, this will follow from Proposition 5.4.9. The details are left to the reader. □

6.2 Cohomology

Starting at Proposition 6.2.11, we assume that $\mathrm{Char} K = 0$.

Definition 6.2.1 *Let*

$$\begin{array}{ccccc} X & \hookrightarrow & Y & \hookrightarrow & P \\ \downarrow & & \downarrow & & \downarrow u \\ C & \hookrightarrow & D & \hookrightarrow & Q \end{array}$$

be a morphism of S-frames. If E is a $j_X^\dagger \mathcal{O}_{]Y[_P}$-module with an integrable connection over S_K, the relative rigid cohomology *of E is the relative de Rham cohomology of E*

$$Ru_{\mathrm{rig}} E := Ru_{K\mathrm{dR}} E$$

with $u_K :]Y[_{P'} \to]C[_Q$. When $Y = X$, we call it the relative convergent cohomology *and denote it by $Ru_{\mathrm{conv}} E$.*

As a particular case, if $(X \subset Y \subset P)$ is an S-frame with structural map $p : P \to S$, the *absolute rigid cohomology* of E is the complex $Rp_{\mathrm{rig}} E$ on S_K

where p is seen as the morphism to the trivial frame ($S_k = S_k \subset S$). Of course, when $Y = X$, we write $Rp_{\text{conv}} E$. We define for each $i \in \mathbf{N}$, the *absolute rigid cohomology* of E

$$\mathcal{H}^i_{\text{rig}}(X \subset Y \subset P/S, E) := R^i p_{\text{rig}} E$$

and the *absolute convergent cohomology* of E

$$\mathcal{H}^i_{\text{conv}}(X \subset P/S, E) := R^i p_{\text{conv}} E.$$

When $S = \mathrm{Spf}\mathcal{V}$, we will write

$$R\Gamma_{\text{rig}}(X \subset Y \subset P, E) := Rp_{\text{rig}} E \quad \text{and} \quad R\Gamma_{\text{conv}}(X \subset P, E) := Rp_{\text{conv}} E$$

as well as

$$H^i_{\text{rig}}(X \subset Y \subset P, E) := R^i p_{\text{rig}} E \quad \text{and} \quad H^i_{\text{conv}}(X \subset P, E) := Rp_{\text{conv}} E.$$

Finally, when E is the trivial module with connection, we drop it in the H-notations.

Let us go back to the general case.

Proposition 6.2.2 *Let*

$$\begin{array}{ccccc} X & \hookrightarrow & Y & \hookrightarrow & P \\ \downarrow & & \downarrow & & \downarrow u \\ C & \hookrightarrow & D & \hookrightarrow & Q \end{array}$$

be a morphism of S-frames and E a $j_X^\dagger \mathcal{O}_{]Y[_P}$-module with an integrable connection over S_K. If W is a strict neighborhood of $]C[_Q$ in $]D[_Q$, V is a strict neighborhood of $]X[_P$ in $u_K^{-1}(W) \cap]Y[_P$ and $u_K : V \to W$ the map induced by u, $Ru_{\text{rig}} E$ is overconvergent and we have

$$(Ru_{\text{rig}} E)_{|W} = Ru_{K\text{dR}} E_{|V}.$$

Proof First of all, it results from Proposition 5.1.17 that $Ru_{\text{rig}} E$ is overconvergent and we may therefore assume that $W =]D[_Q$. It is actually more convenient to call $u_K :]Y[_P \to]D[_Q$ the map induced by u so that, if $j : V \hookrightarrow]Y[_P$ denotes the inclusion map, we have to show that

$$R(u_K \circ j)_{\text{dR}} j^{-1} E = Ru_{K\text{dR}} E.$$

This can be rewritten

$$R(u_K \circ j)_* j^{-1}(E \otimes_{\mathcal{O}_{]Y[_P}} \Omega^\bullet_{]Y[_P/S_K}) = Ru_{K*}(E \otimes_{\mathcal{O}_{]Y[_P}} \Omega^\bullet_{]Y[_P/S_K}).$$

It is therefore sufficient to check that if E is any $j_{X'}^{\dagger}\mathcal{O}_{]Y[_P}$-module, we have

$$Rj_*j^{-1}E = E.$$

But this has been shown in Proposition 5.3.7. □

Definition 6.2.3 *If $(X \subset Y \subset P)$ is an S-frame and E a $j_X^{\dagger}\mathcal{O}_{]Y[_P}$-module with an integrable connection over S_K, the* specialized de Rham complex *of E is the complex*

$$R\underline{\Gamma}_{\mathrm{rig}}E := Rsp_*(E \otimes_{\mathcal{O}_{]Y[_P}} \Omega^{\bullet}_{]Y[_P/S_K})$$

on D. When $Y = X$, we write $R\underline{\Gamma}_{\mathrm{conv}}E$.

Proposition 6.2.4 *Let*

$$\begin{array}{ccccc} X & \hookrightarrow & Y & \hookrightarrow & P \\ \downarrow & & \downarrow g & & \downarrow u \\ C & \hookrightarrow & D & \hookrightarrow & Q \end{array}$$

be a morphism of S-frames and E a $j_X^{\dagger}\mathcal{O}_{]Y[_P}$-module with an integrable connection over S_K. Then

$$Rsp_*Ru_{\mathrm{rig}}E = Rg_*R\underline{\Gamma}_{\mathrm{rig}}E.$$

Proof We have a commutative diagram

$$\begin{array}{ccc}]Y[_P & \xrightarrow{sp} & Y \\ \downarrow u_k & & \downarrow g \\ ]D[_Q & \xrightarrow{sp} & D \end{array}$$

from which we derive

$$\begin{aligned} Rsp_*Ru_{\mathrm{rig}}E &= Rsp_*Ru_{K\mathrm{dR}}E = Rsp_*Ru_{K*}(E \otimes_{\mathcal{O}_{]Y[_P}} \Omega^{\bullet}_{]Y[_P/S_K}) \\ &= Rg_*Rsp_*(E \otimes_{\mathcal{O}_{]Y[_P}} \Omega^{\bullet}_{]Y[_P/S_K}) = Rg_*R\underline{\Gamma}_{\mathrm{rig}}E. \end{aligned}$$

□

Now, we have an immediate consequence of the previous proposition.

Corollary 6.2.5 *Let $(X \subset Y \subset P)$ be a frame (over* Spf$\mathcal{V}$*) and E a $j_X^{\dagger}\mathcal{O}_{]Y[_P}$-module with an integrable connection over S_K. Then*

$$R\Gamma_{\mathrm{rig}}(X \subset Y \subset P, E) = R\Gamma(Y, R\underline{\Gamma}_{\mathrm{rig}}E)$$

Proof In this case, as continuous map, the specialization map

$$\mathrm{Spm} K \to \mathrm{Spec} k$$

is just the identity of the one point topological space. □

Proposition 6.2.6 *Assume that we are given a commutative diagram of morphisms of frames*

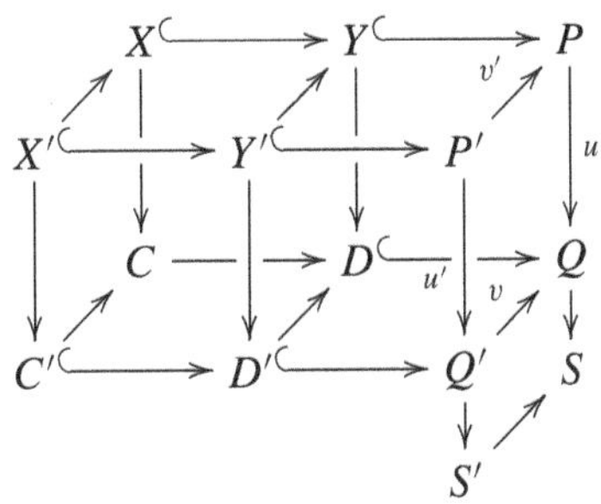

over some morphism of formal schemes $S' \to S$. Then, if E is a $j_X^\dagger \mathcal{O}_{]Y[_P}$-module with an integrable connection over S_K there is a canonical base change map

$$v'* : Lv_K^* Ru_{\mathrm{rig}} E \to Ru'_{\mathrm{rig}} v'^\dagger E$$

with $v_K :]D'[_{Q'} \to]D[_Q$.

Proof If we denote by

$$u_K :]Y[_P \to]D[_P, u'_K :]Y'[_{P'} \to]D'[_{Q'} \quad \text{and} \quad v'_K :]Y'[_{P'} \to]Y[_P$$

the morphisms induced by u, u' and v', respectively, we can consider the base change morphism in de Rham cohomology

$$Lv_K^* Ru_{K\mathrm{dR}} E \to Ru'_{K\mathrm{dR}} v_K'^* E$$

composed with the canonical map

$$Ru'_{K\mathrm{dR}} v_K'^* E \to Ru'_{K\mathrm{dR}} v'^\dagger E.$$

□

Proposition 6.2.7 *Let $(X \subset Y \subset P)$ be an S-frame and E a coherent $j_X^\dagger \mathcal{O}_{]Y[_P}$-module with an integrable connection. If $K \hookrightarrow K'$ is a finite field extension, there is a canonical isomorphism*

$$K' \otimes_K H^q_{\mathrm{rig}}(X \subset Y \subset P, E) \simeq H^q_{\mathrm{rig}}(X_k \subset Y_k \subset P_\mathcal{V}, E_{K'}).$$

Proof Directly follows from Corollary 5.4.16. □

There exists also a general base change morphism for field extensions.

Proposition 6.2.8 *Let*

$$\begin{CD} X @>>> Y @>>> P \\ @VVV @VVV @VV{u}V \\ C @>>> D @>>> Q \end{CD}$$

be a morphism of S-frames and E a $j_X^\dagger \mathcal{O}_{]Y[_P}$-module with an integrable connection over S_K.

If $\sigma : K \hookrightarrow K'$ is an isometric embedding, there is a canonical morphism

$$(Ru_{\mathrm{rig}} E)^\sigma \to Ru^\sigma_{\mathrm{rig}} E^\sigma .$$

Proof If, as usual, we denote by $u_K :]Y[_P \to]D[_Q$ the map induced by u, this is just the base change map in de Rham cohomology

$$(Ru_{K\mathrm{dR}} E)^\sigma \to Ru^\sigma_{K\mathrm{dR}} E^\sigma .$$

□

We now show that rigid cohomology is local downstairs in some sense.

Proposition 6.2.9 *Let*

$$\begin{CD} X @>>> Y @>>> P \\ @V{f}VV @V{g}VV @VV{u}V \\ C @>>> D @>>> Q \end{CD}$$

be a morphism of S-frames,

$$\begin{CD} C' @>>> D' @>>> Q' \\ @VVV @VVV @VVV \\ C @>>> D @>>> Q \end{CD}$$

a cartesian, open or mixed, immersion of frames and

$$\begin{CD} X' @>>> Y' @>>> P' \\ @V{f'}VV @V{g'}VV @VV{u'}V \\ C' @>>> D' @>>> Q' \end{CD}$$

the pullback of the first morphism of frames along this immersion.

If E is a $j_X^\dagger \mathcal{O}_{]Y[_P}$-module with an integrable connection over S_K, then $E_{|]Y'[_{P'}}$ is overconvergent outside $]X'[_{P'}$ and we have

$$(Ru_{\mathrm{rig}}E)_{|]D'[_{Q'}} \simeq Ru'_{\mathrm{rig}}E_{|]Y'[_{P'}}.$$

Proof With our assumptions, we have a cartesian morphism of frames

$$\begin{array}{ccccc} X' & \hookrightarrow & Y' & \hookrightarrow & P' \\ \downarrow & & \downarrow & & \downarrow \\ X & \hookrightarrow & Y & \hookrightarrow & P \end{array}$$

and it follows that $E_{|]Y'[_{P'}}$ is overconvergent outside $]X'[_{P'}$. Moreover, we have $]Y'[_{P'} \subset u_K^{-1}(]D'[_{Q'})$ and our assertion therefore follows from the fact that higher direct images commute to open immersions. □

Our definitions are also local upstairs as we now show. We use the multi-index notation: if $\underline{i} = (i_1, \ldots, i_r)$, then $|\underline{i}| = r$, $X_{\underline{i}} = X_{i_1} \cap \cdots \cap X_{i_r}$, etc.

Proposition 6.2.10 *Let*

$$\begin{array}{ccccc} X & \hookrightarrow & Y & \hookrightarrow & P \\ \downarrow & & \downarrow & & \downarrow u \\ C & \hookrightarrow & D & \hookrightarrow & Q \end{array}$$

be a morphism of S-frames and E a $j_X^\dagger \mathcal{O}_{]Y[_P}$-module with an integrable connection over S_K. Assume that we are given an open or finite mixed covering

$$\begin{array}{ccccc} X_i & \hookrightarrow & Y_i & \hookrightarrow & P_i \\ \downarrow & & \downarrow & & \downarrow \\ X & \hookrightarrow & Y & \hookrightarrow & P \end{array}$$

Then, there is a canonical spectral sequence

$$E_1^{p,q} := \bigoplus_{|\underline{i}|=p} R^q u_{\underline{i},\mathrm{rig}} j_{X_{\underline{i}}}^\dagger E_{|]Y_{\underline{i}}[_{P_{\underline{i}}}} \Rightarrow Ru_{\mathrm{rig}}E.$$

Proof If we denote by $[K^{\bullet,\bullet}]$ the complex associated to a bicomplex $K^{\bullet,\bullet}$, it follows from Proposition 5.2.8 that we have an isomorphism of complexes

$$E \otimes_{\mathcal{O}_{]Y[_P}} \Omega^\bullet_{]Y[_P/S_K} \simeq [\bigoplus_{i_1,\cdots,i_\bullet} j^\dagger_{X_{i_1,\cdots,i_\bullet}} E \otimes_{\mathcal{O}_{]Y[_P}} \Omega^\bullet_{]Y_{i_1,\cdots,i_\bullet}[_{P_{i_1,\cdots,i_\bullet}}/S_K}]$$

We can then consider the corresponding spectral sequence. □

Until the end of the section, we assume that $\mathrm{Char} K = 0$.

Proposition 6.2.11 *If $(X \subset Y \subset P)$ is a smooth frame over $\mathcal{V}$, then $H^0_{\mathrm{rig}}(X \subset Y \subset P)$ is a vector space whose dimension is the number of geometrically connected components of X.*

Proof Using Propositions 6.2.7 and 6.2.10, we may assume that X is connected with a rational point x and we want to show that the natural inclusion

$$K \hookrightarrow H^0_{\mathrm{rig}}(X \subset Y \subset P)$$

is actually bijective. It follows from Lemma 5.4.6, that the map

$$\Gamma(]Y[_P, j_X^\dagger \mathcal{O}_{]Y[_P}) \to \Gamma(]X[_P, \mathcal{O}_{]X[_P})$$

is injective. Since P is smooth in a neighborhood of x, the rational point x lifts to a $\mathcal{V}$-valued point of P that induces a point ξ of $]X[_P$. The map

$$\Gamma(]X[_P, \mathcal{O}_{P_K}) \to \widehat{\mathcal{O}}_{P_K,\xi}$$

is injective because $]X[_P$ is smooth and connected. Finally, we have $\widehat{\mathcal{O}}_{P_K,\xi} \simeq K[[\underline{t}]]$ and the assertion therefore results from the fact that, when $\mathrm{Char} K = 0$, there are no non constant horizontal formal series. □

If A is a formally smooth weakly complete $\mathcal{V}$-algebra and M an A_K-module with an integrable connection, the *de Rham cohomology (or complex) of M* is

$$\Gamma_{\mathrm{dR}}(M) := M \otimes_{A_K} \Omega^\bullet_{A_K}.$$

In other words, we have for each $i \in \mathbf{N}$,

$$H^i_{\mathrm{dR}}(M) := H^i(M \otimes_{A_K} \Omega^\bullet_{A_K}).$$

Proposition 6.2.12 *Let $X = \mathrm{Spec} A$ be a smooth affine $\mathcal{V}$-scheme and Y the closure of X in some projective space for a given presentation of A. If E is a coherent $j_X^\dagger \mathcal{O}_{Y_K^{\mathrm{rig}}}$-module with an integrable connection and $M := \Gamma(Y_K^{\mathrm{rig}}, E)$, we have*

$$R\Gamma_{\mathrm{rig}}(X_k \subset Y_k \subset \widehat{Y}, E) \simeq \Gamma_{\mathrm{dR}}(M).$$

Proof We may consider our cofinal family

$$V_\rho := X_K^{\mathrm{rig}} \cap \mathbf{B}^N(0, \rho^+)$$

of strict neighborhoods of $\widehat{X}_K$ in Y_K^{rig}. Since E is a coherent $j_X^\dagger \mathcal{O}_{Y_K^{\mathrm{rig}}}$-module with an integrable connection, there exists $\rho_0 > 1$ and a coherent module $\mathcal{E}$ on $V := V_{\rho_0}$ with an integrable connection such that $E_{|V} := j_X^\dagger \mathcal{E}$. It follows from Proposition 6.2.2 and the fact that V is quasi-compact and (quasi-) separated that

$$R\Gamma_{\mathrm{rig}}(X_k \subset Y_k \subset \widehat{Y}, E) = R\Gamma_{\mathrm{dR}}(V, j_X^\dagger \mathcal{E})$$
$$= R\Gamma(V, j_X^\dagger \mathcal{E} \otimes_{\mathcal{O}_V} \Omega^\bullet_{V/K}) = \varinjlim_\rho R\Gamma(V, j_{\rho*}(\mathcal{E} \otimes_{\mathcal{O}_{V_\rho}} \Omega^\bullet_{V_\rho/K}))$$

with $j_\rho : V_\rho \hookrightarrow V$ the inclusion map. Since this map is affinoid and each $\mathcal{E} \otimes_{\mathcal{O}_{V_\rho}} \Omega^i_{V_\rho/K}$ is coherent, we have

$$j_{\rho*}(\mathcal{E} \otimes_{\mathcal{O}_{V_\rho}} \Omega^\bullet_{V_\rho/K}) = Rj_{\rho*}(\mathcal{E} \otimes_{\mathcal{O}_{V_\rho}} \Omega^\bullet_{V_\rho/K}).$$

Actually, since each V_ρ is affinoid, it follows that

$$R\Gamma_{\mathrm{rig}}(X_k \subset Y_k \subset \widehat{Y}, E) = \varinjlim_\rho \Gamma(V_\rho, \mathcal{E} \otimes_{\mathcal{O}_{V_\rho}} \Omega^\bullet_{V_\rho/K}).$$

If we write $M_\rho := \Gamma(V_\rho, \mathcal{E})$, we are reduced to checking the following claim.

Claim 6.2.13

$$\Gamma_{\mathrm{dR}}(M) = \varinjlim_\rho \Gamma_{\mathrm{dR}}(M_\rho).$$

This is left to the reader.

□

We may now go back to our old examples of Section 4.2 with Char$K = 0$ and Char$k = p > 0$. Using Claim 6.2.13, we easily see that if $L_\alpha := j_X^\dagger \mathcal{L}_\alpha$, then

$$H^i_{\mathrm{rig}}(\mathbf{A}^1_k \subset \mathbf{P}^1_k \subset \widehat{\mathbf{P}}^1_{\mathcal{V}}, L_\alpha) = 0$$

unless $i = 0$ and $|\alpha| < |p|^{\frac{1}{p-1}}$ in which case we get

$$\dim_K H^0_{\mathrm{rig}}(\mathbf{A}^1_k \subset \mathbf{P}^1_k \subset \widehat{\mathbf{P}}^1_{\mathcal{V}}, L_\alpha) = 1.$$

Also, if $K_\beta := j_X^\dagger \mathcal{K}_\beta$, then

$$H^i_{\mathrm{rig}}(\mathbf{A}^1_k \setminus 0 \subset \mathbf{P}^1_k \subset \widehat{\mathbf{P}}^1_{\mathcal{V}}, K_\beta) = 0$$

unless $i = 0$ and $\beta \in \mathbf{Z}$ in which case,

$$\dim_K H^0_{\mathrm{rig}}(\mathbf{A}^1_k \setminus 0 \subset \mathbf{P}^1_k \subset \widehat{\mathbf{P}}^1_{\mathcal{V}}, K_\beta) = 1,$$

or $i = 1$ and β is Liouville which gives

$$\dim_K H^1_{\mathrm{rig}}(\mathbf{A}^1_k \setminus 0 \subset \mathbf{P}^1_k \subset \widehat{\mathbf{P}}^1_{\mathcal{V}}, K_\beta) = \infty.$$

6.3 Cohomology with support in a closed subset

We need to extend our definition of rigid cohomology in order to take into account cohomology with support in a closed subvariety.

Definition 6.3.1 *Let $(X \subset Y \subset P)$ be an S frame, X' a closed subvariety of X and E a $j_X^\dagger \mathcal{O}_{]Y[_P}$-module with an integrable connection over S_K.*

(i) *If*

$$\begin{array}{ccccc} X & \hookrightarrow & Y & \hookrightarrow & P \\ \downarrow & & \downarrow & & \downarrow u \\ C & \hookrightarrow & D & \hookrightarrow & Q \end{array}$$

is a morphism of S-frames, the relative rigid cohomology with support in X' *is the complex*

$$Ru_{\mathrm{rig},X'}E := Ru_{\mathrm{rig}}\underline{\Gamma}_{X'}^{\dagger}E.$$

(ii) *The* specialized de Rham complex of E with support in X' *is*

$$R\underline{\Gamma}_{\mathrm{rig},X'}E := R\underline{\Gamma}_{\mathrm{rig}}\underline{\Gamma}_{X'}^{\dagger}E.$$

If $p : P \to S$ denotes the structural map, the absolute *rigid cohomology* of E with support in X' is the complex $Rp_{\mathrm{rig},X'}E$. We define for each $i \in \mathbf{N}$, the *absolute rigid cohomology with support in a closed subset* of E

$$\mathcal{H}^i_{\mathrm{rig},X'}(X \subset Y \subset P/S, E) := R^i p_{\mathrm{rig},X'}E.$$

When $S = \mathrm{Spf}\mathcal{V}$, we will write

$$R\Gamma_{\mathrm{rig},X'}(X \subset Y \subset P, E) := Rp_{\mathrm{rig},X'}E$$

and

$$H^i_{\mathrm{rig},X'}(X \subset Y \subset P, E) := R^i p_{\mathrm{rig},X'}E.$$

Finally, when E is the trivial module with connection, we drop it from the H-notations.

Proposition 6.3.2 *Let $(X \subset Y \subset P)$ be a smooth S-frame, X' a closed subvariety of X and E a $j_X^\dagger \mathcal{O}_{]Y[_P}$-module with an integrable connection over S_K Then,*

(i) *There is a canonical map*

$$Ru_{\mathrm{rig},X'}E \to Ru_{\mathrm{rig}}E$$

which is an isomorphism when $X' = X$:

$$Ru_{\mathrm{rig}}E = Ru_{\mathrm{rig},X}E.$$

(ii) *If U is an open neighborhood of X' in X, then*

$$Ru_{\mathrm{rig},X'}E = Ru_{\mathrm{rig},X'}j_U^\dagger E.$$

Proof There is a canonical horizontal map $\Gamma^\dagger_{X'}E \hookrightarrow E$ which is bijective when $X' = X$. The first assertion follows.

The second assertion is a direct consequence of Proposition 5.2.12. □

In particular, we see that rigid cohomology may be seen as a particular case of rigid cohomology with support in a closed subvariety.

In order to compute rigid cohomology with support in a closed subvariety, it is necessary to note the following result:

Proposition 6.3.3 *Let $(X \subset Y \subset P)$ be a smooth S-frame, X' a closed subvariety of X and E a $j_X^\dagger\mathcal{O}_{]Y[_P}$-module with an integrable connection over S_K. If*

$$\begin{array}{ccccc} X & \hookrightarrow & Y & \hookrightarrow & P \\ \downarrow & & \downarrow & & \downarrow{\scriptstyle u} \\ C & \hookrightarrow & D & \hookrightarrow & Q \end{array}$$

is a morphism of S-frames, we have

$$Ru_{\mathrm{rig},X'}E := Ru_{K*}\underline{\Gamma}^\dagger_{X'}(E \otimes_{\mathcal{O}_{]Y[_P}} \Omega^\bullet_{]Y[_P/S_K})$$

and

$$R\underline{\Gamma}_{\mathrm{rig},X'}E := R\underline{\Gamma}_{\mathrm{rig}}\underline{\Gamma}^\dagger_{X'}(E \otimes_{\mathcal{O}_{]Y[_P}} \Omega^\bullet_{]Y[_P/S_K}).$$

Proof Since P is smooth on S in a neighborhood of X, $\Omega^1_{]Y[_P/S_K}$ is locally free on a strict neighborhood of $]X[_P$ in $]Y[_P$ and it follows from Corollary 5.3.3 that

$$\underline{\Gamma}^\dagger_{X'}E_P \otimes_{\mathcal{O}_{]Y[_P}} \Omega^\bullet_{]Y[_P/S_K} = \underline{\Gamma}^\dagger_{X'}(E_P \otimes_{\mathcal{O}_{]Y[_P}} \Omega^\bullet_{]Y[_P/S_K}).$$

□

Proposition 6.3.4 *Let*

$$\begin{array}{ccccc} X & \hookrightarrow & Y & \hookrightarrow & P \\ \downarrow & & \downarrow{\scriptstyle g} & & \downarrow{\scriptstyle u} \\ C & \hookrightarrow & D & \hookrightarrow & Q \end{array}$$

be a morphism of S-frames, X' a closed subvariety of X and E a $j_X^\dagger \mathcal{O}_{]Y[_P}$-module with an integrable connection over S_K. Then

$$Rsp_* Ru_{\mathrm{rig},X'} E = Rg_* R\underline{\Gamma}_{\mathrm{rig},X'} E.$$

Proof Immediate consequence of Proposition 6.2.4. □

Note in particular that if $(X \subset Y \subset P)$ is an S-frame, X' a closed subvariety in X and E a $j_X^\dagger \mathcal{O}_{]Y[_P}$-module with an integrable connection over S_K, we have

$$R\Gamma_{\mathrm{rig},X'}(X \subset Y \subset P, E) = R\Gamma(Y, R\underline{\Gamma}_{\mathrm{rig},X'} E).$$

Rigid cohomology with support in a closed subvariety is functorial:

Proposition 6.3.5 *Assume that we are given a commutative diagram of morphisms of frames*

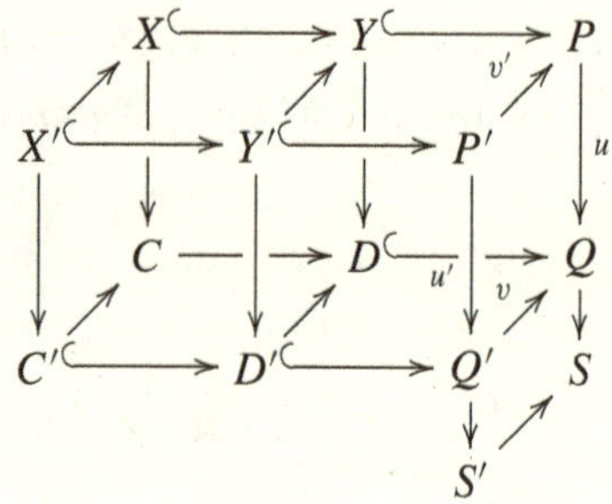

over some morphism of formal schemes $S' \to S$. Let X'' (resp. X''') be a closed subvariety of X (resp. X') with $f^{-1}(X'') \subset X'''$. Then, if E is a $j_X^\dagger \mathcal{O}_{]Y[_P}$-module with an integrable connection over S_K there is a canonical base change map

$$v'*: Lv_K^* Ru_{\mathrm{rig},X''} E \to Ru'_{\mathrm{rig},X'''} v'^\dagger E$$

with $v_K :]D'[_{Q'} \to]D[_Q$.

Proof We have a morphism

$$Lv_K^* Ru_{\mathrm{rig},X''} E = Lv_K^* Ru_{\mathrm{rig}} \underline{\Gamma}_{X''}^\dagger E \to Ru'_{\mathrm{rig},X'''} v'^\dagger \underline{\Gamma}_{X''}^\dagger E$$

and it is therefore sufficient to give a canonical map

$$Ru'_{\mathrm{rig},X'''} v'^\dagger \underline{\Gamma}_{X''}^\dagger E \to Ru'_{\mathrm{rig}} \underline{\Gamma}_{X''}^\dagger v'^\dagger E = Ru'_{\mathrm{rig},X'''} v'^\dagger E$$

and we know from Corollary 5.3.10 that there is a canonical morphism

$$v'^\dagger \underline{\Gamma}_{X''}^\dagger E \to \underline{\Gamma}_{X'''}^\dagger v'^\dagger E.$$

□

We have the following results on the local nature of this notion.

Proposition 6.3.6 *Let*

$$\begin{array}{ccccc} X & \hookrightarrow & Y & \hookrightarrow & P \\ \downarrow{\scriptstyle f} & & \downarrow{\scriptstyle g} & & \downarrow{\scriptstyle u} \\ C & \hookrightarrow & D & \hookrightarrow & Q \end{array}$$

be a morphism of S-frames,

$$\begin{array}{ccccc} C' & \hookrightarrow & D' & \hookrightarrow & Q' \\ \downarrow & & \downarrow & & \downarrow \\ C & \hookrightarrow & D & \hookrightarrow & Q \end{array}$$

a cartesian, open or mixed, immersion of frames and

$$\begin{array}{ccccc} X' & \hookrightarrow & Y' & \hookrightarrow & P' \\ \downarrow{\scriptstyle f'} & & \downarrow{\scriptstyle g'} & & \downarrow{\scriptstyle u'} \\ C' & \hookrightarrow & D' & \hookrightarrow & Q' \end{array}$$

the pullback of the first morphism of frames along the cartesian open immersion. Finally, let X'' be a closed subset of X and $X''' = X'' \cap X'$.

If E is a $j_X^\dagger \mathcal{O}_{]Y[_P}$-module with an integrable connection over S_K, then $E_{|]Y'[_{P'}}$ is overconvergent outside $]X'[_{P'}$ and we have

$$(Ru_{\mathrm{rig},X''}E)_{|]D'[_{Q'}} \simeq Ru'_{\mathrm{rig},X'''}E_{|]Y'[_{P'}}.$$

Proof We know from the case without support that

$$(Ru_{\mathrm{rig},X''}E)_{|]D'[_{Q'}} := (Ru_{\mathrm{rig}}\underline{\Gamma}^\dagger_{X''}E)_{|]D'[_{Q'}} \simeq Ru'_{\mathrm{rig}}(\underline{\Gamma}^\dagger_{X''}E)_{|]Y'[_{P'}}$$

and we only need to check that for any $j_X^\dagger \mathcal{O}_{]Y[_P}$-module E, we have

$$(\underline{\Gamma}^\dagger_{X''}E)_{|]Y'[_{P'}} = \underline{\Gamma}^\dagger_{X'''}E_{|]Y'[_{P'}}.$$

But this follows from Proposition 5.2.5. □

In the next proposition, we use the multi-index notation as explained before.

Proposition 6.3.7 *Let*

$$\begin{array}{ccccc} X & \hookrightarrow & Y & \hookrightarrow & P \\ \downarrow & & \downarrow & & \downarrow{\scriptstyle u} \\ C & \hookrightarrow & D & \hookrightarrow & Q \end{array}$$

be a morphism of S-frames. Assume that we are given an open or finite mixed covering

$$\begin{array}{ccccc} X_i & \hookrightarrow & Y_i & \hookrightarrow & P_i \\ \downarrow & & \downarrow & & \downarrow \\ X & \hookrightarrow & Y & \hookrightarrow & P. \end{array}$$

Let X' be a closed subset of X and for each i, $X'_i := X' \cap X_i$.

If E is a $j_X^\dagger \mathcal{O}_{]Y[_P}$-module with an integrable connection over S_K, there is a canonical spectral sequence

$$E_1^{p,q} := \bigoplus_{|\underline{i}|=p} R^q u_{\underline{i}\mathrm{rig},X'_{\underline{i}}} j_{X_{\underline{i}}}^\dagger E_{|]Y_{\underline{i}}[_{P_{\underline{i}}}} \Rightarrow Ru_{\mathrm{rig},X'} E.$$

Proof From the case without support, we are reduced to check that, for any $j_X^\dagger \mathcal{O}_{]Y[_P}$-module E, we have

$$\underline{\Gamma}^\dagger_{X'_{\underline{i}}} j^\dagger_{X_{\underline{i}}} E_{|]Y_{\underline{i}}[_{P_{\underline{i}}}} = j^\dagger_{X_{\underline{i}}} (\underline{\Gamma}^\dagger_{X'} E)_{|]Y_{\underline{i}}[_{P_{\underline{i}}}}.$$

From Proposition 5.2.5, we have

$$\underline{\Gamma}^\dagger_{X'_{\underline{i}}} E_{|]Y_{\underline{i}}[_{P_{\underline{i}}}} = (\underline{\Gamma}^\dagger_{X'} E)_{|]Y_{\underline{i}}[_{P_{\underline{i}}}}$$

and it is therefore sufficient to check that

$$\underline{\Gamma}^\dagger_{X'_{\underline{i}}} \circ j^\dagger_{X_{\underline{i}}} = j^\dagger_{X_{\underline{i}}} \circ \underline{\Gamma}^\dagger_{X'_{\underline{i}}}.$$

And this follows from Proposition 5.2.6. □

Proposition 6.3.8 *Let*

$$\begin{array}{ccccc} X & \hookrightarrow & Y & \hookrightarrow & P \\ \downarrow & & \downarrow & & \downarrow u \\ C & \hookrightarrow & D & \hookrightarrow & Q \end{array}$$

be a morphism of S-frames, X' be a closed subvariety of X, and E a $j_X^\dagger \mathcal{O}_{]Y[_P}$-module with an integrable connection over S_K.

If $\sigma : K \hookrightarrow K'$ is an isometric embedding, there is a canonical morphism

$$(Ru_{\mathrm{rig},X} E)^\sigma \to Ru^\sigma_{\mathrm{rig},X'^\sigma} E^\sigma.$$

Proof The method is always the same, it reduces to giving a morphism

$$(\Gamma^\dagger_{X''} E)^\sigma \to \Gamma^\dagger_{X'^\sigma} E^\sigma.$$

And it is sufficient to note that, if we set $U := X \setminus X'$, we have a commutative diagram with exact bottom row

$$\begin{array}{ccccccccc} & & (\Gamma^\dagger_{X''} E)^\sigma & \longrightarrow & E^\sigma & \longrightarrow & (j^\dagger_U E)^\sigma & & \\ & & \downarrow & & \| & & \downarrow & & \\ 0 & \longrightarrow & \Gamma^\dagger_{X'^\sigma} E^\sigma & \longrightarrow & E^\sigma & \longrightarrow & j^\dagger_{U^\sigma} E^\sigma & \longrightarrow & 0. \end{array}$$

□

The main reason to introduce rigid cohomology with support in a closed subvariety is the following excision theorem.

Proposition 6.3.9 *Let*

$$\begin{array}{ccccc} X & \hookrightarrow & Y & \hookrightarrow & P \\ \downarrow & & \downarrow & & \downarrow u \\ C & \hookrightarrow & D & \hookrightarrow & Q \end{array}$$

be a morphism of S-frames. Let

$$X'' \hookrightarrow X' \hookrightarrow X$$

be a sequence of closed immersions, $U := X \setminus X''$ and $U' := X' \setminus X''$. If E is a $j^\dagger_X \mathcal{O}_{]Y[_P}$-module with an integrable connection over S_K, there is an exact triangle

$$Ru_{\mathrm{rig},X''} E \to Ru_{\mathrm{rig},X'} E \to Ru_{\mathrm{rig},U'} j^\dagger_U E \to .$$

Proof This immediately follows from Proposition 5.2.11. □

Corollary 6.3.10 *Let $(X \subset Y \subset P)$ be an S-frame, X' a closed subset of X and E a coherent $j^\dagger_X \mathcal{O}_{]Y[_P}$-module with an integrable connection. If $K \hookrightarrow K'$ is a finite field extension, there is a canonical isomorphism*

$$K' \otimes_K H^q_{\mathrm{rig},X'}(X \subset Y \subset P, E) \simeq H^q_{\mathrm{rig},X'_k}(X_k \subset Y_k \subset P_\mathcal{V}, E_{K'}).$$

Proof Since we have a natural long exact sequence

$$\cdots \to H^q_{\mathrm{rig},X'}(X \subset Y \subset P, E) \to H^q_{rig}(X \subset Y \subset P, E)$$
$$\to H^q_{\mathrm{rig}}(U \subset Y \subset P, E) \to H^{q+1}_{\mathrm{rig},X'}(X \subset Y \subset P, E) \to \cdots,$$

this is a consequence of Proposition 6.2.7. □

It is not difficult to describe rigid cohomology with support in a hypersurface in the Monsky–Washnitzer setting.

Proposition 6.3.11 *Let $X = \mathrm{Spec} A$ be a connected smooth affine $\mathcal{V}$-scheme and Y the closure of X in some projective space for a given presentation of A. Let X' be the hypersurface defined by some non zero $f \in A$. Let E be a locally free $j_X^\dagger \mathcal{O}_{Y_K^{\mathrm{rig}}}$-module with an integrable connection and $M := \Gamma(Y_K^{\mathrm{rig}}, E)$. If $M_f := (A_f)_K^\dagger \otimes_{A_K^\dagger} M$, then $M^f := M_f/M$ is an $A_K^\dagger$-module with an integrable connection and we have*

$$R\Gamma_{\mathrm{rig},X'_k}(X_k \subset Y_k \subset \widehat{Y}, E) \simeq \Gamma_{\mathrm{dR}}(M^f)[-1].$$

Proof By definition, there is a short exact sequence of complexes

$$0 \to \Gamma_{\mathrm{dR}}(M) \to \Gamma_{\mathrm{dR}}(M_f) \to \Gamma_{\mathrm{dR}}(M^f) \to 0$$

and we finish with Proposition 6.3.9. □

6.4 Cohomology with compact support

Unlike its analog without support, rigid cohomology with compact support requires some coherence hypothesis in order to be defined. Also, rigid cohomology with compact support has never been defined in the literature in full generality. Nobuo Tsuzuki has done some work on this question but there is no written material available. Since our aim is not to introduce new results, we will therefore stick to the particular case $C = D$.

Starting at Proposition 6.4.15, we will assume $\mathrm{Char} K = 0$.

Proposition 6.4.1 *Let*

$$\begin{array}{ccccc} X & \hookrightarrow & Y & \hookrightarrow & P \\ \downarrow & & \downarrow & & \downarrow u \\ D & = & D & \hookrightarrow & Q \end{array}$$

be a quasi-compact morphism of S-frames.

Let V (resp. V') be a strict neighborhood of $]X[_P$ in $]Y[_P$, $u_K : V \to]D[_Q$ (resp. $u'_K : V' \to]D[_Q$) the map induced by u and $\mathcal{E}$ (resp. $\mathcal{E}'$) a coherent $\mathcal{O}_V$-module (resp. $\mathcal{O}_{V'}$-module) with an integrable connection.

If $j_X^\dagger \mathcal{E}' = j_X^\dagger \mathcal{E}$, then

$$Ru'_{K*} R\underline{\Gamma}_{]X[_P}(\mathcal{E}' \otimes_{\mathcal{O}_{V'}} \Omega^\bullet_{V'/S_K}) = Ru_{K*} R\underline{\Gamma}_{]X[_P}(\mathcal{E} \otimes_{\mathcal{O}_V} \Omega^\bullet_{V/S_K}).$$

Proof The question is local on $]D[_Q$ and we may therefore assume that P is quasi-compact. Moreover, using Theorem 5.4.4, we may assume that $V' \subset V$ and $\mathcal{E}' = \mathcal{E}_{|V'}$. Thus, it remains to show that if $j : V' \hookrightarrow V$ denotes the inclusion map, we have

$$Ru_{K*}R\underline{\Gamma}_{]X[_P}(\mathcal{E} \otimes_{\mathcal{O}_V} \Omega^\bullet_{V/S_K}) = R(u_K \circ j)_* R\underline{\Gamma}_{]X[_P}(\mathcal{E}_{|V'} \otimes_{\mathcal{O}_{V'}} \Omega^\bullet_{V'/S_K}).$$

This immediately follows from Proposition 5.2.15. □

We may therefore give the following definition.

Definition 6.4.2 *Let*

$$\begin{array}{ccccc} X & \hookrightarrow & Y & \hookrightarrow & P \\ \downarrow & & \downarrow & & \downarrow u \\ D & = & D & \hookrightarrow & Q \end{array}$$

be a quasi-compact morphism of S-frames. Let V be a strict neighborhood of $]X[_P$ in $]Y[_P$, $\mathcal{E}$ a coherent $\mathcal{O}_V$-module with an integrable connection and $E = j_X^\dagger \mathcal{E}$. If $u_K : V \to]D[_Q$ denotes the induced map, the relative rigid cohomology with compact support *of E is the complex*

$$Ru_{\mathrm{rig},c}E := Ru_{K*}R\underline{\Gamma}_{]X[_P}(\mathcal{E} \otimes_{\mathcal{O}_V} \Omega^\bullet_{V/S_K}).$$

As a particular case, if $(X \subset Y \subset P)$ is a quasi-compact S-frame and if $p : P \to S$ denotes the structural map, the absolute *rigid cohomology* of E with compact support is the complex $Rp_{\mathrm{rig},c}E$. We define for each $i \in \mathbf{N}$, the *absolute rigid cohomology with compact support* of E

$$\mathcal{H}^i_{\mathrm{rig},c}(X \subset Y \subset P/S, E) := R^i p_{\mathrm{rig},c}E.$$

When $S = \mathrm{Spf}\mathcal{V}$, we will write

$$R\Gamma_{\mathrm{rig},c}(X \subset Y \subset P, E) := Rp_{\mathrm{rig},c}E$$

and

$$H^i_{\mathrm{rig},c}(X \subset Y \subset P, E) := R^i p_{\mathrm{rig},c}E.$$

Finally, when E is the trivial module with connection, we drop it from the H-notations.

In the next proposition, we denote by $[K^{\bullet,\bullet}]$ the simple complex associated to a bicomplex $K^{\bullet,\bullet}$.

Proposition 6.4.3 *Let*

$$\begin{array}{ccccc} X & \hookrightarrow & Y & \hookrightarrow & P \\ \downarrow & & \downarrow & & \downarrow u \\ D & = & D & \hookrightarrow & Q \end{array}$$

be a morphism of quasi-compact S-frames.

Let V be a strict neighborhood of $]X[_P$ in $]Y[_P$, $\mathcal{E}$ a coherent $\mathcal{O}_V$-module with an integrable connection and $E = j_X^\dagger \mathcal{E}$. If we denote as usual by $u_K : V \to]D[_Q$ the induced map, we have

(i) *If $h : T := V \setminus]X[_P \hookrightarrow V$ denotes the inclusion map, then*

$$Ru_{\mathrm{rig},c}E = Ru_{K*}[\mathcal{E} \otimes_{\mathcal{O}_V} \Omega^{\bullet}_{V/S_K} \to h_*(\mathcal{E}_{|T} \otimes_{\mathcal{O}_T} \Omega^{\bullet}_{T/S_K})].$$

(ii) *If $u'_K : T \to]D[_Q$ denotes the induced map, we have an exact triangle*

$$Ru_{\mathrm{rig},c}E \to Ru_{K\mathrm{dR}}\mathcal{E} \to Ru'_{K\mathrm{dR}}\mathcal{E}_{|T} \to .$$

Proof This follows easily from Proposition 5.4.17. More precisely, we have an isomorphism

$$R\underline{\Gamma}_{]X[_P}(\mathcal{E} \otimes_{\mathcal{O}_V} \Omega^{\bullet}_{V/S_K}) = [\mathcal{E} \otimes_{\mathcal{O}_V} \Omega^{\bullet}_{V/S_K} \to h_*(\mathcal{E}_{|T} \otimes_{\mathcal{O}_T} \Omega^{\bullet}_{T/S_K})]$$

which we may also see as an exact triangle

$$R\underline{\Gamma}_{]X[_P}(\mathcal{E} \otimes_{\mathcal{O}_V} \Omega^{\bullet}_{V/S_K}) \to \mathcal{E} \otimes_{\mathcal{O}_V} \Omega^{\bullet}_{V/S_K} \to h_*(\mathcal{E}_{|T} \otimes_{\mathcal{O}_T} \Omega^{\bullet}_{T/S_K}) \to .$$

Using one or the other of these two interpretations give the results since h is quasi-Stein. □

Proposition 6.4.4 *Let $(X \subset Y \subset P)$ be an S-frame, V be a strict neighborhood of $]X[_P$ in $]Y[_P$, $\mathcal{E}$ a coherent $\mathcal{O}_V$-module with an integrable connection. Then, the complex*

$$Rsp_*R\underline{\Gamma}_{*]X[_P}(\mathcal{E} \otimes_{\mathcal{O}_V} \Omega^{\bullet}_{V/S_K}),$$

with $sp : V \hookrightarrow]Y[_P \to Y$, only depends on $E := j_X^\dagger \mathcal{E}$ and not on $\mathcal{E}$ or V.

Proof The question is local on Y and we may therefore assume that P is quasi-compact. Then, it follows from Theorem 5.4.4 and Proposition 5.2.15 that this is independent of the choice of $\mathcal{E}$ and V. □

Definition 6.4.5 *Let $(X \subset Y \subset P)$ be an S-frame, V be a strict neighborhood of $]X[_P$ in $]Y[_P$ and $\mathcal{E}$ a coherent $\mathcal{O}_V$-module with an integrable connection.*

Then the specialized de Rham complex with compact support *of $E := j_X^\dagger \mathcal{E}$ is*

$$R\underline{\Gamma}_{\mathrm{rig},c}E := Rsp_*R\underline{\Gamma}_{*]X[_P}(\mathcal{E} \otimes_{\mathcal{O}_V} \Omega^{\bullet}_{V/S_K})$$

with $sp : V \hookrightarrow]Y[_P \to Y$.

And we have as before:

Proposition 6.4.6
Let

$$\begin{array}{ccccc} X & \hookrightarrow & Y & \hookrightarrow & P \\ \downarrow & & \downarrow g & & \downarrow u \\ D & = & D & \hookrightarrow & Q \end{array}$$

be a quasi-compact morphism of S-frames and E be a coherent $j_X^\dagger \mathcal{O}_{]Y[_P}$-module with an integrable connection over S_K. Then

$$Rsp_* Ru_{\mathrm{rig},c} E = Rg_* R\underline{\Gamma}_{\mathrm{rig},c} E.$$

Proof The question is local on D and we may therefore assume that P is quasi-compact. We can find a strict neighborhood V of $]X[_P$ in $]Y[_P$ and a coherent $\mathcal{O}_V$-module $\mathcal{E}$ with an integrable connection such that $E = j_X^\dagger \mathcal{E}$.

We consider the following commutative diagram

$$\begin{array}{ccc} V & \xrightarrow{sp} & Y \\ \downarrow u_K & & \downarrow g \\ ]D[_Q & \xrightarrow{sp} & D. \end{array}$$

Then, we have

$$\begin{aligned} Rsp_* Ru_{\mathrm{rig},c} E &= Rsp_* Ru_{K*} R\underline{\Gamma}_{]X[_P}(\mathcal{E} \otimes_{\mathcal{O}_V} \Omega^\bullet_{V/S_K}) \\ &= Rg_* Rsp_* R\underline{\Gamma}_{]X[_P}(\mathcal{E} \otimes_{\mathcal{O}_V} \Omega^\bullet_{V/S_K}) = Rg_* R\underline{\Gamma}_{\mathrm{rig},c} E. \end{aligned}$$

□

Note in particular that if $(X \subset Y \subset P)$ is a quasi-compact S-frame and E a coherent $\mathcal{O}_V$-module with an integrable connection, we have

$$R\Gamma_{\mathrm{rig},c}(X \subset Y \subset P, E) = R\Gamma(Y, R\underline{\Gamma}_{\mathrm{rig},c} E).$$

Proposition 6.4.7 *Assume that we are given a commutative diagram of morphisms of frames*

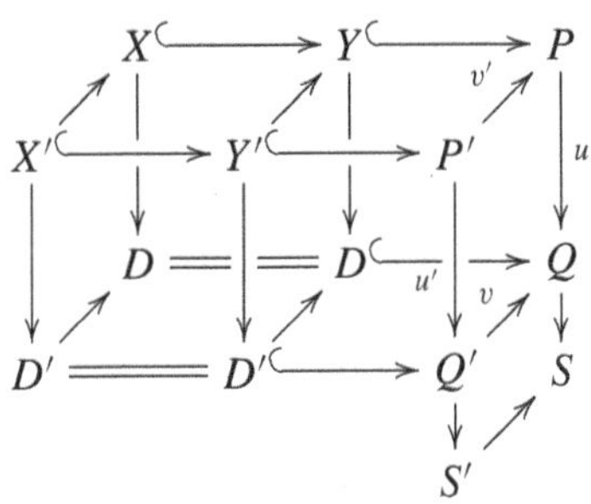

over some morphism $S' \to S$, with u and u' quasi-compact and v' cartesian.

If E is a coherent $j_X^\dagger \mathcal{O}_{]Y[_P}$-module with an integrable connection over S_K, there is a canonical base change map

$$v'^* : Lv_K^* Ru_{\mathrm{rig},c} E \to Ru'_{\mathrm{rig},c} v'^\dagger E$$

with $v_K :]D'[_{Q'} \to]D[_Q$.

Proof First of all note that it is sufficient to define a morphism

$$v_K^{-1} Ru_{\mathrm{rig},c} E \to Ru'_{\mathrm{rig},c} v'^\dagger E$$

and extend it with the functor $\mathcal{O}_{]D'[_{Q'}} \otimes^L_{v_K^{-1}\mathcal{O}_{]D[_Q}} -$.

Now, let V be a strict neighborhood of $]X[_P$ in $]Y[_P$ and $\mathcal{E}$ a coherent $\mathcal{O}_V$-module with an integrable connection such that $E = j_X^\dagger \mathcal{E}$. If $u_K : V \to]D[_Q$ is the map induced by u, we have by definition

$$v_K^{-1} Ru_{\mathrm{rig},c} E = v_K^{-1} Ru_{K*} R\underline{\Gamma}_{]X[_P}(\mathcal{E} \otimes_{\mathcal{O}_V} \Omega^\bullet_{V/S_K}).$$

Let V' be a strict neighborhood of $]X'[_{P'}$ in $u_K^{-1}(V) \cap]Y'[_{P'}$ and $v'_K : V' \to V$ the map induced by v'. Then, we may consider the adjunction map

$$v_K^{-1} Ru_{K*} R\underline{\Gamma}_{]X[_P}(\mathcal{E} \otimes_{\mathcal{O}_V} \Omega^\bullet_{V/S_K}) \to Ru'_{K*} {v'_K}^{-1} R\underline{\Gamma}_{]X[_P}(\mathcal{E} \otimes_{\mathcal{O}_V} \Omega^\bullet_{V/S_K})$$

where $u'_K : V' \to]D'[_{Q'}$ is the map induced by u'. Since v' is a cartesian morphism of frames, we also have thanks to Proposition 5.2.17, a canonical map

$${v'_K}^{-1} R\underline{\Gamma}_{]X[_P}(\mathcal{E} \otimes_{\mathcal{O}_V} \Omega^\bullet_{V/S_K}) \to R\underline{\Gamma}_{]X[_P} {v'_K}^{-1}(\mathcal{E} \otimes_{\mathcal{O}_V} \Omega^\bullet_{V/S_K}).$$

And we also have a canonical morphism

$${v'_K}^{-1}(\mathcal{E} \otimes_{\mathcal{O}_V} \Omega^\bullet_{V/S_K}) \to {v'_K}^* \mathcal{E} \otimes_{\mathcal{O}_{V'}} \Omega^\bullet_{V'/S'_K}.$$

Finally, since $v'^\dagger E = j_X^\dagger {v'_K}^* \mathcal{E}$, we have

$$Ru'_{\mathrm{rig},c} v'^\dagger E = Ru'_{K*} R\underline{\Gamma}_{]X[_P}({v'_K}^* \mathcal{E} \otimes_{\mathcal{O}_{V'}} \Omega^\bullet_{V'/S'_K}).$$

Composing everything gives what we want. □

Note that it is also possible to derive this proposition from the base change theorem in de Rham cohomology.

Proposition 6.4.8 *Let $(X \subset Y \subset P)$ be a quasi-compact S-frame and E a coherent $j_X^\dagger \mathcal{O}_{]Y[_P}$-module with an integrable connection. If $K \hookrightarrow K'$ is a finite* *field extension, there is a canonical isomorphism*

$$K' \otimes_K H^q_{\mathrm{rig},c}(X \subset Y \subset P, E) \simeq H^q_{\mathrm{rig},c}(X_k \subset Y_k \subset P_{\mathcal{V}}, E_{K'}).$$

Proof Using Proposition 6.4.3, this easily follows from Proposition 5.4.16. □

We can also show that rigid cohomology with compact support is local downstairs.

Proposition 6.4.9 *Let*

$$\begin{array}{ccccc} X & \hookrightarrow & Y & \hookrightarrow & P \\ \downarrow{\scriptstyle f} & & \downarrow{\scriptstyle g} & & \downarrow{\scriptstyle u} \\ D & = & D & \hookrightarrow & Q \end{array}$$

be a quasi-compact morphism of S-frames,

$$\begin{array}{ccc} D' & \hookrightarrow & Q' \\ \downarrow & & \downarrow \\ D & \hookrightarrow & Q \end{array}$$

an open or mixed immersion of formal embeddings and

$$\begin{array}{ccccc} X' & \hookrightarrow & Y' & \hookrightarrow & P' \\ \downarrow{\scriptstyle f'} & & \downarrow{\scriptstyle g'} & & \downarrow{\scriptstyle u'} \\ D' & = & D' & \hookrightarrow & Q' \end{array}$$

the pullback of the first morphism of frames along

$$\begin{array}{ccccc} D' & = & D' & \hookrightarrow & Q' \\ \downarrow & & \downarrow & & \downarrow \\ D & = & D & \hookrightarrow & Q. \end{array}$$

If E is a coherent $j_X^\dagger \mathcal{O}_{]Y[_P}$-module with an integrable connection over S_K, then

$$(Ru_{\mathrm{rig},c}E)_{|]D'[_{Q'}} \simeq Ru'_{\mathrm{rig},c}E_{|]Y'[_{P'}}.$$

Proof Let V be a strict neighborhood of $]X[_P$ in $]Y[_P$ and $\mathcal{E}$ a coherent $\mathcal{O}_V$-module with an integrable connection such that $E = j_X^\dagger \mathcal{E}$. If

$$u_K : V \to]D[_Q$$

denotes the induced map and

$$h : T := V\backslash]X[_P \hookrightarrow V$$

the inclusion map, we know that

$$Ru_{\mathrm{rig},c}E = Ru_{K*}[\mathcal{E} \otimes_{\mathcal{O}_V} \Omega^{\bullet}_{V/S_K} \to h_*(\mathcal{E}_{|T} \otimes_{\mathcal{O}_T} \Omega^{\bullet}_{T/S_K})].$$

Moreover, if $V' = V \cap]Y'[_{P'}$, it follows from Proposition 5.1.5 that $E_{|]Y'[_{P'}} = j^{\dagger}_{X'}\mathcal{E}_{|V'}$. Therefore, if $u'_K : V' \to]D'[_{Q'}$ denotes the induced map and

$$h' : T' := V'\backslash]X'[_{P'} \hookrightarrow V'$$

the inclusion map and it follows that

$$Ru'_{\mathrm{rig},c}E_{|]Y'[_{P'}} = Ru'_{K*}[\mathcal{E}_{|V'} \otimes_{\mathcal{O}'_V} \Omega^{\bullet}_{V'/S_K} \to h'_*(\mathcal{E}_{|T'} \otimes_{\mathcal{O}'_T} \Omega^{\bullet}_{T'/S_K})].$$

Since higher direct images commute with open immersions, it is therefore sufficient to verify that if $\mathcal{E}$ is any module on V, we have

$$h'_*\mathcal{E}_{|T'} = (h_*\mathcal{E}_{|T})_{V'}.$$

This follows from the fact that the following diagram is cartesian

$$\begin{array}{ccc} T' & \xhookrightarrow{h'} & V'\backslash]X'[_{P'} \\ \downarrow & & \downarrow \\ T & \xrightarrow{h} & V\backslash]X[_P. \end{array}$$

□

Our definitions are local upstairs for cartesian coverings as we now show:

Proposition 6.4.10 *Let*

$$\begin{array}{ccccc} X & \hookrightarrow & Y & \hookrightarrow & P \\ \downarrow & & \downarrow & & \downarrow u \\ D & = & D & \hookrightarrow & Q \end{array}$$

be a quasi-compact morphism of S-frames and E a coherent $j^{\dagger}_X\mathcal{O}_{]Y[_P}$-module with an integrable connection over S_K. Assume that we are given a cartesian, *open or finite mixed, covering*

$$\begin{array}{ccccc} X_{\underline{i}} & \hookrightarrow & Y_{\underline{i}} & \hookrightarrow & P_{\underline{i}} \\ \downarrow & & \downarrow & & \downarrow \\ X & \hookrightarrow & Y & \hookrightarrow & P. \end{array}$$

Then, there is a canonical spectral sequence

$$E_1^{p,q} := \bigoplus_{\underline{i}} R^q u_{\underline{i}\mathrm{rig},c} E_{|]Y_{\underline{i}}[_{P_{\underline{i}}}} \Rightarrow Ru_{\mathrm{rig},c}E.$$

Proof First of all, we can write $E = j_X^\dagger \mathcal{E}$ with $\mathcal{E}$ on V as usual. We may then consider the spectral sequence for u_{K*} and $R\underline{\Gamma}_{]X[_P}\mathcal{E} \otimes_{\mathcal{O}_V} \Omega^\bullet_{V/S_K}$ with respect to the covering $V = \cup V_i$ with $V_i := V \cap]Y_i[_{P_i}$:

$$E_1^{p,q} := \bigoplus_{\underline{i}} R^q u_{\underline{i}K*}(R\underline{\Gamma}_{]X[_P}\mathcal{E} \otimes_{\mathcal{O}_V} \Omega^\bullet_{V/S_K})_{|V_{\underline{i}}} \Rightarrow Ru_{\mathrm{rig},c}E.$$

It is therefore sufficient to recall from Proposition 5.2.17 that if $\mathcal{E}$ is any sheaf on V, we have

$$(R\underline{\Gamma}_{]X[_P}\mathcal{E})_{|V_{\underline{i}}} = R\underline{\Gamma}_{]X_i[_{P_i}}\mathcal{E}_{|V_{\underline{i}}}.$$

□

Proposition 6.4.11 *Let*

$$\begin{array}{ccccc} X & \hookrightarrow & Y & \hookrightarrow & P \\ \downarrow & & \downarrow & & \downarrow u \\ D & = & D & \hookrightarrow & Q \end{array}$$

be a quasi-compact morphism of S-frames and E be a coherent $j_X^\dagger \mathcal{O}_{]Y[_P}$-module with an integrable connection over S_K.

If $\sigma : K \hookrightarrow K'$ is an isometric embedding, there is a canonical morphism

$$(Ru_{\mathrm{rig},c}E)^\sigma \to Ru^\sigma_{\mathrm{rig},c}E^\sigma.$$

Proof Let V be a strict neighborhood of $]X[_P$ in $]Y[_P$ and $\mathcal{E}$ a coherent $\mathcal{O}_V$-module with an integrable connection such that $E = j_X^\dagger \mathcal{E}$. If we denote by $\varpi : \widetilde{V}^\sigma \to \widetilde{V}$ the morphism of toposes, it is sufficient to describe a morphism

$$\varpi^{-1} Ru_{K*} R\underline{\Gamma}_{]X[_P}(\mathcal{E} \otimes_{\mathcal{O}_V} \Omega^\bullet_{V/S_K}) \to Ru^\sigma_{K*} R\underline{\Gamma}_{]X^\sigma[_{P^\sigma}}(\mathcal{E}^\sigma \otimes_{\mathcal{O}_{V'^\sigma}} \Omega^\bullet_{V'^\sigma/S^\sigma_{K'}}).$$

We use the adjunction map

$$\varpi^{-1} Ru_{K*} R\underline{\Gamma}_{]X[_P}(\mathcal{E} \otimes_{\mathcal{O}_V} \Omega^\bullet_{V/S_K})) \to Ru^\sigma_{K*}\varpi^{-1}(R\underline{\Gamma}_{]X[_P}(\mathcal{E} \otimes_{\mathcal{O}_V} \Omega^\bullet_{V/S_K})),$$

the natural morphism

$$\varpi^{-1} R\underline{\Gamma}_{]X[_P}(\mathcal{E} \otimes_{\mathcal{O}_V} \Omega^{\bullet}_{V/S_K})) \to R\underline{\Gamma}_{]X^{\sigma}[_{P^{\sigma}}} \varpi^{-1}(\mathcal{E} \otimes_{\mathcal{O}_V} \Omega^{\bullet}_{V/S_K})$$

that comes from Proposition 5.2.18 and the canonical map

$$\varpi^{-1}(\mathcal{E} \otimes_{\mathcal{O}_V} \Omega^{\bullet}_{V/S_K}) \to \mathcal{E}^{\sigma} \otimes_{\mathcal{O}_{V^{\sigma}}} \Omega^{\bullet}_{V^{\sigma}/S^{\sigma}_K}.$$

□

Proposition 6.4.12 *Let*

$$\begin{array}{ccccc} X & \hookrightarrow & Y & \hookrightarrow & P \\ \downarrow & & \downarrow & & \downarrow u \\ D & = & D & \hookrightarrow & Q \end{array}$$

be a quasi-compact morphism of S-frames and E be a coherent $j^{\dagger}_X \mathcal{O}_{]Y[_P}$-module with an integrable connection over S_K. Then, there is a canonical morphism

$$Ru_{\mathrm{rig},c} E \to Ru_{\mathrm{rig}} E$$

which is an isomorphism when $Y = X$.

Proof The composite map

$$\underline{\Gamma}_{]X[_P} \to \mathrm{Id} \to j^{\dagger}_X$$

is obviously bijective when $Y = X$. It induces a morphism $R\underline{\Gamma}_{]X[_P} \to j^{\dagger}_X$ which is also bijective when $Y = X$. Our assertion follows. □

Proposition 6.4.13 *Let*

$$\begin{array}{ccccc} X & \hookrightarrow & Y & \hookrightarrow & P \\ \downarrow & & \downarrow & & \downarrow u \\ D & = & D & \hookrightarrow & Q \end{array}$$

be a quasi-compact morphism of S-frames and E be a coherent $j^{\dagger}_X \mathcal{O}_{]Y[_P}$-module with an integrable connection over S_K. Let $X' \hookrightarrow X$ be a closed immersion, $U := X \setminus X'$ and Y' a closed subvariety of Y such that $X' = Y \cap X$.

Then, there is an exact triangle

$$Ru_{\mathrm{rig},c} j^{\dagger}_U E \to Ru_{\mathrm{rig},c} E \to Ru_{\mathrm{rig},c} E_{|]Y'[} \to .$$

Proof Using the definition, this is an immediate consequence of Proposition 5.2.19. □

We finish with Poincaré duality.

Proposition 6.4.14 *Let*

$$\begin{array}{ccccc} X & \hookrightarrow & Y & \hookrightarrow & P \\ \downarrow & & \downarrow & & \downarrow u \\ D & = & D & \hookrightarrow & Q \end{array}$$

be a quasi-compact morphism of S-frames and E be a coherent $j_X^\dagger \mathcal{O}_{]Y[_P}$-module with an integrable connection over S_K and dual module $E^\vee$. Then, there is a canonical morphism

$$Ru_{\mathrm{rig}} E^\vee \otimes^L_{\mathcal{O}_{S_K}} Ru_{\mathrm{rig},c} E \to Ru_{\mathrm{rig},c} j_X^\dagger \mathcal{O}_{]Y[_P}.$$

More generally, if X' is a closed subset of X, there is a canonical morphism

$$Ru_{\mathrm{rig},X'} E^\vee \otimes^L_{\mathcal{O}_{S_K}} Ru_{\mathrm{rig},c} E_{|]Y'[_P} \to Ru_{\mathrm{rig},c} j_X^\dagger \mathcal{O}_{]Y[_P}$$

where Y' is a closed subvariety of Y with $X' = Y' \cap X$.

Proof Let $\mathcal{E}$ be a coherent module with an integrable connection on a strict neighborhood V of $]X[_P$ in $]Y[_P$ such that $E = j_X^\dagger \mathcal{E}$. We may consider the map induced by multiplication

$$(\mathcal{E}^\vee \otimes_{\mathcal{O}_V} \Omega^\bullet_V) \otimes_{\mathcal{O}_{S_K}} (\mathcal{E} \otimes_{\mathcal{O}_V} \Omega^\bullet_V) \to \Omega^\bullet_V.$$

It follows from Corollary 5.3.6 that this map induces a map

$$(j_X^\dagger \mathcal{E}^\vee \otimes_{\mathcal{O}_V} \Omega^\bullet_V) \otimes^L_{\mathcal{O}_{S_K}} (R\Gamma_{]X[_P} \mathcal{E} \otimes_{\mathcal{O}_V} \Omega^\bullet_V) \to R\Gamma_{]X[_P} \Omega^\bullet_V.$$

Recall that if $\mathcal{F}$ and $\mathcal{G}$ are two $\mathcal{O}_{S_K}$-modules on V, there is a canonical map

$$u_{K*}\mathcal{F} \otimes_{\mathcal{O}_{S_K}} u_{K*}\mathcal{G} \to u_{K*}(\mathcal{F} \otimes_{\mathcal{O}_{]D[_Q}} \mathcal{G}).$$

We may therefore apply Ru_{K*} on each factor and obtain

$$Ru_{\mathrm{rig}} E^\vee \otimes^L_{\mathcal{O}_{S_K}} Ru_{\mathrm{rig},c} E \to Ru_{\mathrm{rig},c} j_X^\dagger \mathcal{O}_{]Y[_P}.$$

Of course, if X' is a closed subset of X, we may consider the induced map

$$(\underline{\Gamma}^\dagger_{X'} j_X^\dagger \mathcal{E}^\vee \otimes_{\mathcal{O}_V} \Omega^\bullet_V) \otimes^L_{\mathcal{O}_{S_K}} (R\Gamma_{]X'[_P} \mathcal{E} \otimes_{\mathcal{O}_V} \Omega^\bullet_V) \to R\Gamma_{]X[_P} \Omega^\bullet_V$$

and get the expected pairing as before. □

Until the end of the section, we assume that $\mathrm{Char} K = 0$.

There is a down-to-earth description of rigid cohomology with compact support in the Monsky–Washnitzer setting.

Proposition 6.4.15 *Let $X = \operatorname{Spec} A$ be a smooth affine $\mathcal{V}$-scheme of pure dimension d and Y the closure of X in some projective space for a given presentation of A. If E is a coherent $j_X^\dagger \mathcal{O}_{Y_K^{\mathrm{rig}}}$-module with an integrable connection over S_K and $M := \Gamma(Y_K^{\mathrm{rig}}, E)$, we have*

$$R\Gamma_{\mathrm{rig,c}}(X_k \subset Y_k \subset \widehat{Y}, E) \simeq \Gamma_{\mathrm{dR}}(M_c)[-d].$$

Proof By definition, if $\mathcal{E}$ is a coherent module with an integrable connection on some strict neighborhood V of $\widehat{X}_K$ in X_K^{rig} such that $E = j^\dagger \mathcal{E}$, we have

$$R\Gamma_{\mathrm{rig,c}}(X_k \subset Y_k \subset \widehat{Y}, E)$$
$$= R\Gamma(V, R\underline{\Gamma}_{\widehat{X}_K}\mathcal{E} \otimes_{\mathcal{O}_V} \Omega_V^\bullet) = R\Gamma_{\widehat{X}_K}(V, \mathcal{E} \otimes_{\mathcal{O}_V} \Omega_V^\bullet).$$

It follows from Proposition 5.2.21 that

$$H^i_{\widehat{X}_K}(V, \mathcal{E} \otimes_{\mathcal{O}_V} \Omega^j_V) = 0$$

unless $i = d$ and that

$$H^d_{\widehat{X}_K}(V, \mathcal{E} \otimes_{\mathcal{O}_V} \Omega^j_V) = M_c \otimes_{A_K^\dagger} \Omega^j_{A_K^\dagger}.$$

□

It is time to say a few more words about the Robba ring $\mathcal{R}$ and study its relation to rigid cohomology. First of all, recall that the *de Rham cohomology (or complex)* of a coherent $\mathcal{R}$-module M of finite rank with a connection ∇ is

$$\Gamma_{\mathrm{dR}} M := [M \xrightarrow{\nabla} M dt/t].$$

In other words, we have an exact sequence

$$0 \to H^0_{\mathrm{dR}}(M) \to M \xrightarrow{\nabla} M dt/t \to H^1_{\mathrm{dR}}(M) \to 0$$

Proposition 6.4.16 *Let $\mathcal{Y}$ be a flat formal $\mathcal{V}$-scheme whose special fiber Y is a smooth connected curve and X a dense open subset of Y such that all points of $Y \setminus X$ are rational.*

Let E be a coherent $j_X^\dagger \mathcal{O}_{\mathcal{Y}_K}$-module with a (integrable) connection. Then, there is a long exact sequence

$$0 \to H^0_{\mathrm{rig}}(E) \to \oplus_{x \notin X} H^0_{\mathrm{dR}}(E_x) \to H^1_{\mathrm{rig},c}(E) \to$$
$$H^1_{\mathrm{rig}}(E) \to \oplus_{x \notin X} H^1_{\mathrm{dR}}(E_x) \to H^2_{\mathrm{rig},c}(E) \to 0$$

where E_x denotes the stalk of E on the Robba ring at x.

Proof We first choose a strict neighborhood V of $]X[_{\mathcal{Y}}$ in $\mathcal{Y}_K$ and a coherent $\mathcal{O}_V$-module with an integrable connection such that $E = j^\dagger \mathcal{E}$. If

$h : V \setminus]X[_\mathcal{Y} \hookrightarrow V$ denotes the inclusion map, we saw in Proposition 5.4.17 that there is an exact triangle

$$R\underline{\Gamma}_{]X[}(\mathcal{E} \otimes_{\mathcal{O}_V} \Omega^\bullet_{V/S_K}) \to (\mathcal{E} \otimes_{\mathcal{O}_V} \Omega^\bullet_{V/S_K}) \to h_* h^{-1}(\mathcal{E} \otimes_{\mathcal{O}_V} \Omega^\bullet_{V/S_K}) \to .$$

Since $j_X^\dagger$ is exact and $R\underline{\Gamma}_{]X[}\mathcal{E}$ is always overconvergent, we deduce another exact triangle

$$R\underline{\Gamma}_{]X[}(\mathcal{E} \otimes_{\mathcal{O}_V} \Omega^\bullet_{V/S_K}) \to j_X^\dagger(\mathcal{E} \otimes_{\mathcal{O}_V} \Omega^\bullet_{V/S_K}) \to j_X^\dagger h_* h^{-1}(\mathcal{E} \otimes_{\mathcal{O}_V} \Omega^\bullet_{V/S_K}) \to$$

and therefore also an exact triangle

$$R\Gamma_{\mathrm{rig},c} E \to R\Gamma_{\mathrm{rig}} E \to R\Gamma(V, j_X^\dagger h_* h^{-1}(\mathcal{E} \otimes_{\mathcal{O}_V} \Omega^\bullet_{V/S_K})) \to .$$

This gives a long exact sequence. Now, it easily follows from Proposition 5.4.9 that

$$R\Gamma(V, j_X^\dagger h_* h^{-1}(\mathcal{E} \otimes_{\mathcal{O}_V} \Omega^\bullet_{V/S_K})) = \oplus_{x \notin X} \Gamma_{\mathrm{dR}} E_x .$$

□

Note that we may always assume that all points of $Y \setminus X$ are rational after a finite extension of K.

In the Monsky–Washnitzer setting, we can improve a little bit on Poincaré duality. First of all, we may define the trace map.

Proposition 6.4.17 *Let $X = \operatorname{Spec} A$ be a smooth affine $\mathcal{V}$-scheme of dimension d and $X \hookrightarrow \mathbf{A}^N_\mathcal{V}$ a closed embedding. Then, the trace map*

$$\int_A : \omega_{A,c} \to K$$

on higher differentials with compact support factors through $H^d_{\mathrm{dR}}(A_c)$ and induces a trace map

$$\mathrm{tr} : H^d_{\mathrm{dR}}(A_c) \to K.$$

Proof Since the Gysin map commutes with the differential, we may assume that $X = \mathbf{A}^N_\mathcal{V}$. If $f \in K[\underline{t}]_c$, we have

$$\int d(f dt_1 \wedge \cdots \wedge \widehat{dt_k} \wedge \cdots \wedge dt_N) = \int \partial f/\partial t_k \underline{dt}.$$

But, if $f = \sum_{\underline{i}<0} a_{\underline{i}} \underline{t}^{\underline{i}}$, then $\partial f/\partial t_k = \sum_{\underline{i}<0} i_k a_{\underline{i}} \underline{t}^{\underline{i}-1_k}$ whose first coefficient is zero. □

Proposition 6.4.18 *Let $X = \operatorname{Spec} A$ be a smooth affine $\mathcal{V}$-scheme of pure dimension d. If M is a locally free $A_K^\dagger$-module of finite rank with an integrable* *connection, Serre duality induces a topologically perfect paring of complexes*

$$\Gamma_{\mathrm{dR}} M^{\vee} \times \Gamma_{\mathrm{dR}} M_c \to K[-d].$$

Proof For each $i = 1, \ldots, d$, we apply Serre duality to

$$N_i := M \otimes_{A_K^\dagger} \Omega^{d-i}_{A_K^\dagger} \otimes_{A_K^\dagger} (\Omega^d_{A_K^\dagger})^{\vee}.$$

We have

$$N_i^{\vee} \simeq M^{\vee} \otimes_{A_K^\dagger} \Omega^i_{A_K^\dagger}$$

and

$$N^i \otimes \omega_c \simeq M_c \otimes_{A_K^\dagger} \Omega^{d-i}_{A_K^\dagger}.$$

Thus, we get a perfect paring

$$(M^{\vee} \otimes_{A_K^\dagger} \Omega^i_{A_K^\dagger}) \times (M_c \otimes_{A_K^\dagger} \Omega^{d-i}_{A_K^\dagger}) \to K.$$

It remains to show that they are compatible with the differentials. But this easily follows from the construction. □

We should stress at this point the fact that this perfect duality is just topological Serre duality and only holds at the cochain level.

The next proposition gives a characterization of the trace map on cohomology. More precisely, the second assertion tells us that it is unique up to a constant and the first one tells us what this constant is.

Proposition 6.4.19 *Let $X = \operatorname{Spec} A$ be a smooth affine $\mathcal{V}$-scheme of pure dimension d. Then,*

(i) *If x is a point of $\widehat{X}_K$ and*

$$g_x : K(x) \to \omega_{A,c},$$

denotes the Gysin map coming from the inclusion $x \hookrightarrow \hat{X}_K$, then the trace map

$$\mathrm{tr}_{K(x)/K} : K(x) \to K$$

sends α to $\int_A g_x(\alpha)$.

(ii) *If X is geometrically connected, the trace map*

$$\mathrm{tr}_X : H^d_{\mathrm{dR}}(A_c) \to K$$

is, up to a constant, the unique non-trivial continuous linear form on $H^d_{\mathrm{dR}}(A_c)$.

Proof The first assertion follows from the second part of Proposition 5.4.20.

We turn now to the second assertion. From the first part, we know that the trace map is non trivial and it is clearly linear and continuous. Since topological duality is left exact, the topological dual of $H^d_{\mathrm{dR}}(A_c)$ is $H^0_{\mathrm{dR}}(A^\dagger_K)$ which is one dimensional by Proposition 6.2.11. It exactly means that, up to multiplication by a constant, there exists a unique non trivial continuous linear map $H^d_{\mathrm{dR}}(A_c) \to K$. □

We finish with Poincaré duality on the Robba ring.

Proposition 6.4.20 *If $\mathcal{R}$ denotes the Robba ring on K and M is a free $\mathcal{R}$-module of finite rank with a connection, Serre duality induces a topologically perfect paring of complexes*

$$\Gamma_{\mathrm{dR}} M^\vee \times \Gamma_{\mathrm{dR}} M \to K[-1].$$

Proof This follows from Serre duality on the Robba ring that was shown in Proposition 5.4.22. □

6.5 Comparison theorems

We assume in this section that $\mathrm{Char} K = 0$ and fix a formal $\mathcal{V}$-scheme S.

We start with a technical lemma.

Lemma 6.5.1 *Let $(X \subset Y \subset P)$ be an S-frame, $X \hookrightarrow Y'$ an open immersion and $h : Y' \to Y$ be a projective morphism which induces the identity on X. Then, locally on $(X \subset Y \subset P)$, there exists a closed subvariety Y'' of Y' containing X such that the morphism induced by h extends to a proper étale morphism of frames*

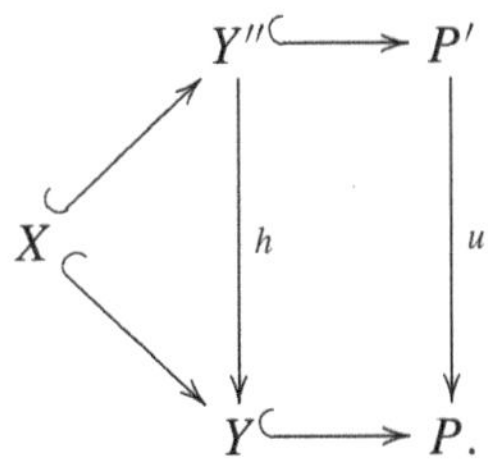

Proof Since the question is local, we may assume that P is affine. Since h is projective, there exists a commutative diagram

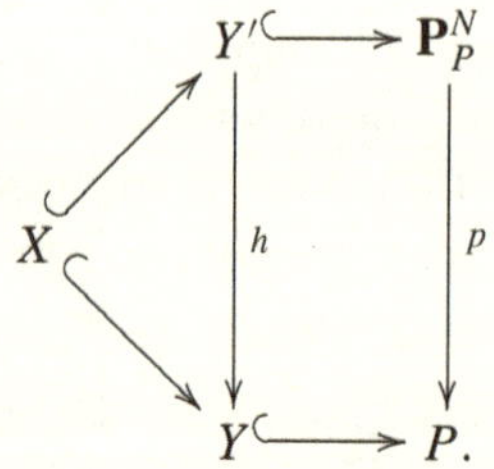

The closed immersion $X \hookrightarrow p^{-1}(X) = \mathbf{P}^N_X$ is a section of the canonical projection which is smooth. It follows that, locally on X, there exists an open subset U of $\mathbf{P}^N_X$ such that X is defined in U by a regular sequence. We may of course assume that $U = D^+(s) \cap p^{-1}(Y)$ with $s \in \Gamma(\mathbf{P}^N_P, \mathcal{O}(m))$ for some m and that the regular sequence is induced by $t_1, \ldots, t_d \in \Gamma(\mathbf{P}^N_P, \mathcal{O}(n))$ for some n. We put $P' := V(t_1, \ldots, t_d)$ and $Y'' = Y' \cap P'$. □

Theorem 6.5.2 *Let*

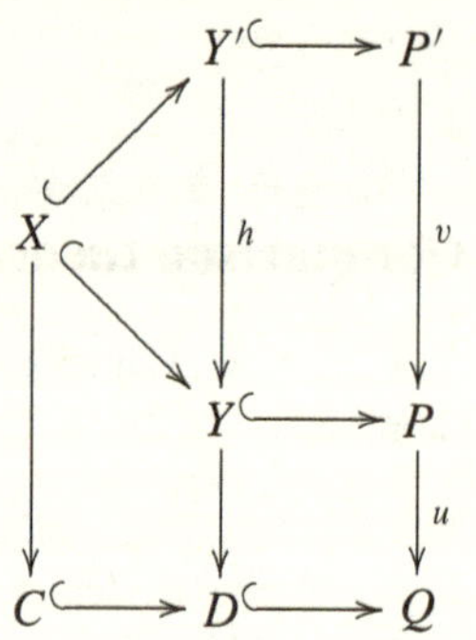

be a sequence of morphisms of S-frames, the upper one being a proper smooth morphism of smooth S-frames.

If E is a coherent $j_X^\dagger \mathcal{O}_{]Y[_P}$*-module with an integrable connection over* S_K*, the base change morphism is an isomorphism*

$$v^* : Ru_{\mathrm{rig}}E \simeq R(u \circ v)_{\mathrm{rig}} v^\dagger E.$$

Recall that our assumptions mean that h is proper and v is smooth in a neighborhood of X.

Proof We show here that the theorem can be deduced from Proposition 6.5.3 below which is simply the case where the lower morphism is the identity.

Clearly, if we have

$$E \otimes_{\mathcal{O}_{]Y[_P}} \Omega^{\bullet}_{]Y[_P/S_K} \simeq Rv_{K*}(v_K^* E \otimes_{\mathcal{O}_{]Y[_{P'}}} \Omega^{\bullet}_{]Y[_{P'}/S_K})$$

and we apply Ru_{K*}, we obtain the isomorphism of the theorem. □

Proposition 6.5.3 *Let*

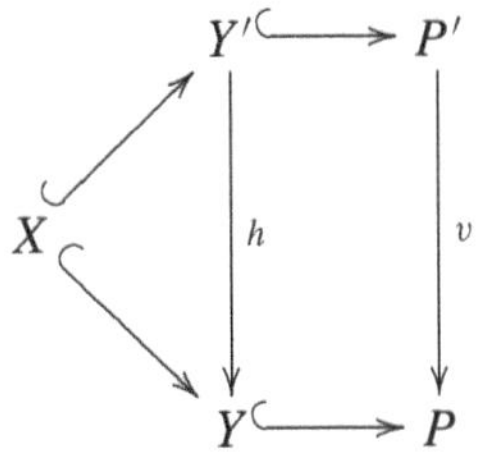

be a proper smooth morphism of smooth S-frames. If E is a coherent $j_X^\dagger \mathcal{O}_{]Y[_P}$-module with an integrable connection over S_K, the base change map is an isomorphism

$$v_* : E \otimes_{\mathcal{O}_{]Y[_P}} \Omega^{\bullet}_{]Y[_P/S_K} \simeq Rv_{K*} v^\dagger E \otimes_{\mathcal{O}_{]Y'[_{P'}}} \Omega^{\bullet}_{]Y'[_{P'}/S_K}.$$

Proof We do not prove directly the proposition but only show that it can be deduced from Lemma 6.5.4 below which is the special case $h = \mathrm{Id}_Y$. Note that in this lemma, we have $v_K^* E = v^\dagger E$ because the morphism of frames is trivially cartesian.

We first reduce to the projective case. Since the question is local on $(X \subset Y \subset P)$, we may assume that X is quasi-projective. Then, Chow's lemma (Corollary 5.7.14 of [75]), tells us that we may blow up a closed subvariety of Y' outside X in P' and obtain a projective morphism

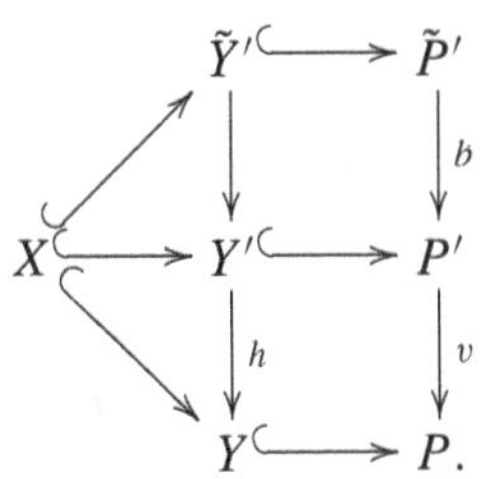

Then, Proposition 3.1.13 tells us that our blowing up induces an isomorphism on strict neighborhoods and therefore (see Proposition 6.2.2) an isomorphism

$$b^* : v^\dagger E \otimes_{\mathcal{O}_{]Y'[_P}} \Omega^{\bullet}_{]Y'[_P/S_K} \simeq Rb_{K*}(b^\dagger v^\dagger E \otimes_{\mathcal{O}_{]\tilde{Y}'[_{\tilde{P}'}}} \Omega^{\bullet}_{]\tilde{Y}'[_{\tilde{P}'}/S_K}).$$

Thus the proposition directly results from the projective case that gives us

$$(v \circ b)^* : E \otimes_{\mathcal{O}_{]Y[_P}} \Omega^\bullet_{]Y[_P/S_K} \simeq Rv_{K*}Rb_{K*}b^\dagger v^\dagger E \otimes_{\mathcal{O}_{]\tilde{Y}'[_{\tilde{P}'}}} \Omega^\bullet_{]\tilde{Y}'[_{\tilde{P}'}/S_K}.$$

Thus, we may assume that h is projective. Thanks to Lemma 6.5.1, since the question is local on $(X \subset Y \subset P)$ and only depends on a closed subset of Y containing X, we can extend h to a proper étale map

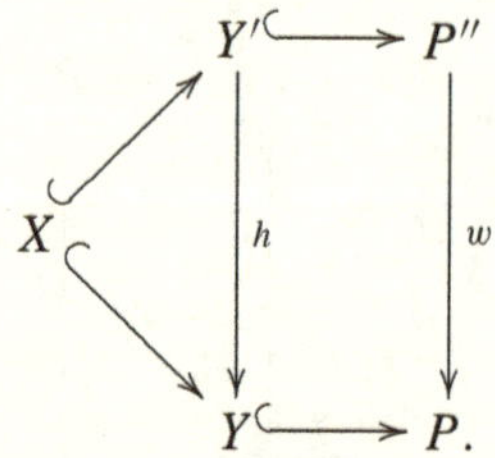

It induces an isomorphism on strict neighborhoods by Theorem 3.4.12 and it follows from Proposition 6.2.2 that

$$w^* : E \otimes_{\mathcal{O}_{]Y[_P}} \Omega^\bullet_{]Y[_P/S_K} \simeq Rw_{K*}(w^\dagger E \otimes_{\mathcal{O}_{]Y'[_{P''}}} \Omega^\bullet_{]Y'[_{P''}/S_K}).$$

We may also consider the diagonal embedding $Y' \hookrightarrow P''' := P' \times_P P''$ and the projections p_1 and p_2. Thanks to Lemma 6.5.4, we have

$$w^\dagger E \otimes_{\mathcal{O}_{]Y'[_{P''}}} \Omega^\bullet_{]Y'[_{P''}/S_K} \simeq Rp_{2K*}(p_2^\dagger w^\dagger E \otimes_{\mathcal{O}_{]Y'[_{P'''}}} \Omega^\bullet_{]Y'[_{P'''}/S_K})$$

and

$$v^\dagger E \otimes_{\mathcal{O}_{]Y'[_{P'}}} \Omega^\bullet_{]Y'[_{P''}/S_K} \simeq Rp_{1K*}(p_1^\dagger v^\dagger E \otimes_{\mathcal{O}_{]Y'[_{P'''}}} \Omega^\bullet_{]Y'[_{P'''}/S_K}).$$

Using the fact that $w \circ p_1 = v \circ p_2$, we get

$$E \otimes \Omega_{]Y[_P/S_K} \simeq Rw_{K*}Rp_{2K*}(p_2^\dagger w^\dagger E \otimes_{\mathcal{O}_{]Y'[_{P'''}}} \Omega^\bullet_{]Y'[_{P'''}/S_K})$$

$$\simeq Rv_{K*}Rp_{1K*}(p_1^\dagger v^\dagger E \otimes_{\mathcal{O}_{]Y'[_{P'''}}} \Omega^\bullet_{]Y'[_{P'''}/S_K}) \simeq Rv_{K*}v^\dagger E \otimes_{\mathcal{O}_{]Y'[_{P'}}} \Omega^\bullet_{]Y'[_{P''}/S_K}.$$

□

Thus, we have to prove the following result:

Lemma 6.5.4 *Let*

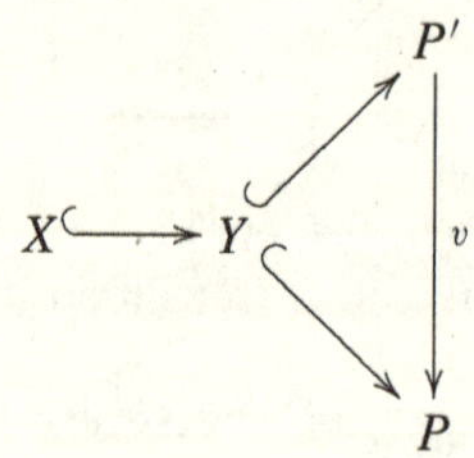

be a smooth morphism of smooth S-frames. If E is a coherent $j_X^\dagger \mathcal{O}_{]Y[_P}$-module with an integrable connection over S_K, the base change map is an isomorphism

$$v* : E \otimes_{\mathcal{O}_{]Y[_P}} \Omega^\bullet_{]Y[_P/S_K} \simeq Rv_{K*}(v_K^* E \otimes_{\mathcal{O}_{]Y[_{P'}}} \Omega^\bullet_{]Y[_{P'}/S_K})$$

with $v_K :]Y[_{P'} \to]Y[_P$.

Proof Again, we do not really prove the lemma but we show that it results from the *Global Poincaré Lemma* 6.5.5 below which is simply the case $S = P$.

Actually, this is a consequence of the Gauss–Manin construction. More precisely, the complex

$$K^\bullet := v_K^* E \otimes_{\mathcal{O}_{]Y[_{P'}}} \Omega^\bullet_{]Y[_{P'}/S_K}$$

comes with its Gauss–Manin filtration and we have

$$Gr^k Rv_{K*}K^\bullet = Rv_{K*}L^\bullet[-k] \otimes_{\mathcal{O}_{]Y[_P}} \Omega^k_{]Y[_P/S_K}$$

with

$$L^\bullet := v_K^* E \otimes_{\mathcal{O}_{]Y[_{P'}}} \Omega^\bullet_{]Y[_{P'}/]Y[_P}.$$

The Global Poincaré Lemma tells us that $E = Rv_{K*}L^\bullet$ and therefore

$$Gr^k Rv_{K*}K^\bullet = E[-k] \otimes_{\mathcal{O}_{]Y[_{P'}}} \Omega^k_{]Y[_P/S_K}.$$

which implies that

$$Rv_{K*}K^\bullet = E \otimes_{\mathcal{O}_{]Y[_{P'}}} \Omega^\bullet_{]Y[_P/S_K}$$

as expected. □

So now, it remains to prove the Global Poincaré Lemma.

Lemma 6.5.5 *(Global Poincaré Lemma) Let*

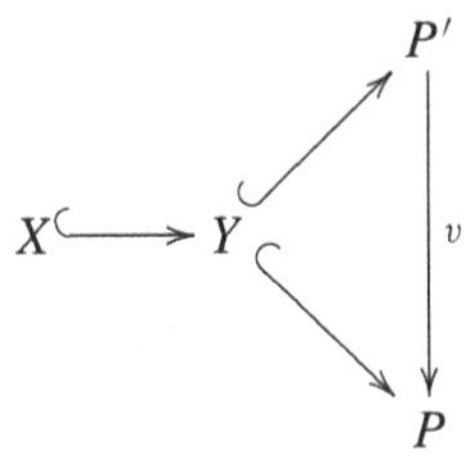

a smooth morphism of frames. If E is a coherent $j_X^\dagger \mathcal{O}_{]Y[_P}$-module, there is a canonical isomorphism

$$E \simeq Rv_{K*}(v_K^* E \otimes_{\mathcal{O}_{]Y[_{P'}}} \Omega^\bullet_{]Y[_{P'}/]Y[_P}).$$

In order prove this, we will use induction on the relative dimension. We will need the following result:

Lemma 6.5.6 *If we are given another smooth morphism*

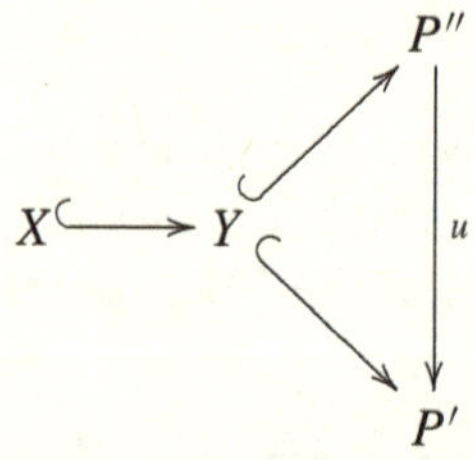

of frames and both u and v satisfy the Global Poincaré Lemma, so does $v \circ u$.

Proof The proof goes exactly as in the previous lemma. With

$$K^\bullet := (v \circ u)_K^* E \otimes_{\mathcal{O}_{]Y[_{P'''}}} \Omega^\bullet_{]Y[_{P''}/]Y[_P}$$

and

$$L^\bullet := (v \circ u)_K^* E \otimes_{\mathcal{O}_{]Y[_{P''}}} \Omega^\bullet_{]Y[_{P'}/]Y[_{P'}}$$

we have

$$Gr^k Ru_{K*} K^\bullet = Ru_{K*} L^\bullet[-k] \otimes_{\mathcal{O}_{]Y[_{P'}}} \Omega^k_{]Y[_{P'}/]Y[_P}.$$

Since we assume that v satisfies the Global Poincaré Lemma, we get

$$Gr^k Ru_{K*} K^\bullet = v_K^* E[-k] \otimes_{\mathcal{O}_{]Y[_{P'}}} \Omega^k_{]Y[_P/]Y[_P}$$

and therefore,

$$Ru_{K*} K^\bullet = v_K^* E \otimes_{\mathcal{O}_{]Y[_{P'}}} \Omega^\bullet_{]Y[_P/]Y[_P}.$$

Since we also assume that u satisfies the Global Poincaré Lemma, we get

$$R(v \circ u)_{K*} K^\bullet = Rv_{K*}(v_K^* E \otimes_{\mathcal{O}_{]Y[_{P'}}} \Omega^\bullet_{]Y[_P/]Y[_P}) = E.$$

□

Proof (of the Global Poincaré Lemma 6.5.5) Again, this is going to be a reduction. More precisely, we will show that it is sufficient to consider the projection of the relative affine line. The question is local on $]Y[_P$. We may therefore assume that P and P' are affine. It follows from Proposition 5.2.8 that our assertion is local on X also. We may therefore assume that there exists $t_1, \ldots, t_d$ in the ideal I' of Y in P' that give a basis $(\bar{t}_1, \ldots, \bar{t}_d)$ of the conormal sheaf $\omega_{X'/X}$ of X in $X' := u^{-1}(X)$. Then, we know from Proposition 3.3.13

that if we embed Y into $\widehat{\mathbf{A}}^d_P$ using the zero section, then $(t_1, \ldots, t_d)$ defines an étale morphism of frames

$$\begin{array}{ccccc} & & & & P' \\ & & & \nearrow & \downarrow \\ X & \hookrightarrow & Y & & \downarrow \\ & & & \searrow & \downarrow \\ & & & & \widehat{\mathbf{A}}^d_P. \end{array}$$

Since the case of a quasi-compact étale morphism obviously results from Theorem 3.4.12, we may assume thanks to Lemma 6.5.6 that $P' = \widehat{\mathbf{A}}^d_P$. Using the same lemma, we may even assume that $d = 1$. In other words, we are reduced to the Local Poincaré Lemma 6.5.7 below. □

Lemma 6.5.7 *(Local Poincaré Lemma*[†]*) Let $(X \subset Y \subset P)$ be a strictly local S-frame and*

$$p : \widehat{\mathbf{A}}^1_P \to P$$

the projection. If E is a coherent $j_X^\dagger \mathcal{O}_{]Y[_P}$-module and W an affinoid subset of $]Y[_P$, there is a canonical isomorphism

$$\Gamma(W, E) \simeq R\Gamma(W \times \mathbf{D}(0, 1^-), p_K^* E \overset{\partial/\partial t}{\to} p_K^* E).$$

Proof It follows from Theorem 5.4.4 that there exists a strict neighborhood V of $]X[_P$ in $]Y[_P$ and a coherent $\mathcal{O}_V$-module $\mathcal{E}$ on V such that $E = j_X^\dagger \mathcal{E}$. We fix a closed complement Z for X in Y and we let as usual $V^\lambda :=]Y[_P \backslash]Z[_{P\lambda}$ for $\lambda < 1$. Since W is quasi-compact, there exists $\lambda_0 < 1$ such that $W \cap V_\lambda \subset V$ when $\lambda_0 \leq \lambda < 1$. Then, it follows from the second assertion of Proposition 5.1.12 that, if $M := \Gamma(W, E)$, we have

$$M = \varinjlim_{\lambda} M^\lambda$$

with $M^\lambda := \Gamma(W \cap V^\lambda, \mathcal{E})$.

Now, we choose a sequence $\eta_k \overset{<}{\to} 1$ and we define

$$M_k := \Gamma(W \times \mathbf{D}(0, \eta_k^+), p_K^* E).$$

If we let $A^\lambda = \Gamma(W \cap V^\lambda, \mathcal{E})$ and

$$M_k^\lambda := M^\lambda \otimes_{A^\lambda} A^\lambda\{t/\eta_k\},$$

[†] this lemma implies that assertion (*) in the proof of Proposition 8.3.5 of [26] is valid but their proof is not correct.

Proposition 5.1.12 again tells us that for each $k \in \mathbf{N}$, we have

$$M_k := \varinjlim_{\lambda} M_k^{\lambda}.$$

It follows from Proposition 5.4.14 that we can use the affinoid covering

$$W \times \mathbf{D}(0, 1^+) = \cup_k \left(W \times \mathbf{D}(0, \eta_k^+)\right)$$

in order to compute cohomology of coherent $j_X^\dagger \mathcal{O}_{]Y[_P}$-sheaves. Thus, Lemma 6.5.10 below tells us that the right-hand side of our isomorphism is the complex associated to the bicomplex

$$\begin{array}{ccc} \prod_k M_k & \xrightarrow{\partial} & \prod_k M_k \\ \downarrow d & & \downarrow d \\ \prod_k M_k & \xrightarrow{\partial} & \prod_k M_k \end{array}$$

with

$$d(s_k) = (s_{k+1} - s_k) \qquad \text{and} \qquad \partial(s_k) = (\partial/\partial t(s_k)).$$

Note that, if we still denote by $d : M^{\mathbf{N}} \to M^{\mathbf{N}}$ the map induced on constant terms, there is a short exact sequence

$$0 \to M \xrightarrow{\delta} M^{\mathbf{N}} \xrightarrow{d} M^{\mathbf{N}} \to 0.$$

Exactness on the left should be clear and one can check that the map

$$\begin{array}{rcl} M^{\mathbf{N}} & \xrightarrow{h} & M^{\mathbf{N}} \\ (s_k) & \longmapsto & (-\sum_{i=0}^{k-1} s_i) \end{array}$$

is a section for d:

$$(d \circ h)(s_k) = d(-\sum_{i=0}^{k-1} s_i) = (-\sum_{i=0}^{k-1} s_i + \sum_{i=0}^{k} s_i) = (s_k).$$

We will consider the "inclusion" map $i : M^{\lambda} \hookrightarrow M_k^{\lambda}$, the "constant" map

$$\begin{array}{rcl} M_k^{\lambda} & \xrightarrow{c} & M^{\lambda} \\ t^i & \longmapsto & \begin{cases} 0 & \text{if} \quad i > 0 \\ 1 & \text{if} \quad i = 0 \end{cases} \end{array}$$

and the "integration" map

$$\begin{aligned} M_k^\lambda &\xrightarrow{\quad I \quad} M_{k-1}^\lambda \\ t^i &\longmapsto t^{i+1}/i+1. \end{aligned}$$

Note that this last map is well defined, because we choose a (strictly) increasing sequence $\eta_k \overset{<}{\to} 1$ and therefore

$$\left(|a_i|\eta_k^i \to 0\right) \Rightarrow \left(|\frac{a_{i-1}}{i}|\eta_{k-1}^i \to 0\right).$$

Of course, when $k=0$, we set I to be the 0 map.

Taking limits, we can extend i, c and I to maps

$$i : M \hookrightarrow M_k, \quad c : M_k \to M \text{ and } \quad I : M_k \to M_{k-1}.$$

Taking products, we obtain maps

$$i : M^{\mathbf{N}} \hookrightarrow \prod_k M_k,$$

$$c : \prod_k M_k \to M^{\mathbf{N}}$$

and

$$I : \prod_k M_k \to \prod_k M_k.$$

Claim 6.5.8 *We have the following formulas:*

$$d \circ I - I \circ d = 0, \quad \partial \circ I - I \circ \partial = i \circ c \quad \text{ and } \quad \partial \circ I - d = \mathrm{Id}.$$

The first equality follows from the fact that I is compatible with the inclusions $M_{k+1} \subset M_k$. The second one can be checked on each M_k. Clearly, t^i is sent to itself both by $\partial \circ I$ and $I \circ \partial$ unless $i=0$ and we have

$$(\partial \circ I)(1) = 1 \quad \text{and} \quad (I \circ \partial)(1) = 0.$$

This shows the second assertion. But we also have $(\partial \circ I)(s_k) = (s_{k+1})$ from which the last formula follows.

Now, the lemma is equivalent to the following claim:

Claim 6.5.9 *The sequence*

$$0 \to M \overset{i\circ\delta}{\to} \prod_k M_k \overset{(d,\partial)}{\longrightarrow} \prod_k M_k \oplus \prod_k M_k \overset{\partial - d}{\longrightarrow} \prod_k M_k \to 0$$

is exact.

It is clearly a complex because $d \circ \delta \circ i = \partial \circ \delta \circ i$ and the first map is injective. Moreover, since we always have $s = \partial(I(s)) - d(s)$, we see that the last map is surjective. But it also follows that when $d(s) = \partial(s) = 0$, we have

$$s = (\partial \circ I)(s) = (I \circ \partial)(s) + (i \circ c)(s) = (i \circ c)(s).$$

Thus, we see that

$$i(d(c(s))) = d(i(c(s))) = d(s) = 0.$$

Since i is injective, we obtain $d(c(s)) = 0$ and therefore $c(s) = \delta(m)$. In other words, we proved that if $d(s) = \partial(s) = 0$, we have $s = (\delta \circ i) - (m)$ and it follows that our sequence is exact in the second term.

It remains to show exactness in the third term. In other words, we assume that $\partial(s) = d(s')$ and we hunt for an s'' such that $d(s'') = s$ and $\partial(s'') = s'$. It is not difficult to find a good candidate for s'' because

$$\partial(s) = d(s') = (\partial \circ I)(s') - s'$$

and therefore $s' = \partial(I(s') - s)$. But when we compute

$$\begin{aligned} d(I(s') - s) &= (d \circ I)(s') - d(s) = (I \circ d)(s') - d(s) \\ &= (I \circ \partial)(s) - [(\partial \circ I)(s) - s] = s - (i \circ c)(s), \end{aligned}$$

we do not get exactly s. This is not really serious because

$$(i \circ c)(s) = (i \circ (d \circ h) \circ c)(s) = (d \circ i \circ h \circ c)(s) = d((i \circ h \circ c)(s))$$

and

$$\partial((i \circ h \circ c)(s)) = 0$$

since $\partial \circ i = 0$. Thus, we may take

$$s'' = I(s') - s + (i \circ h \circ c)(s).$$

□

Lemma 6.5.10 *Let $V = \cup_{k \in \mathbf{N}} V_k$ an increasing admissible covering of a rigid analytic variety which is acyclic for all the terms of a complex of abelian sheaves $E^\bullet$ on V:*

$$\forall q > 0, k \geq 0, i \geq 0, \quad H^q(V_k, E^i) = 0.$$

Then we have

$$R\Gamma(V, E^\bullet) \simeq \left[\prod_{k \in \mathbf{N}} \Gamma(V_k, E^\bullet) \xrightarrow{d} \prod_{k \in \mathbf{N}} \Gamma(V_k, E^\bullet) \right]$$

with $d(s_k) = (s_{k+1|V_k} - s_k)$.

Proof It is of course sufficient to do the case a sheaf $E := E^0$. Note also that for any sheaf E on V, we have

$$\Gamma(V, E) = \varprojlim \Gamma(V_k, E).$$

It follows that

$$R\Gamma(V, E) = R\varprojlim R\Gamma(V_k, E) = R\varprojlim \Gamma(V_k, E).$$

In particular,

$$H^q(V, E) = R^q \varprojlim \Gamma(V_k, E) = 0$$

for $q > 1$. Moreover, with our assumptions, we may use Cech cohomology so that

$$R\Gamma(V, E) \simeq \left[\prod_k \Gamma(V_k, E) \xrightarrow{d} \prod_{k<l} \Gamma(V_k, E) \to \cdots\right]$$

with $d(s_k) = (s_{l|V_k} - s_k)$. Thus, it only remains to show that the canonical morphism of complexes

$$\begin{array}{ccccc} \prod_k \Gamma(V_k, E) & \longrightarrow & \prod_k \Gamma(V_k, E) & & \\ \| & & \uparrow & & \\ \prod_k \Gamma(V_k, E) & \longrightarrow & \prod_{k<l} \Gamma(V_k, E) & \longrightarrow & \cdots \end{array}$$

induces an isomorphism on the H^q for $q = 0, 1$. Clearly the conditions

$$\forall l < k, s_{l|V_k} = s_k \qquad \text{and} \qquad \forall k, s_{k+1|V_k} = s_k$$

are equivalent. This settles the case $q = 0$.

Now a family $(s_{k,l}) \in \prod_{l<k} \Gamma(V_k, E)$ is a cocycle if and only if

$$\forall m < l < k, \quad s_{m,k} = s_{m,l} + s_{l,k}$$

which just means that

$$\forall k, l, s_{k,l} = \sum_{i=k+1}^{l} s_{k,k+1}.$$

It follows that we get an isomorphism on 1-cocycles; and therefore also on cohomology. □

Of course, this lemma is purely of topological nature and holds on more general topological spaces. We can now derive some corollaries from our theorem. First of all, the theorem is still valid for cohomology with support in a closed subvariety.

Corollary 6.5.11 *Let*

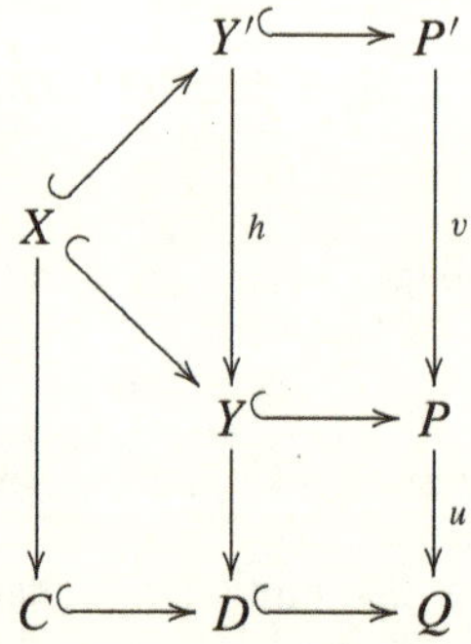

be a sequence of morphisms of S-frames where v is a proper smooth morphism of smooth S-frames. Let X' be a closed subvariety of X. If E is a coherent $j_X^\dagger \mathcal{O}_{]Y[_P}$-module with an integrable connection over S_K and $u' := u \circ v$, there is a canonical isomorphism

$$v^* : Ru_{\mathrm{rig},X'}E \simeq Ru'_{\mathrm{rig},X'}v^\dagger E.$$

Proof If $U := X \setminus X'$, the morphism of triangles

$$\begin{array}{ccccccc}
Ru_{\mathrm{rig},X'}E & \longrightarrow & Ru_{\mathrm{rig}}E & \longrightarrow & Ru_{\mathrm{rig}}j_U^\dagger E & \longrightarrow & \\
\downarrow & & \downarrow & & \downarrow & & \\
Ru'_{\mathrm{rig},X'}v^\dagger E & \longrightarrow & Ru'_{\mathrm{rig}}v^\dagger E & \longrightarrow & Ru'_{\mathrm{rig}}v^\dagger j_U^\dagger E & \longrightarrow &
\end{array}$$

is an isomorphism on the last two columns. It is therefore also an isomorphism on the first one. □

We can also derive a comparison theorem for the specialized de Rham complexes.

Corollary 6.5.12 *Let*

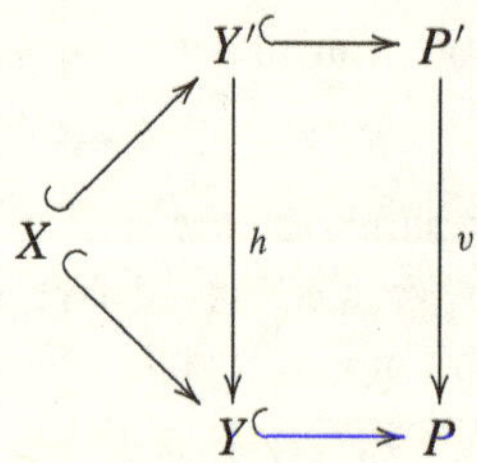

be a proper smooth morphism of smooth S-frames. If E is a coherent $j_X^\dagger \mathcal{O}_{]Y[_P}$*-module with an integrable connection over* S_K*, we have*

$$R\underline{\Gamma}_{\mathrm{rig}} E \simeq R\underline{\Gamma}_{\mathrm{rig}} u^\dagger E.$$

Actually, if X' *is a closed subvariety of* X*, we have*

$$R\underline{\Gamma}_{\mathrm{rig},X'} E \simeq R\underline{\Gamma}_{\mathrm{rig},X'} u^\dagger E.$$

Proof It is of course sufficient to prove the second assertion. The case $u = \mathrm{Id}_P$ of the previous corollary reads

$$\underline{\Gamma}_{X'}^\dagger (E \otimes_{\mathcal{O}_{]Y[_P}} \Omega^\bullet_{]Y[_P/S_K}) \simeq Rv_{K*}\underline{\Gamma}_{X'}^\dagger (v_K^* E \otimes_{\mathcal{O}_{]Y[_{P'}}} \Omega^\bullet_{]Y[_{P'}/S_K})$$

and it follows that

$$\begin{aligned} R\underline{\Gamma}_{\mathrm{rig},X'} E &= Rsp_* \underline{\Gamma}_{X'}^\dagger (E \otimes_{\mathcal{O}_{]Y[_P}} \Omega^\bullet_{]Y[_P/S_K}) \\ &\simeq Rsp_* Rv_{K*} \underline{\Gamma}_{X'}^\dagger (v_K^* E \otimes_{\mathcal{O}_{]Y[_{P'}}} \Omega^\bullet_{]Y[_{P'}/S_K}) \\ &= Rsp_* \underline{\Gamma}_{X'}^\dagger (v_K^* E \otimes_{\mathcal{O}_{]Y[_{P'}}} \Omega^\bullet_{]Y[_{P'}/S_K}) = R\underline{\Gamma}_{\mathrm{rig}} u^\dagger E. \end{aligned}$$

□

We are going now to state and prove the comparison theorem for cohomology with compact support. I have never seen this in print. Anyway, it is much easier than the other case.

Theorem 6.5.13 *Let*

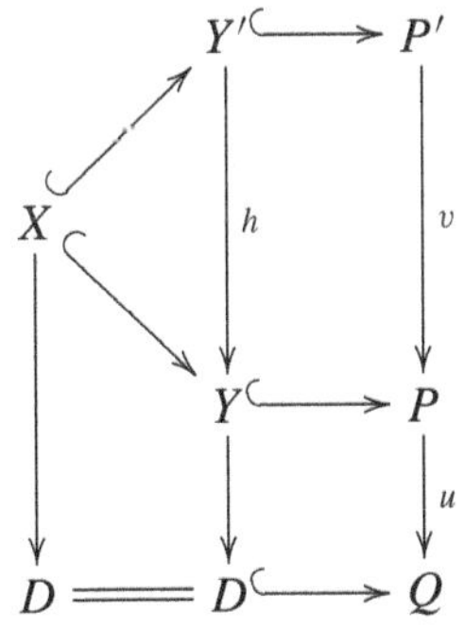

be a sequence of morphisms of S-frames. We assume that v *is a proper smooth morphism of smooth S-frames, that* u *is quasi-compact and we let* $u' := u \circ v$*.*

Then, if E is a coherent $j_X^\dagger \mathcal{O}_{]Y[_P}$*-module with an integrable connection over* S_K*, the base change map is an isomorphism*

$$v^* : Ru_{\mathrm{rig},c} E \simeq Ru'_{\mathrm{rig},c} v^\dagger E.$$

Proof As before, the theorem reduces to the next lemma. More precisely, we may clearly assume that u is the identity and $D := Y$, but we can also assume that $h = \mathrm{Id}_Y$. For this second step, we want to use Lemma 6.5.1. We can localize on Y. We may then blow up and then shrink Y' and assume that h extends to a proper étale map

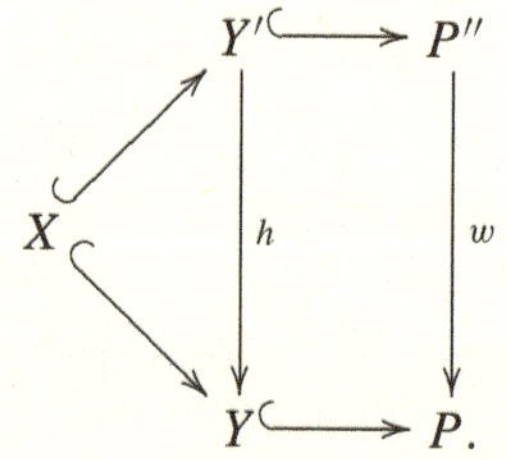

It induces an isomorphism $w_K : V'' \simeq V$ between strict neighborhoods, and in particular on the tubes $]X[_{P''} \simeq]X[_P$. Shrinking V if necessary, we may assume that $E = j_X^\dagger \mathcal{E}$ where $\mathcal{E}$ is a coherent $\mathcal{O}_V$-module with an integrable connection. Thus, we have

$$w^* : R\Gamma_{]X[_P}\mathcal{E} \otimes_{\mathcal{O}_V} \Omega^\bullet_{V/S_K} \simeq R w_{K*} R\Gamma_{]X[_{P''}} w_K^* \mathcal{E} \otimes_{\mathcal{O}_{V''}} \Omega^\bullet_{V''/S_K}.$$

We may also consider the diagonal embedding $Y' \hookrightarrow P''' := P' \times_P P''$ and the projections p_1 and p_2. Thanks to Lemma 6.5.14, we have

$$R\underline{\Gamma}_{]X[_{P''}} w_K^* \mathcal{E} \otimes_{\mathcal{O}_{V''}} \Omega^\bullet_{V''/S_K} \simeq R p_{2K*} R\underline{\Gamma}_{]X[_{P'''}} p_{2K}^* w_K^* \mathcal{E} \otimes_{\mathcal{O}_{V'''}} \Omega^\bullet_{V'''/S_K}$$

and

$$R\underline{\Gamma}_{]X[_{P'}} v_K^* \mathcal{E} \otimes_{\mathcal{O}_{V'}} \Omega^\bullet_{V'/S_K} \simeq R p_{1K*} R\underline{\Gamma}_{]X[_{P'''}} p_{1K}^* v_K^* \mathcal{E} \otimes_{\mathcal{O}_{V'''}} \Omega^\bullet_{V'''/S_K}.$$

Using the fact that $w \circ p_1 = v \circ p_2$, we get

$$R\underline{\Gamma}_{]X[_P}\mathcal{E} \otimes_{\mathcal{O}_V} \Omega^\bullet_{V/S_K} \simeq R v_{K*} R\underline{\Gamma}_{]X[_{P'}} v_K^* \mathcal{E} \otimes_{\mathcal{O}_{V'}} \Omega^\bullet_{V'/S_K}$$

as expected. □

Lemma 6.5.14 *Let*

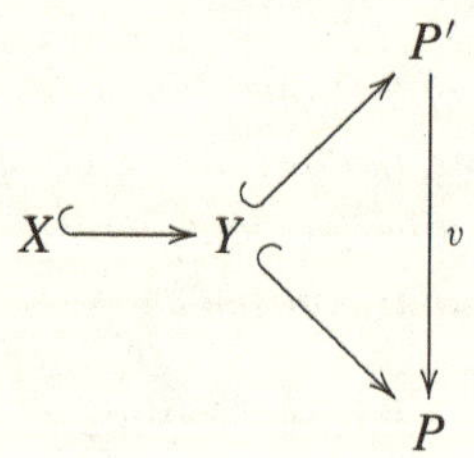

be a smooth morphism of smooth S-frames. Let V be a strict neighborhood of $]X[_P$ in $]Y[_P$, V' a strict neighborhood of $]X[_{P'}$ in $u_K^{-1}(V)\cap]Y[_{P'}$ and $\mathcal{E}$ a coherent $\mathcal{O}_V$-module with an integrable connection.

Then, there is a canonical isomorphism

$$R\underline{\Gamma}_{]X[_P}(\mathcal{E}\otimes_{\mathcal{O}_V}\Omega^\bullet_{V/S_K})\simeq Rv_{K*}R\underline{\Gamma}_{]X[_{P'}}(v_K^*\mathcal{E}\otimes_{\mathcal{O}_{V'}}\Omega^\bullet_{V'/S_K})$$

with $v_K : V' \to V$.

Proof We keep the same strategy as before and prove that the lemma results from the Global Poincaré Lemma with compact support (Lemma 6.5.15 below) which is simply the case $S = P$.

We will denote by

$$h : T := V\backslash]X[_P \hookrightarrow V \quad \text{and} \quad h' : T' := V'\backslash]X[_{P'} \hookrightarrow V'$$

the inclusion maps.

We need a refinement of the Gauss–Manin construction. We consider the de Rham complex

$$M^\bullet := v_K^*\mathcal{E}\otimes_{\mathcal{O}_{V'}}\Omega^\bullet_{V'/S_K}$$

with its Gauss–Manin filtration. Since for each k, $Fil^kM^\bullet$ has coherent terms and that h' is a quasi-Stein morphism, $h'_*h'^{-1}M^\bullet$ inherits a filtration such that

$$Gr^k h'_*h'^{-1}M^\bullet = h'_*h'^{-1}Gr^kM^\bullet.$$

It follows that

$$K^\bullet := [M^\bullet \to h'_*h'^{-1}M^\bullet]$$

is canonically endowed with a filtration such that

$$Gr^kK^\bullet := [Gr^kM^\bullet \to h'_*h'^{-1}Gr^kM^\bullet].$$

But we know that

$$Gr^kM^\bullet = N^\bullet[-k]\otimes_{\mathcal{O}_{V'}} u_K^*\Omega^k_{V/S_K}$$

with

$$N^\bullet := v_K^*\mathcal{E}\otimes_{\mathcal{O}_{V'}}\Omega^\bullet_{V'/V_K}.$$

Thus, if we set

$$L^\bullet := [N^\bullet \to h'_*h'^{-1}N^\bullet]$$

we get as usual

$$Gr^kRv_{K*}K^\bullet = Rv_{K*}L^\bullet[-k]\otimes_{\mathcal{O}_V}\Omega^k_{V/S_K}.$$

But the Global Poincaré Lemma with compact support tells us that

$$Rv_{K*}L^\bullet = [\mathcal{E} \to h_*h^{-1}\mathcal{E}]$$

and it follows that

$$Rv_{K*}K^\bullet = [\mathcal{E} \otimes_{\mathcal{O}_{V'}} \Omega^\bullet_{V/S_K} \to h_*h^{-1}\mathcal{E} \otimes_{\mathcal{O}_V} \Omega^\bullet_{V/S_K}]$$

as expected. □

Lemma 6.5.15 *(Global Poincaré Lemma with compact support) Let*

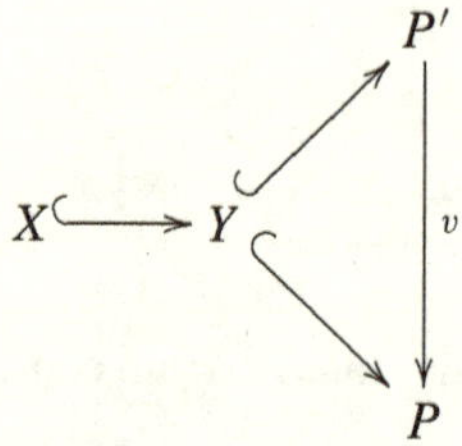

be a smooth morphism of frames. Let V be a strict neighborhood of $]X[_P$ in $]Y[_P$, V' a strict neighborhood of $]X[_{P'}$ in $u_K^{-1}(V)\cap]Y[_{P'}$ and $\mathcal{E}$ a coherent $j_X^\dagger\mathcal{O}_V$-module.

Then, there is a canonical isomorphism

$$R\underline{\Gamma}_{]X[_P}\mathcal{E} \simeq Rv_{K*}R\underline{\Gamma}_{]X[_{P'}}(v_K^*\mathcal{E} \otimes_{\mathcal{O}_{V'}} \Omega^\bullet_{V'/V}).$$

Again, we will use induction on the relative dimension and we will therefore need the following lemma.

Lemma 6.5.16 *If we are given another quasi-compact smooth morphism*

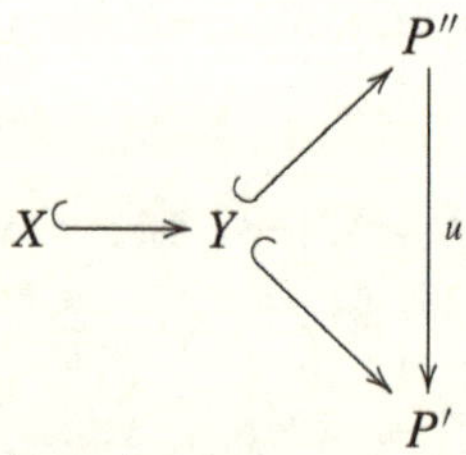

of smooth frames over S and both u and v satisfy the Global Poincaré Lemma, so does $v \circ u$.

Proof This relies again on the Gauss–Manin construction. Let V be a strict neighborhood of $]X[_P$ in $]Y[_P$, V' a strict neighborhood of $]X[_{P'}$ in $u_K^{-1}(V)\cap]Y[_{P'}$, V'' a strict neighborhood of $]X[_{P''}$ in $u_K^{-1}(V')\cap]Y[_{P''}$ and $\mathcal{E}$ a coherent $j_X^\dagger \mathcal{O}_V$-module with an integrable connection. We will denote by

$$h : V\backslash]X[_P \hookrightarrow V, \quad h' : V'\backslash]X[_{P'} \hookrightarrow V' \quad \text{and} \quad h'' : V''\backslash]X[_{P''} \hookrightarrow V''$$

the inclusion maps.

We let

$$M^\bullet := (v \circ u)_K^* \mathcal{E} \otimes_{\mathcal{O}_{V''}} \Omega^\bullet_{V''/V}$$

and

$$K^\bullet := [M^\bullet \to h''_* h''^{-1} M^\bullet]$$

so that

$$Gr^k K^\bullet = L^\bullet[-k] \otimes_{\mathcal{O}_{V''}} u_K^* \Omega^k_{V'/V}$$

with

$$L^\bullet := [N^\bullet \to h''_* h''^{-1} N^\bullet]$$

and

$$N^\bullet := (v \circ u)_K^* \mathcal{E} \otimes_{\mathcal{O}_{V''}} \Omega^\bullet_{V''/V'}.$$

We have

$$\begin{aligned} Gr^k Ru_{K*} K^\bullet &= Ru_{K*} L^\bullet[-k] \otimes_{\mathcal{O}_{V'}} \Omega^k_{V'/V} \\ &= [v_K^* \mathcal{E} \to h'_* h'^{-1} v_K^* \mathcal{E}][-k] \otimes_{\mathcal{O}_{V'}} \Omega^k_{V'/V} \end{aligned}$$

and therefore,

$$Ru_{K*} K^\bullet = [v_K^* \mathcal{E} \otimes_{\mathcal{O}_{V'}} \Omega^\bullet_{V'/V} \to h'_* h'^{-1} v_K^* \mathcal{E} \otimes_{\mathcal{O}_{V'}} \Omega^\bullet_{V'/V}].$$

Finally, we get

$$R(v \circ u)_{K*} K^\bullet = Rv_{K*}[v_K^* \mathcal{E} \otimes_{\mathcal{O}_{V'}} \Omega^\bullet_{V'/V} \to h'_* h'^{-1} v_K^* \mathcal{E} \otimes_{\mathcal{O}_{V'}} \Omega^\bullet_{V'/V}] = E.$$

□

Proof (of the Global Poincaré Lemma with compact support) Again, this is going to be a reduction. We use induction on the dimension of X. First of all, the question is local on $]Y[_P$ and we may therefore assume that P and P' are both affine. By induction, using Proposition 6.4.13, we may replace X by any dense open subset. We may therefore assume that, if we embed Y into $\widehat{\mathbf{A}}^d_P$ using the zero section, there is an étale morphism of frames

$$\begin{array}{ccccc} & & & & P' \\ & & & \nearrow & \downarrow \\ X & \hookrightarrow & Y & & \downarrow \\ & & & \searrow & \downarrow \\ & & & & \widehat{\mathbf{A}}^d_P. \end{array}$$

Using Lemma 6.5.16 (and of course Theorem 3.4.12), we are reduced to the case of the projection $p : \widehat{\mathbf{A}}^1_P \to P$ and $V' = V \times \mathbf{D}(0, 1^-)$.

Let us denote by

$$h : T := V \backslash]X[_P \hookrightarrow V \quad \text{and} \quad h' : T \times \mathbf{D}(0, 1^-) \hookrightarrow V \times \mathbf{D}(0, 1^-)$$

the inclusion maps and by $p'_K : T' \to T$ the map induced by p. It follows from Proposition 6.4.3 that there is a morphism of triangles

$$\begin{array}{ccccccc} \underline{\Gamma}_X \mathcal{E} & \longrightarrow & \mathcal{E} & \longrightarrow & Rh_* \mathcal{E}_{|T} & \longrightarrow & \\ \downarrow & & \downarrow & & \downarrow & & \\ Rv_{\mathrm{rig},c}\, p^\dagger \mathcal{E} & \longrightarrow & Rp_{K\mathrm{dR}} \mathcal{E} & \longrightarrow & Rh_* Rp'_{K\mathrm{dR}} \mathcal{E}_{|T} & \longrightarrow & . \end{array}$$

It is therefore sufficient to show that both right arrows are bijective. In other words, we are reduced to the following lemma:

□

Lemma 6.5.17 *Let W be an affinoid variety, $\mathcal{E}$ be a coherent $\mathcal{O}_W$-module and*

$$p : W \times \mathbf{D}(0, 1^-) \to W$$

the first projection. Then, we have

$$H^q(W \times \mathbf{D}(0, 1^-), p^* \mathcal{E}) = 0$$

for $q > 1$ and the sequence

$$0 \to \Gamma(W, E) \to \Gamma(W \times \mathbf{D}(0, 1^-), p^* \mathcal{E}) \stackrel{\partial/\partial t}{\to} \Gamma(W \times \mathbf{D}(0, 1^-), p^* \mathcal{E}) \to 0$$

is exact.

Proof This is completely standard and left to the reader. □

Corollary 6.5.18 *Let*

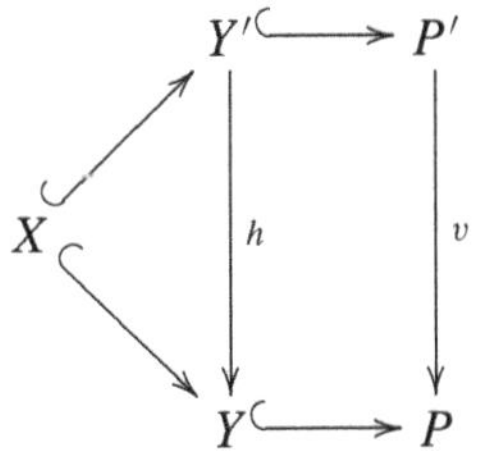

be a proper smooth morphism of smooth S-frames and E a coherent $j_X^\dagger \mathcal{O}_{]Y[_P}$*-module with an integrable connection over* S_K. *We have*

$$R\underline{\Gamma}_{\mathrm{rig},c} E \simeq R\underline{\Gamma}_{\mathrm{rig},c} u^\dagger E.$$

Proof It is sufficient to apply Rsp_* to the case $u = \mathrm{Id}_P$ of the theorem. □

7
Overconvergent isocrystals

We fix a complete ultrametric field K with $\mathcal{V}$, k and π as usual and we denote by S a formal $\mathcal{V}$-scheme.

7.1 Overconvergent isocrystals on a frame

Definition 7.1.1 *A (finitely presented)* overconvergent isocrystal *on an S-frame $(X \subset Y \subset P)$ is*

(i) *a family of (coherent) $j^\dagger_{X'}\mathcal{O}_{]Y'[_{P'}}$-modules $E_{P'}$ for each morphism of S-frames*

$$\begin{array}{ccccc} X' & \hookrightarrow & Y' & \hookrightarrow & P' \\ \downarrow & & \downarrow & & \downarrow \\ X & \hookrightarrow & Y & \hookrightarrow & P \end{array}$$

(ii) *a family of isomorphisms*

$$\varphi_u : u^\dagger E_{P'} \simeq E_{P''}$$

for each commutative diagram

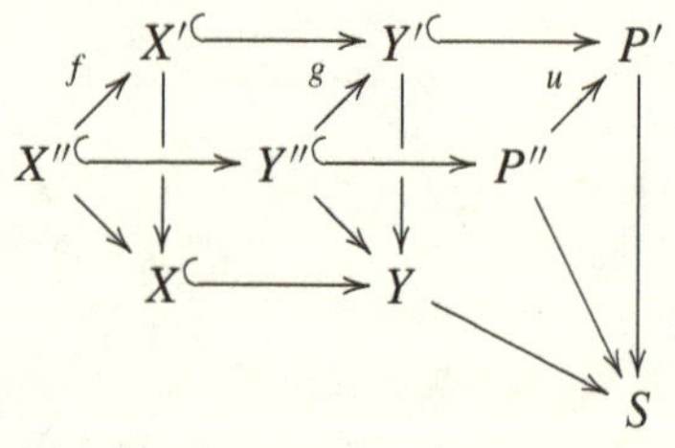

subject to the cocycle condition

$$\varphi_{v\circ u} = \varphi_v \circ u'^* \varphi_u.$$

We call $E_{P'}$ the realization *of E on* $(X' \subset Y' \subset P'/S')$.

A morphism of overconvergent isocrystals *is a family of compatible morphisms of $j^{\dagger}_{X'}\mathcal{O}_{]Y'[_{P'}}$-modules.*

When $Y = X$, we simply say convergent isocrystal.

We point out that the transition morphisms are not required to be compatible with the structural morphism to P, but only to S. This is very important. In other words, $E_{P'}$ depends only on P' and not on a particular morphism $P' \to P$.

We will denote by $\mathrm{Isoc}^{\dagger}(X \subset Y \subset P/S)$ the category of finitely presented overconvergent isocrystals on $(X \subset Y \subset P/S)$ and by $\mathrm{Isoc}(X \subset P/S)$ the category of finitely presented convergent isocrystals on $(X \subset P/S)$.

Proposition 7.1.2 *Overconvergent isocrystals on an S-frame $(X \subset Y \subset P)$ form an abelian category with faithful exact functor $E \mapsto E_P$. In particular,* $\mathrm{Isoc}^{\dagger}(X \subset Y \subset P/S)$ *is an abelian category.*

Proof By definition, if

$$\begin{array}{ccccc} X' & \hookrightarrow & Y' & \hookrightarrow & P' \\ \downarrow{\scriptstyle f} & & \downarrow{\scriptstyle g} & & \downarrow{\scriptstyle u} \\ X & \hookrightarrow & Y & \hookrightarrow & P \end{array}$$

is a morphism of S-frames and E an overconvergent isocrystal on $(X \subset Y \subset P)$, then $E_{P'} = u^{\dagger}E_P$. It immediately follows that the functor $E \mapsto E_P$ is faithful. It is also clear that we get an abelian category where kernels and cokernels are obtained by first considering the realization on P and then pulling back by $u^{\dagger}$. In particular, the functor $E \mapsto E_P$ is also exact. □

Proposition 7.1.3 *Let*

$$\begin{array}{ccccc} X' & \hookrightarrow & Y' & \hookrightarrow & P' \\ \downarrow{\scriptstyle f} & & \downarrow{\scriptstyle g} & & \downarrow{\scriptstyle u} \\ X & \hookrightarrow & Y & \hookrightarrow & P \end{array}$$

be a morphism from an S'-frame to an S-frame over some morphism $v : S' \to S$. Then, any overconvergent isocrystal E on $(X \subset Y \subset P/S)$ defines by restriction an overconvergent isocrystal

$$u^*E := E_{|(X' \subset Y' \subset P')}$$

on $(X' \subset Y' \subset P'/S')$. And this is functorial in E.

In particular, there is a pullback functor

$$u^* : \mathrm{Isoc}^\dagger(X \subset Y \subset P/S) \to \mathrm{Isoc}^\dagger(X' \subset Y' \subset P'/S').$$

Proof Looking at the definition, this is really a triviality. □

Note that, by definition, we have $(u^*E)_{P'} = u^\dagger E_P$. In practice, when both u and g are the identity we will simply write $E_{|X'}$. We should be careful with this notation because $(E_{|X'})_P = j^\dagger_{X'} E_P$. When only u is the identity but the morphism is cartesian, we will also write $E_{|Y'}$. In this case, there is no ambiguity because then g is necessarily a closed immersion and $(E_{|Y'})_P = (E_P)_{|]Y'[_P}$.

Proposition 7.1.4 *If*

$$\begin{array}{ccccc} X' & \hookrightarrow & Y' & \hookrightarrow & P' \\ \downarrow{\scriptstyle f} & & \downarrow{\scriptstyle g} & & \downarrow{\scriptstyle u} \\ X & \hookrightarrow & Y & \hookrightarrow & P \end{array}$$

is a flat morphism from an S'-frame to an S-frame over some morphism $v : S' \to S$, then u^* is exact on overconvergent isocrystals.

Proof Since the realization functors $E \mapsto E_P$ and $u^*E \mapsto u^\dagger E_P$ are faithfully exact, our assertion results from Proposition 5.3.13. □

In the next proposition, we show that the category $\mathrm{Isoc}^\dagger(X \subset Y \subset P/S)$ is local on $(X \subset Y \subset P)$.

Proposition 7.1.5 *Let*

$$\begin{array}{ccccc} X_i & \hookrightarrow & Y_i & \hookrightarrow & P_i \\ \downarrow & & \downarrow & & \downarrow \\ X & \hookrightarrow & Y & \hookrightarrow & P \end{array}$$

be an open covering or a finite mixed covering of an S-frame. Then, there is an equivalence of categories

$$\mathrm{Isoc}^\dagger(X \subset Y \subset P) \simeq$$
$$\varprojlim \Big[\prod_i \mathrm{Isoc}^\dagger(X_i \subset Y_i \subset P_i) \rightrightarrows \prod_{i,j} \mathrm{Isoc}^\dagger(X_{ij} \subset Y_{ij} \subset P_{ij})$$
$$\rightrightarrows\!\!\!\to \prod_{i,j,k} \mathrm{Isoc}^\dagger(X_{ijk} \subset Y_{ijk} \subset P_{ijk})\Big].$$

Proof This is an immediate consequence of Proposition 5.4.12. □

Proposition 7.1.6 *Let*

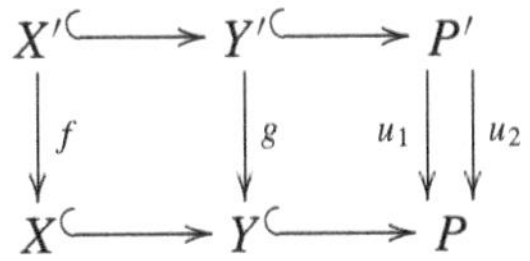

be two morphisms of frames over the same $S' \to S$ and E an overconvergent isocrystal on $(X \subset Y \subset P/S)$. Then, there exists a canonical isomorphism of overconvergent isocrystals

$$u_2^* E \simeq u_1^* E.$$

Proof This is a formal (and immediate) consequence of the definitions. □

Corollary 7.1.7 *Assume that there exists two morphisms of S-frames*

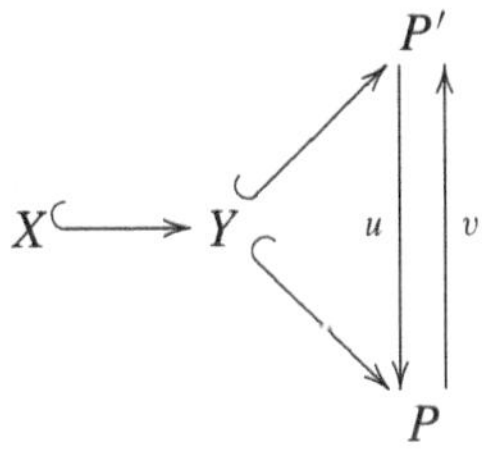

Then u^* and v^* induce an equivalence of categories between overconvergent isocrystals on $(X \subset Y \subset P/S)$ and $(X \subset Y \subset P'/S)$.

Proof It follows from Proposition 7.1.6 that $u^* \circ v^* = \mathrm{Id}$ and $v^* \circ u^* = \mathrm{Id}$. □

Theorem 7.1.8 *Let S be a formal $\mathcal{V}$-scheme and*

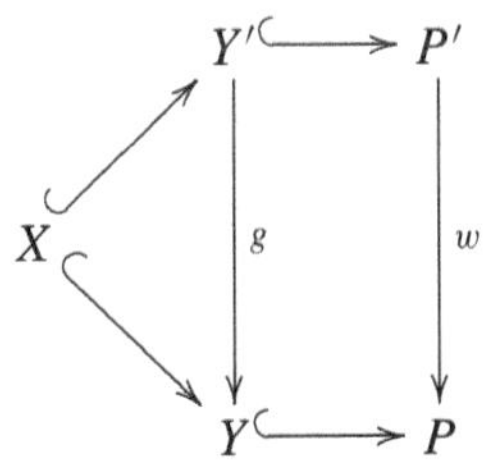

a proper smooth morphism of frames over S. Then the pullback functor is an equivalence of categories:

$$u^* : \mathrm{Isoc}^\dagger(X \subset Y \subset P/S) \simeq \mathrm{Isoc}^\dagger(X \subset Y' \subset P'/S).$$

Proof We assume first that $g = \mathrm{Id}_Y$. It follows from Proposition 7.1.5 that this question is local on $(X \subset Y \subset P)$. We may assume that the frames are affine and that there exists $t_1, \ldots, t_d$ in the ideal I' of Y in P' that give a basis $(\bar{t}_1, \ldots, \bar{t}_d)$ of the conormal sheaf $\omega_{X'/X}$ of X in $X' := u^{-1}(X)$. Then, we know from Proposition 3.3.13 that if we embed Y into $\widehat{\mathbf{A}}^d_P$ using the zero section, then $(t_1, \ldots, t_d)$ defines an étale morphism of frames

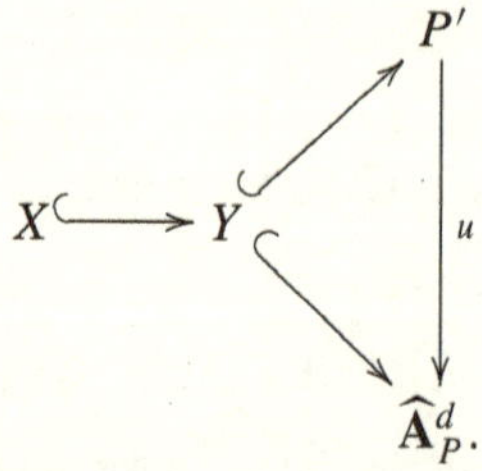

Since the case of a quasi-compact étale morphism obviously results from Theorem 3.4.12, we may assume that $P' = \widehat{\mathbf{A}}^d_P$. We may then consider the morphism of frames

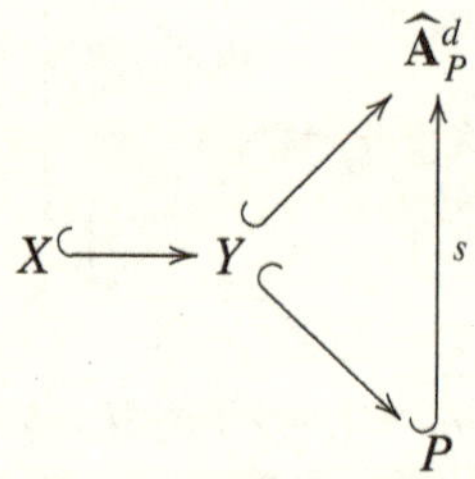

induced by the zero section and our assertion follows from Corollary 7.1.7.

We now turn to the general case. The assertion is local on $(X \subset Y \subset P)$ and only depends on a closed subset of Y containing X. In particular, we may assume that Y is quasi-projective and Chow's lemma (Corollary 5.7.14 of [75]) reduces to the case g projective. Thanks to Lemma 6.5.1, we may therefore assume that g extends to a proper étale map

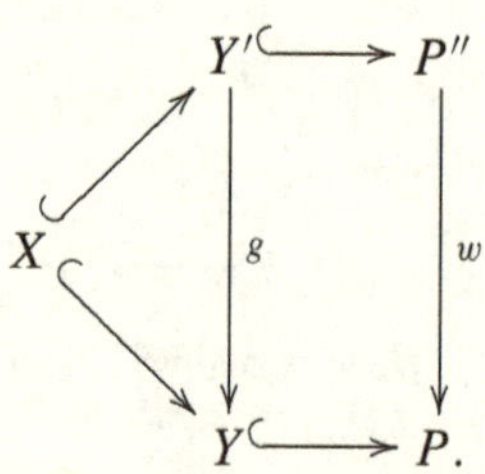

It induces an isomorphism on strict neighborhoods by Theorem 3.4.12 again, and it follows that

$$w^* : \mathrm{Isoc}^\dagger(X \subset Y \subset P/S) \simeq \mathrm{Isoc}^\dagger(X \subset Y' \subset P''/S).$$

We may also consider the diagonal embedding $Y' \hookrightarrow P''' := P' \times_P P''$ and the projections p_1 and p_2 and it follows from the first case that

$$p_1^* : \mathrm{Isoc}^\dagger(X \subset Y' \subset P'/S) \simeq \mathrm{Isoc}^\dagger(X \subset Y' \subset P'''/S)$$

and

$$p_2^* : \mathrm{Isoc}^\dagger(X \subset Y' \subset P''/S) \simeq \mathrm{Isoc}^\dagger(X \subset Y' \subset P'''/S).$$

Our assertion now follows from the fact that $w \circ p_2 = v \circ p_1$. □

Proposition 7.1.9 *Let $(X \subset Y \subset P)$ be an S-frame and $j : U \hookrightarrow X$ the inclusion of an open subset. Then, any overconvergent isocrystal E on $(U \subset Y \subset P/S)$ defines also an overconvergent isocrystal j_*E on $(X \subset Y \subset P/S)$. More precisely, if*

$$\begin{array}{ccccc} X' & \hookrightarrow & Y' & \hookrightarrow & P' \\ \downarrow f & & \downarrow g & & \downarrow u \\ X & \hookrightarrow & Y & \hookrightarrow & P \end{array}$$

is a morphism of S-frames and $E_{P'}$ denotes the realization of E on $(f^{-1}(U) \subset Y' \subset P')$, we set $(j_*E)_{P'} = E_{P'}$.

Proof The point is to show that there is a canonical isomorphism $j^\dagger_{X'} u_K^* E_P \simeq E_{P'}$ knowing that we already have a canonical isomorphism $j^\dagger_{U'} u_K^* E_P \simeq E_{P'}$ with $U' := f^{-1}(U)$. This can be split into two cases as usual, the cartesian case and the open immersion case. In the cartesian case, there is nothing to do. We may therefore assume that $Y' = Y$, $P' = P$ and that X' is simply an open subset in X so that $U' = X' \cap U$. Then, we have

$$j^\dagger_{X'} E_P = j^\dagger_{X'} j^\dagger_U E_P = j^\dagger_{X' \cap U} E_P = j^\dagger_{U'} E_P$$

as expected. □

Of course, when if E is finitely presented, j_*E is not likely to be so!

Definition 7.1.10 *Let $(X \subset Y \subset P)$ be an S-frame, X' a closed subvariety of X and $j : U := X \setminus X' \hookrightarrow X$ the inclusion of the open complement. If E is an overconvergent isocrystal on $(X \subset Y \subset P/S)$, then*

$$\underline{\Gamma}_{X'} E := \ker(E \to j_* E_{|U})$$

is the overconvergent sub-isocrystal of E with support in X'.

Proposition 7.1.11 *Let $(X \subset Y \subset P)$ be an S-frame, X' a closed subvariety of X and $j : U := X \setminus X' \hookrightarrow X$ the inclusion of the open complement. If E*

is a an overconvergent isocrystal on $(X \subset Y \subset P/S)$*, there is a short exact sequence*

$$0 \to \underline{\Gamma}_{X'} E \to E \to j_* E_{|U} \to 0$$

of overconvergent isocrystals on $(X \subset Y \subset P/S)$ *and*

$$(\underline{\Gamma}_{X'} E)_P = \underline{\Gamma}^\dagger_{X'} E_P.$$

Proof It follows from Proposition 5.2.11 that we have an exact sequence

$$0 \to \underline{\Gamma}^\dagger_{X'} E_P \to E_P \to j^\dagger_U E_P \to 0.$$

Since the realization functor on P is exact and faithful, both assertions follow. □

We can do a little better:

Proposition 7.1.12 *Let* $(X \subset Y \subset P)$ *be an S-frame. Let*

$$X'' \hookrightarrow X' \hookrightarrow X$$

be a sequence of closed immersions, $U := X \setminus X''$ *and* $U' := X' \setminus X''$*. If* E *is an overconvergent isocrystal on* $(X \subset Y \subset P/S)$*, there is an exact sequence*

$$0 \to \underline{\Gamma}_{X''} E \to \underline{\Gamma}_{X'} E \to j_{U*} \underline{\Gamma}_{U'} E_{|U} \to 0.$$

Proof This follows also from Proposition 5.2.11. □

7.2 Overconvergence and calculus

The next step consists in building overconvergence into the definition of the stratification. Starting at Proposition 7.2.11, we assume that $\mathrm{Char} K = 0$.

If $(X \subset Y \subset P)$ is an S-frame, we can consider the projections

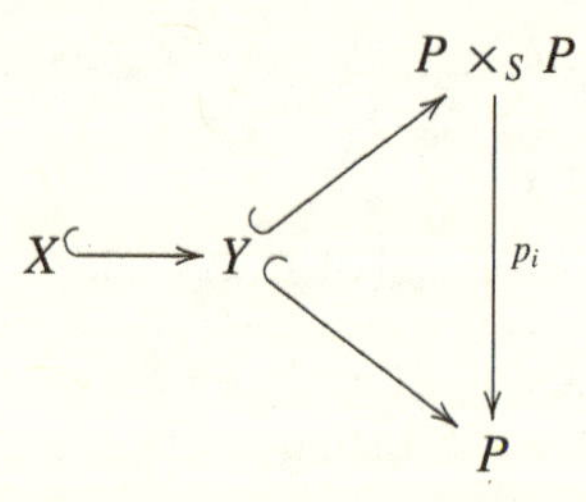

for $i = 1, 2$, and also the projections

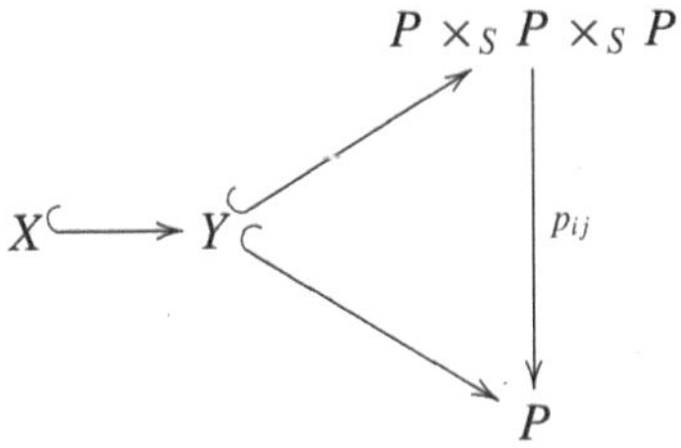

for $i, j = 1, 2, 3$, and we will still use the same letters for the morphisms induced on the tubes on X or Y.

Definition 7.2.1 *Let* $(X \subset Y \subset P)$ *be an* S*-frame and* E *a* $j_X^\dagger \mathcal{O}_{]Y[_P}$*-module. An* overconvergent stratification on E *is an isomorphism of* $j_X^\dagger \mathcal{O}_{]Y[_{P\times_S P}}$*-modules, also called the* (Taylor) isomorphism of E,

$$\epsilon : p_2^* E \simeq p_1^* E$$

such that

$$p_{13}^*(\epsilon) = p_{12}^*(\epsilon) \circ p_{23}^*(\epsilon)$$

on $]Y[_{P\times_S P\times_S P}$.

When $Y = X$, *we simply say* convergent.

A morphism of $j_X^\dagger \mathcal{O}_{]Y[_P}$*-modules with overconvergent stratification is a morphism of* $j_X^\dagger \mathcal{O}_{]Y[_P}$*-modules compatible with the Taylor isomorphisms.*

Note that, in this definition, we could use $p_i^\dagger$ instead of p_i^*. This is the same thing since the projections induce cartesian morphisms of frames.

Proposition 7.2.2 *The category of (finitely presented) overconvergent isocrystals on an* S*-frame* $(X \subset Y \subset P)$ *is naturally equivalent to the category of (coherent)* $j_X^\dagger \mathcal{O}_{]Y[_P}$*-modules with an overconvergent stratification.*

Proof Given an overconvergent isocrystal E on an S-frame $(X \subset Y \subset P)$, we have on one hand a $j_X^\dagger \mathcal{O}_{]Y[_P}$-module E_P corresponding to $(X \subset Y \subset P)$ and on the other hand a $j_X^\dagger \mathcal{O}_{]Y[_{P\times_S P}}$-module $E_{P\times_S P}$ corresponding to the diagonal embedding $(X \subset Y \subset P \times_S P)$. And we can consider the composite isomorphism

$$\epsilon : p_2^* E_P \simeq E_{P\times_S P} \simeq p_1^* E_P.$$

Clearly, this defines an overconvergent stratification on E_P.

Conversely, if we are given a $j_X^\dagger \mathcal{O}_{]Y[_P}$-module E with an overconvergent stratification and a morphism of frames

$$\begin{array}{ccccc} X' & \hookrightarrow & Y' & \hookrightarrow & P' \\ \downarrow & & \downarrow & & \downarrow u \\ X & \hookrightarrow & Y & \hookrightarrow & P \end{array}$$

we can define $E_{P'} := u^\dagger E_P$. Now, assume that we are given another morphism of frames

$$\begin{array}{ccccc} X'' & \hookrightarrow & Y'' & \hookrightarrow & P'' \\ \downarrow & & \downarrow & & \downarrow u' \\ X & \hookrightarrow & Y & \hookrightarrow & P \end{array}$$

so that $E_{P''} := u'^\dagger E_P$ and a commutative diagram

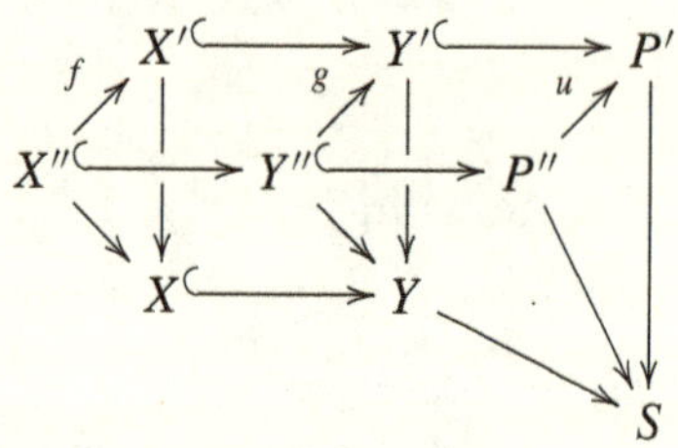

as in the definition of overconvergent isocrystals. We may consider the morphism of frames

$$\begin{array}{ccccc} X'' & \hookrightarrow & Y'' & \hookrightarrow & P'' \\ \downarrow & & \downarrow & & \downarrow (u', u\circ v) \\ X & \hookrightarrow & Y & \hookrightarrow & P \times_S P \end{array}$$

and pullback the Taylor isomorphism in order to get an isomorphism

$$\begin{aligned} v^\dagger E_{P'} &= v^\dagger u^\dagger E_P = (u', u \circ v)^\dagger p_2^* E_P \\ &\simeq (u', u \circ v)^\dagger p_1^* E_P = u'^\dagger E_P = E_{P''}. \end{aligned}$$

It only remains to check that the cocycle condition holds and that we do get a quasi-inverse to our functor. This is left to the reader. □

Corollary 7.2.3 *In the definition of finitely presented overconvergent isocrystals, we get an equivalent category by considering only*

(i) *frames of the form* $X \subset Y \subset P \times_S \cdots \times_S P$ *and projections (we may even stop at triple products), or*
(ii) *frames of the form* $(X \subset Y \subset P')$ *with any* P', *or else*
(iii) *strictly local frames* $(X' \subset Y' \subset P')$ *with* Y' *open in* Y *(and consequently* X' *open in* X*).*

Proof The first case is just a reformulation of Proposition 7.2.2. The second one results immediately from the first. And the last one, then results from Proposition 5.4.12. □

Proposition 7.2.4 *Let* $(X \subset Y \subset P)$ *be an* S*-frame and* $\sigma : K \hookrightarrow K'$ *be an isometric embedding. If* E *is an overconvergent isocrystal on* $(X \subset Y \subset P)$, *there exists a unique overconvergent isocrystal* E^σ *on* $(X^\sigma \subset Y^\sigma \subset P^\sigma)$ *such that* $E^\sigma_{P'^\sigma} = (E_{P'})^\sigma$ *for any frame* $(X' \subset Y' \subset P')$ *over* $(X \subset Y \subset P)$. *And this is functorial in* E..

In particular, there is a pullback functor

$$\begin{array}{ccc} \mathrm{Isoc}^\dagger(X \subset Y \subset P/S) & \longrightarrow & \mathrm{Isoc}^\dagger(X^\sigma \subset Y^\sigma \subset P^\sigma/S^\sigma) \\ E & \longmapsto & E^\sigma \end{array}$$

Proof We simply use the functoriality of the pullback along σ of the Taylor isomorphism in order to get an overconvergent stratification. Then we know from Proposition 7.2.2 that this overconvergent stratification extends uniquely to a structure of overconvergent isocrystal. □

Proposition 7.2.5 *If* E *is an overconvergent isocrystal on an* S*-frame* $(X \subset Y \subset P)$, *then* E_P *has a natural stratification and this is functorial in* E. *In particular,* E_P *has an integrable connection and there is a forgetful functor*

$$\mathrm{Isoc}^\dagger(X \subset Y \subset P/S) \to \mathrm{MIC}(X \subset Y \subset P/S).$$

Proof It is sufficient to pull back the overconvergent stratification of E along the morphism of frames

$$(X \subset Y \subset P^{(n)}) \hookrightarrow (X \subset Y \subset P \times_S P)$$

for each n. □

Definition 7.2.6 *Let*

$$\begin{array}{ccccc} X & \hookrightarrow & Y & \hookrightarrow & P \\ \downarrow{\scriptstyle f} & & \downarrow{\scriptstyle g} & & \downarrow{\scriptstyle u} \\ C & \hookrightarrow & D & \hookrightarrow & Q \end{array}$$

be a morphism of S-frames and E an overconvergent isocrystal on $(X \subset Y \subset P/S)$. *The* rigid cohomology *of E is*

$$Ru_{\mathrm{rig}}E = Ru_{\mathrm{rig}}E_P.$$

If X' is a closed subset of X, the rigid cohomology of E with support in X' *is*

$$Ru_{\mathrm{rig},X'}E = Ru_{\mathrm{rig},X'}E_P.$$

Finally, if $C = D$, u is quasi-compact and E is finitely presented, the rigid cohomology of E with compact support *is*

$$Ru_{\mathrm{rig},c}E = Ru_{\mathrm{rig},c}E_P.$$

We will not review here all the properties of these functors. They translate directly from their counterpart on modules with connection.

Note that the restriction functor of Proposition 7.2.5 is not fully faithful a priori and has no reason to be so in general. This is however the case when we restrict to finitely presented overconvergent isocrystals on smooth frames as we shall see shortly.

Lemma 7.2.7 *Let $(X \subset Y \subset P)$ be a smooth S-frame. If E and F are two finitely presented $j_X^\dagger \mathcal{O}_{]Y[_P}$-modules, then the canonical map*

$$p_{1*}\mathcal{H}om(p_2^*E, p_1^*F) \to \varprojlim \mathcal{H}om(p_2^{(n)*}E, p_1^{(n)*}F)$$

is injective. More generally, if If E' is a coherent $j_X^\dagger \mathcal{O}_{]Y[_{P\times_S P}}$-module, then the canonical map

$$p_{1*}\mathcal{H}om(E', p_1^*F) \to \varprojlim \mathcal{H}om(E'_{|]Y[_P^{(n)}}, p_1^{(n)*}F)$$

is injective.

Proof The question is local on $]Y[_P$ and therefore also on Y thanks to Proposition 2.2.15. We may therefore assume that $(X \subset Y \subset P)$ is local with coordinates $t_1, \ldots, t_d$. Actually, using Corollary 5.4.7, we may even assume that $Y = X$, so that in particular $j_X^\dagger = \mathrm{Id}$ and $F = \mathcal{F}$ is simply a coherent $\mathcal{O}_{]X[_P}$-module.

Since E' has locally a finite presentation and our functors are left exact, we may assume that E' is trivial. Thus, we have to show that the canonical map

$$p_{1*}p_1^*\mathcal{F} \to \varprojlim p_1^{(n)*}\mathcal{F}$$

is injective.

The effective version of the weak fibration theorem (Corollary 2.3.17) tells us that the morphism

$$(\tau_1, \ldots, \tau_d) : P \times_S P \to \widehat{\mathbf{A}}^d_P$$

with

$$\tau_i := p_2^*(t_i) - p_1^*(t_i),$$

induces an isomorphism

$$]X[_{P\times_S P} \simeq]X[_P \times \mathbf{B}^d(0, 1^-).$$

Our assertion therefore follows from Proposition 4.3.6. □

Proposition 7.2.8 *The functor that associates, to a finitely presented overconvergent isocrystal on a smooth S-frame $(X \subset Y \subset P)$, the corresponding stratified module is fully faithful.*

Proof Our functor is clearly faithful and we have to show that it is full. In other words, we want to prove that if E and F are two $j_X^\dagger \mathcal{O}_{]Y[_P}$-modules with an overconvergent stratification and $\varphi : E \to F$ is a morphism of $j_X^\dagger \mathcal{O}_{]Y[_P}$-modules compatible with the stratifications $\epsilon_E^{(n)}$ and $\epsilon_F^{(n)}$ for all $n \in \mathbf{N}$, then it is also compatible with the overconvergent stratifications ϵ_E and ϵ_F. Thus, the point is to show that the morphism

$$s := p_1^*(\varphi) \circ \epsilon_E - \epsilon_F \circ p_2^*(\varphi) : p_2^* E \to p_1^* F$$

is equal to 0. But our hypothesis tells us that for each $n \in \mathbf{N}$, the morphism

$$s_n := p_1^{(n)*}(\varphi) \circ \epsilon_E^{(n)} - \epsilon_F^{(n)} \circ p_2^{(n)*}(\varphi) : p_2^{(n)*} E \to p_1^{(n)*} F$$

is zero.

Our assertion therefore follows from Lemma 7.2.7. □

Proposition 7.2.9 *Let*

$$\begin{array}{ccccc} X_i & \hookrightarrow & Y_i & \hookrightarrow & P_i \\ \downarrow & & \downarrow & & \downarrow \\ X & \hookrightarrow & Y & \hookrightarrow & P \end{array}$$

be an open covering of a smooth S-frame. Let E be a coherent stratified $j_X^\dagger \mathcal{O}_{]Y[_P}$-module. If the stratification of $E_{|(X_i \subset Y_i \subset P_i)}$ is overconvergent for each i, then the stratification of E is already overconvergent.

Proof This follows from the fact that both categories are local and that the forgetful functor is fully faithful. □

Definition 7.2.10 *Let $(X \subset Y \subset P)$ be a smooth S-frame. An integrable connection on a coherent $j_X^\dagger \mathcal{O}_{]Y[_P}$-module E is* overconvergent *if there exists a strict neighborhood V of $]X[_P$ in $]Y[_P$ and a coherent $\mathcal{O}_V$-module $\mathcal{E}$ with an overconvergent integrable connection (see Definition 4.3.4) such that $E = j_X^\dagger \mathcal{E}$ (and ∇ too comes from the connection of $\mathcal{E}$).*

We will denote by $\mathrm{MIC}^\dagger(X \subset Y \subset P/S)$ the full subcategory of coherent $j_X^\dagger \mathcal{O}_{]Y[_P}$-modules with an overconvergent integrable connection.

Until the end of this section, we assume that $\mathrm{Char}K = 0$.

Proposition 7.2.11 *If $(X \subset Y \subset P)$ is a smooth quasi-compact S-frame, we have an equivalence of categories*

$$\varinjlim_V \mathrm{MIC}^\dagger(V/S_K) \simeq \mathrm{MIC}^\dagger(X \subset Y \subset P/S).$$

Proof This follows immediately from Proposition 6.1.15 and the definition of an overconvergent connection. □

Proposition 7.2.12 *Let*

$$\begin{array}{ccccc} X_i & \hookrightarrow & Y_i & \hookrightarrow & P_i \\ \downarrow & & \downarrow & & \downarrow \\ X & \hookrightarrow & Y & \hookrightarrow & P \end{array}$$

be an open covering of a smooth S-frame. Let E be a coherent $j_X^\dagger \mathcal{O}_{]Y[_P}$-module with an integrable connection. If the connection of $E_{|(X_i \subset Y_i \subset P_i)}$ is overconvergent for each i, so is the connection of E.

Proof This follows from the analogous result on stratifications. □

Proposition 7.2.13 *If $(X \subset Y \subset P)$ is a smooth quasi-compact S-frame, there is an equivalence of categories*

$$\mathrm{Isoc}^\dagger(X \subset Y \subset P/S) \simeq \mathrm{MIC}^\dagger(X \subset Y \subset P/S).$$

Proof It follows from Propositions 7.2.8 and 6.1.10 that the functor

$$\mathrm{Isoc}^\dagger(X \subset Y \subset P/S) \to \mathrm{MIC}(X \subset Y \subset P/S)$$

is fully faithful. It remains to show that its image is exactly $\mathrm{MIC}^\dagger(X \subset Y \subset P/S)$.

We can identify $\mathrm{Isoc}^\dagger(X \subset Y \subset P/S)$ with the category of coherent $j_X^\dagger \mathcal{O}_{]Y[_P}$-modules E with an overconvergent stratification. If E is such an object, there exists a strict neighborhood V of $]X[_P$ in $]Y[_P$ and a coherent $\mathcal{O}_V$-module $\mathcal{E}$ with an integrable connection such that $E = j_X^\dagger \mathcal{E}$. Now, if

$$\epsilon : p_2^* E \simeq p_1^* E$$

denotes the Taylor isomorphisms of E, there exists a strict neighborhood V' of $]X[_{P\times_S P}$ in $]Y[_{P\times_S P}$ and an isomorphism

$$\epsilon' : p_2^* \mathcal{E}_{|V'} \simeq p_1^* \mathcal{E}_{|V'}$$

such that $\epsilon = j_X^\dagger \epsilon'$. It follows that the connection of $\mathcal{E}$ is overconvergent. Conversely, if $\mathcal{E}$ is a coherent $\mathcal{O}_V$-module with an overconvergent integrable connection, it is clear that $E := j_X^\dagger \mathcal{E}$ is a coherent $j_X^\dagger \mathcal{O}_{]Y[_P}$-modules with an overconvergent stratification. □

Let us consider again the Monsky–Washnitzer situation.

Definition 7.2.14 *Let $X = \mathrm{Spec}\, A$ be an affine $\mathcal{V}$-scheme and Y denotes the closure of X in some projective space for a given presentation of A. An integrable connection on a coherent $A_K^\dagger$-module M is* overconvergent *if M corresponds to an object of $\mathrm{MIC}^\dagger(X_k \subset Y_k \subset \widehat{Y}/\mathcal{V})$ under the equivalence of categories*

$$\mathrm{MIC}(X_k \subset Y_k \subset \widehat{Y}/\mathcal{V}) \simeq \mathrm{MIC}(A_K^\dagger).$$

We will denote by $\mathrm{MIC}^\dagger(A_K^\dagger)$ the full subcategory of coherent $A_K^\dagger$-modules with an overconvergent connection. Since we already defined overconvergence with respect to a given set of étale coordinates, we need to check that our new definition is compatible with the old one.

Proposition 7.2.15 *Let $X = \mathrm{Spec}\, A$ be a smooth affine $\mathcal{V}$-scheme with étale coordinates $t_1, \ldots, t_d$ and M be a coherent $A_K^\dagger$-module with an integrable connection. Then, M is overconvergent if and only if it is overconvergent with respect to $t_1, \cdots, t_d$.*

Recall that our condition means that M is η-convergent for any $\eta < 1$ which may also be written in terms of radius of convergence as

$$\lim_{\rho \stackrel{<}{\to} 1} R(M, \rho) = 1.$$

Proof Using Propositions 3.3.9 and 6.1.17, our assertion is a consequence of Proposition 4.4.10 but we need to be a little careful because the ambient space is not affine and the coordinates are not everywhere defined. More precisely, we must consider the closure Y of X. Then, as in Proposition 4.3.8, we can find an affine covering $\widehat{Y} = \cup \mathcal{Y}_i$ and for each i, coordinates defined on $\mathcal{Y}_i$ and étale on $\mathcal{X}_i := \widehat{X} \cap \mathcal{Y}_i$. We may then apply Proposition 4.4.10 to $\mathcal{Y}_i$ which is affine and $\mathcal{X}_i$ which is the smooth complement of a hypersurface. Moreover, using the second assertion of Proposition 4.4.4, one can show that we may actually use our given étale coordinates on X. Since overconvergence is a local notion, it only remains to verify that our conditions on the radii of convergence are of local nature in our situation also. This immediately follows from the fact that $V_\rho = \cup(\mathcal{Y}_{iK} \cap V_\rho)$ is a finite affinoid covering. Details are left to the reader. □

Note in particular, that our former definition of overconvergence does not depend on the choice of étale coordinates. Using the second assertion of Proposition 4.4.4, it would be possible to give a direct proof of this result.

Corollary 7.2.16 *Let $X = \operatorname{Spec} A$ be a smooth affine $\mathcal{V}$-scheme with étale coordinates $t_1, \ldots, t_d$. If $f \in A_K^\dagger$ and $\eta < 1$, there exists $\rho_0 > 1$ such that whenever $1 \leq \rho \leq \rho_0$, we have $f \in A_\rho$ and f is η-convergent in A_ρ.*

Proof This follows from the fact that $A_K^\dagger$ itself is overconvergent. □

Recall that, if θ denotes the Taylor series, this assertion says that $\theta(f) \in A_\rho\{\underline{\tau}/\eta\}$. This result may also be derived from Corollary 3.4.14.

The next proposition shows that the overconvergence of a connection can be checked on the Robba rings.

Proposition 7.2.17 *Let $\mathcal{Y}$ be a flat formal $\mathcal{V}$-scheme whose special fiber Y is a smooth connected curve and X a dense open subset of Y such that all points of $Y \setminus X$ are rational. Let E be a coherent $j_X^\dagger \mathcal{O}_{\mathcal{Y}_K}$-module with a connection. Then, E is overconvergent if for each $x \in Y \setminus X$, the connection of E_x is overconvergent.*

Proof The question is local and we may therefore assume that $\mathcal{Y}$ is affine with an étale coordinate $t : \mathcal{Y} \to \widehat{\mathbf{A}}^1_{\mathcal{V}}$ such that

$$Y \setminus X = \{x\} = t^{-1}(0).$$

We write for $\lambda < 1$, $V^\lambda :=]Y[_P \setminus]x[_{P\lambda}$ and $E = j_X^\dagger \mathcal{E}$ where $\mathcal{E}$ is a coherent $\mathcal{O}_{V^{\lambda_0}}$-module with connection. We know from Proposition 4.4.10 that E is overconvergent if and only if

$$\lim_{\lambda \xrightarrow{<} 1} R(\mathcal{E}, V^\lambda) = 1.$$

And, by definition, E_x is overconvergent if and only if

$$\lim_{\lambda \overset{<}{\to} 1} R(E_x, \lambda) = 1.$$

We endow $\mathcal{E}$ with a quotient norm induced by the spectral norm on V^{λ_0}. Note now that for any $s \in \Gamma(V^\lambda, \mathcal{E})$, we have

$$\|s\|_{V^\lambda} = \sup(\|s\|_\lambda, \|s\|_{]X]_y}),$$

where $\| - \|_\lambda$ denotes the norm on the annulus $A(0, \lambda^+, \lambda^+)$ under the isomorphism $]x[_y \simeq \mathbf{D}(0, 1^-)$. It immediately follows that our condition is necessary. This equality also reduces the converse assertion to the proof that $E_{|(X \subset \mathcal{Y})}$ is convergent when E_x is overconvergent. Thus we have to show that $R(\mathcal{E},]X[_Y) = 1$. This is a consequence of the continuity of the radius of convergence (see for example [27]) which we will not prove here because it would lead us too far. Let us just say that it follows from the fact that $\|s_\lambda\| \to \|s\|_{]X[_Y}$ when $\lambda \to 1$. □

7.3 Virtual frames

Roughly speaking, a virtual frame is a frame where the third term is missing. Alternately, it might be seen as a smooth frame whose third term is not specified.

Definition 7.3.1 *A* virtual frame $(X \subset Y)$ *is an open immersion* $X \hookrightarrow Y$ *of algebraic k-varieties. A* morphism of virtual frames *from* $(X' \subset Y')$ *to* $(X \subset Y)$ *is a commutative diagram*

$$\begin{array}{ccc} X' & \hookrightarrow & Y' \\ \downarrow{\scriptstyle f} & & \downarrow{\scriptstyle g} \\ X & \hookrightarrow & Y. \end{array}$$

Unless necessary, we will only mention g. The morphism is said to be cartesian *if the square is cartesian.*

The standard situation is the following: starting with a quasi-projective variety X over k, we embed it into some projective space $\mathbf{P}^n_k$ and we take $Y = \overline{X}$, the algebraic closure of X. It is also worth mentioning the case of a smooth curve over k because such a curve has a unique smooth projective compactification.

Proposition 7.3.2 *Virtual frames form a category with finite inverse limits and the forgetful functors are left exact.*

Proof It is clear that we obtain a category. The second part of the assertion tells us that finite inverse limits exist and can be computed term by term. This follows from the fact that finite inverse limits of open immersions are still open immersions. □

Of course, if we fix some virtual frame, we may always consider the category of virtual frames over the fixed one. As a particular case, if S is a formal $\mathcal{V}$-scheme, we may consider the category of virtual frames over S, meaning over $(S_k = S_k)$.

Proposition 7.3.3 *If $v : S' \to S$ is a morphism of formal $\mathcal{V}$-schemes, there is an obvious forgetful functor from virtual S'-frames to virtual S-frames as well as a pullback functor in the other direction. Also, if $\sigma : K \hookrightarrow K'$ is an isometric embedding, there is an obvious extension functor*

$$(X \subset Y) \mapsto (X^\sigma \subset Y^\sigma).$$

Proof Pulling back an open immersion gives an open immersion. □

Proposition 7.3.4

(i) *Given a virtual frame $(X \hookrightarrow Y)$, any morphism of algebraic k-varieties $g : Y' \to Y$ extends uniquely to a cartesian morphism of virtual frames.*
(ii) *Any morphism of virtual frames can be split into a morphism with $g = \mathrm{Id}_Y$ (and f an open immersion), followed by a cartesian morphism.*

Proof The first assertion is completely trivial. For the second one, simply note that any morphism splits as

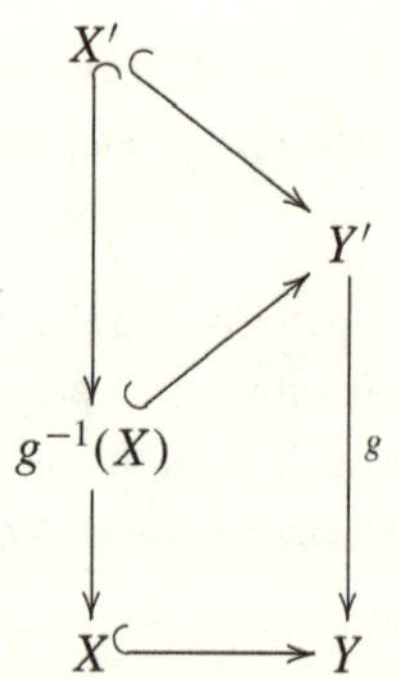

□

Definition 7.3.5 *A* morphism from an S-frame $(X' \subset Y' \subset P')$ to a virtual S-frame $(X \subset Y)$ *is a commutative diagram*

$$\begin{array}{ccccc} X' & \hookrightarrow & Y' & \hookrightarrow & P' \\ \downarrow{\scriptstyle f} & & \downarrow{\scriptstyle g} & & \downarrow \\ X & \hookrightarrow & Y & \longrightarrow & S \end{array}$$

We will also say that $(X' \subset Y' \subset P')$ is a frame over $(X \subset Y/S)$. *A* morphism of S-frames over the virtual frame $(X \subset Y/S)$ *is a commutative diagram*

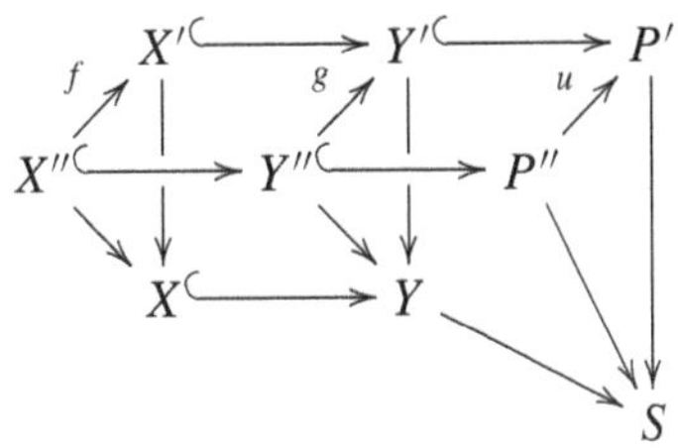

Proposition 7.3.6 *If $(X \subset Y/S)$ is a virtual frame over S, then S-frames over $(X \subset Y/S)$ form a category with non empty finite inverse limits.*

Proof This is easily checked. □

Note however that this category has no final object in general.

Definition 7.3.7 *A* (finitely presented) overconvergent isocrystal *on a virtual S-frame $(X \subset Y)$ is a family of (coherent) $j^\dagger_{X'}\mathcal{O}_{]Y'[_{P'}}$-modules $E_{P'}$ for each frame $(X' \subset Y' \subset P')$ over $(X \subset Y/S)$ and, for each morphism of S-frames*

$$\begin{array}{ccccc} X'' & \hookrightarrow & Y'' & \hookrightarrow & P'' \\ \downarrow{\scriptstyle f} & & \downarrow{\scriptstyle g} & & \downarrow{\scriptstyle u} \\ X' & \hookrightarrow & Y' & \hookrightarrow & P' \end{array}$$

over $(X \subset Y/S)$, an isomorphism

$$\varphi_u : u^\dagger E_{P'} \simeq E_{P''}.$$

Moreover, these isomorphisms are subject to the cocycle condition

$$\varphi_{v \circ u} = \varphi_v \circ u'^* \varphi_u.$$

We call $E_{P'}$ the realization *of E on* $(X' \subset Y' \subset P'/S')$.

A morphism of overconvergent isocrystals *is a family of compatible morphisms of* $j^\dagger_{X'}\mathcal{O}_{]Y'[_{P'}}$*-modules.*

When $Y = X$*, we simply say* convergent isocrystal.

We will denote by $\mathrm{Isoc}^\dagger(X \subset Y/S)$ the category of finitely presented overconvergent isocrystals on $(X \subset Y/S)$ and by $\mathrm{Isoc}(X/S)$ the category of finitely presented convergent isocrystals on (X/S).

Proposition 7.3.8 *Let*

$$\begin{array}{ccc} X' & \hookrightarrow & Y' \\ \downarrow{\scriptstyle f} & & \downarrow{\scriptstyle g} \\ X & \hookrightarrow & Y \end{array}$$

be a morphism from a virtual S'*-frame to a virtual* S*-frame over some morphism of formal* $\mathcal{V}$*-schemes* $S' \to S$*. Then, any overconvergent isocrystal* E *on* $(X \subset Y/S)$ *defines by restriction an isocrystal*

$$g^*E := E_{|(X' \subset Y'/S')}$$

on $(X' \subset Y'/S')$*. And this is functorial in* E*.*

In particular, there is a pullback functor

$$g^* : \mathrm{Isoc}^\dagger(X \subset Y/S) \to \mathrm{Isoc}^\dagger(X' \subset Y'/S').$$

Proof As usual, looking at the definition, this is a triviality. □

In practice, when g is the identity, we will simply write $E_{|X'}$ and when the morphism is cartesian, we will simply write $E_{|Y'}$.

In the next proposition, we show the local nature of the category $\mathrm{Isoc}^\dagger(X \subset Y/S)$. Before that, we need a definition.

Definition 7.3.9 *An* (open, closed) immersion *of virtual frames is a morphism of virtual frames*

$$\begin{array}{ccc} X' & \hookrightarrow & Y' \\ \downarrow & & \downarrow \\ X & \hookrightarrow & Y \end{array}$$

where both vertical arrows are (open, closed) immersions.

An (open, closed) covering *is a family of (open, closed) immersions*

$$\begin{array}{ccc} X_i & \hookrightarrow & Y_i \\ \downarrow & & \downarrow \\ X & \hookrightarrow & Y \end{array}$$

which is a covering at each step. It is said to be cartesian if each immersion of virtual frames is cartesian.

Proposition 7.3.10 *Let*

$$\begin{array}{ccc} X_i & \hookrightarrow & Y_i \\ \downarrow & & \downarrow \\ X & \hookrightarrow & Y \end{array}$$

be an open covering or a finite closed covering of a virtual S-frame. Then, there is an equivalence of categories

$$\mathrm{Isoc}^\dagger(X \subset Y/S) \simeq \varprojlim \Big[\textstyle\prod_i \mathrm{Isoc}^\dagger(X_i \subset Y_i/S)$$
$$\rightrightarrows \textstyle\prod_{\{i,j\}} \mathrm{Isoc}^\dagger(X_{ij} \subset Y_{ij}/S) \,\substack{\rightarrow\\\rightarrow\\\rightarrow}\, \textstyle\prod_{\{i,j,k\}} \mathrm{Isoc}^\dagger(X_{ijk} \subset Y_{ijk}/S)\Big].$$

Proof This is an immediate consequence of Proposition 5.4.12. □

Proposition 7.3.11 *Let $(X \subset Y \subset P)$ be a smooth S-frame. Then, the restriction functor*

$$\mathrm{Isoc}^\dagger(X \subset Y/S) \to \mathrm{Isoc}^\dagger(X \subset Y \subset P/S)$$

is an equivalence of categories.

Proof Let $E \in \mathrm{Isoc}^\dagger(X \subset Y \subset P/S)$ and $(X' \subset Y' \subset P')$ be a frame over $(X \subset Y/S)$. Consider the following commutative diagram:

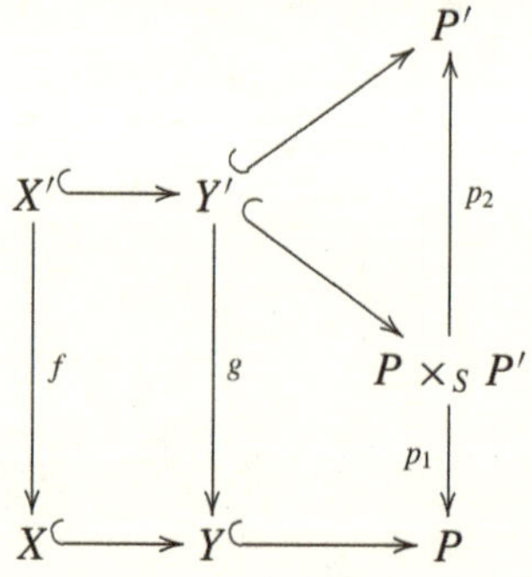

Since P is smooth over S in the neighborhood of X, it follows from Proposition 7.1.8 that p_2 induces an equivalence of categories

$$p_2^\dagger : \mathrm{Isoc}^\dagger(X' \subset Y' \subset P'/S) \simeq \mathrm{Isoc}^\dagger(X' \subset Y' \subset P \times_S P'/S).$$

In particular, there exists $E' \in \mathrm{Isoc}^\dagger(X' \subset Y' \subset P'/S)$ such that

$$p_2^\dagger E' = p_1^\dagger E$$

and one can set $E_{P'} := E'_{P'}$. It is not difficult to see that this defines an element of $\mathrm{Isoc}^\dagger(X \subset Y/S)$ and that we do get a quasi-inverse to the forgetful functor. The details are left to the reader. □

Corollary 7.3.12 *Let $X \subset Y$ be a virtual S-frame. If $Y \hookrightarrow P$ is a closed embedding into a formal S-scheme which is smooth in a neighborhood of X, then the category* $\mathrm{Isoc}^\dagger(X \subset Y \subset P/S)$ *does not depend on P, up to natural equivalence.*

Proof This is an immediate consequence of the proposition. □

Corollary 7.3.13 *In the definition of finitely presented overconvergent isocrystals over $(X \subset Y/S)$, we get an equivalent category by considering only strictly local frames $(X' \subset Y' \subset P')$ with Y' open in Y (and consequently X' open in X).*

Proof Using Proposition 7.3.10, we may assume that Y is affine and consider the smooth frame $X \subset Y \subset \widehat{\mathbf{A}}_S^N$. Then, our assertion follows from Propositions 7.3.11 and 7.2.3. □

Corollary 7.3.14 *If $(X \subset Y)$ is a virtual S-frame, then the category* $\text{Isoc}^\dagger(X \subset Y/S)$ *is abelian with tensor product and internal Hom. Moreover, if*

$$\begin{array}{ccc} X' & \hookrightarrow & Y' \\ \downarrow f & & \downarrow g \\ X & \hookrightarrow & Y \end{array}$$

is a morphism from a virtual S'-frame to a virtual S-frame over some morphism of formal $\mathcal{V}$-schemes $S' \to S$, the pullback functor

$$g^* : \text{Isoc}^\dagger(X \subset Y/S) \to \text{Isoc}^\dagger(X' \subset Y'/S')$$

is exact.

Proof Using localization and comparison, the first assertion follows from Proposition 7.1.2 and the second one from Proposition 7.1.4. □

Proposition 7.3.15 *Let $(X \subset Y)$ be a virtual S-frame and $\sigma : K \hookrightarrow K'$ an isometric embedding. Then there is a pullback functor*

$$\begin{array}{ccc} \text{Isoc}^\dagger(X \subset Y/S) & \longrightarrow & \text{Isoc}^\dagger(X^\sigma \subset Y^\sigma/S^\sigma) \\ E & \longmapsto & E^\sigma \end{array}$$

such that for any frame $(X' \subset Y' \subset P')$ over $(X \subset Y)$, we have

$$E^\sigma_{P'^\sigma} = (E_{P'})^\sigma .$$

Proof Using Proposition 7.3.11, this directly follows from Proposition 7.2.4 since this assertion is local and that we may embed Y in $\widehat{\mathbf{A}}^N_S$. □

7.4 Cohomology of virtual frames

We assume in this section that $\text{Char} K = 0$. Also, as soon as the definition is given we will always assume that the virtual S-frames and the morphisms from virtual frames to frames are all realizable.

Definition 7.4.1 *If $(X \subset Y \subset P)$ is a smooth S-frame, we call it a* realization *of the virtual S-frame $(X \subset Y)$. And we will say that the virtual frame is* realizable.

A virtual morphism *from the virtual S-frame* $(X \subset Y)$ *to an S-frame* $(C \subset D \subset Q)$ *is just a morphism to the underlying virtual S-frame:*

$$\begin{array}{ccccc} X & \hookrightarrow & Y & & \\ \downarrow{\scriptstyle f} & & \downarrow{\scriptstyle g} & & \\ C & \hookrightarrow & D & \hookrightarrow & Q \end{array}$$

over S. *A* realization *on* $(X \subset Y \subset P)$ *of this virtual morphism is a morphism of S-frames:*

$$\begin{array}{ccccc} X & \hookrightarrow & Y & \hookrightarrow & P \\ \downarrow{\scriptstyle f} & & \downarrow{\scriptstyle g} & & \downarrow{\scriptstyle u} \\ C & \hookrightarrow & D & \hookrightarrow & Q \end{array}$$

Note that a virtual S-frame $(X \subset Y)$ is always realizable when Y is quasi-projective over S: write Y as a formal subscheme of $\widehat{\mathbf{P}}^N_{\mathcal{V}}$ and choose an open subset P with Y closed in P. Also, if $(X \subset Y)$ is a realizable S-frame, then any morphism to a smooth S-frame $(C \subset D \subset Q)$ is realizable : first, embed Y in some P and then use the diagonal embedding into the product $P \times_S Q$. Finally, note that when g is quasi-compact, we may always assume that u is also quasi-compact.

Unless otherwise specified, we will only consider realizable virtual frames and realizable morphisms.

Proposition 7.4.2 *Let*

$$\begin{array}{ccccc} X & \hookrightarrow & Y & \hookrightarrow & P \\ \downarrow{\scriptstyle f} & & \downarrow{\scriptstyle g} & & \downarrow{\scriptstyle u} \\ C & \hookrightarrow & D & \hookrightarrow & Q \end{array} \quad \text{and} \quad \begin{array}{ccccc} X & \hookrightarrow & Y & \hookrightarrow & P' \\ \downarrow{\scriptstyle f} & & \downarrow{\scriptstyle g} & & \downarrow{\scriptstyle u'} \\ C & \hookrightarrow & D & \hookrightarrow & Q \end{array}$$

be two realizations of the same virtual morphism of S-frames.

If $E \in \mathrm{Isoc}^\dagger(X \subset Y/S)$, *there exists an isomorphism*

$$\psi_{u,u'} : Ru_{\mathrm{rig}} E_P \simeq Ru'_{\mathrm{rig}} E_{P'}.$$

Moreover, if

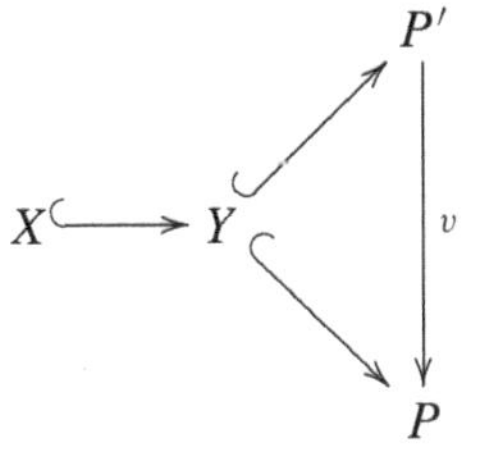

is any morphism of frames such that $u' = u \circ v$, *the base change map is an isomorphism*

$$v^* : Ru_{\mathrm{rig}} E_P \simeq Ru'_{\mathrm{rig}} v^\dagger E_P$$

and $\psi_{u,u'} = \varphi_v \circ v^*$ *where* φ_v *comes from the canonical isomorphism* $v^\dagger E_P \simeq E_{P'}$.

The analogous results for $Ru_{\mathrm{rig},X'} E$ *when* X' *is a subvariety of* X *and for* $Ru_{\mathrm{rig},c} E$ *when* u *is quasi-compact and* $C = D$ *are also true.*

Proof We do only the case without support. The other cases are strictly analogous and left to the reader.

First of all, with v as in the second part, we denote by

$$\psi_v : Ru_{\mathrm{rig}} E_P \to Ru'_{\mathrm{rig}} v^\dagger E_P \simeq Ru'_{\mathrm{rig}} E_{P'}$$

the composite map. Note that, if

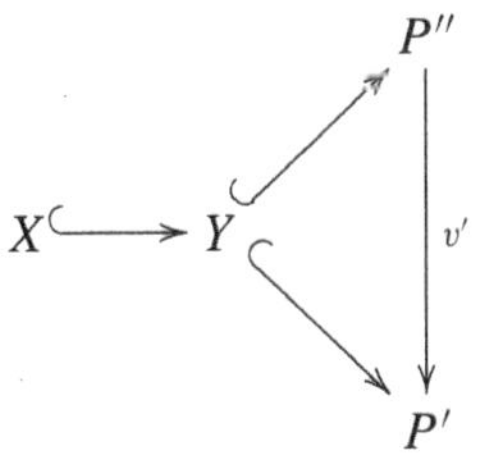

is a morphism of S-frames such that $u'' = u' \circ v'$, then by functoriality $\psi_{v \circ v'} = \psi_{v'} \circ \psi_v$. On the other hand, we know from Theorem 6.5.2 that when v is smooth, then ψ_v is an isomorphism. It follows that, if v is smooth and v' is a section of v, then $\psi_{v'}$ is also an isomorphism and $\psi_{v'} = \psi_v^{-1}$.

We now turn to the definition of $\psi_{u,u'}$. Since u and u' are smooth, so are the projections $p_1 : P \times_Q P' \to P$ and $p_2 : P \times_Q P' \to P'$ and we let

$$\psi_{u,u'} := \psi_{p_2}^{-1} \circ \psi_{p_1}.$$

Back to our morphism v, its graph $\gamma : P' \to P \times_Q P'$ is a section of p_2 and it follows that

$$\psi_{u,u'} = \psi_\gamma \circ \psi_{p_1} = \psi_{p_1 \circ \gamma} = \psi_v.$$

□

We may reformulate this proposition in the following form:

Corollary 7.4.3 *Let*

$$\begin{array}{ccccc} X & \hookrightarrow & Y & & \\ \downarrow & & \downarrow g & & \\ C & \hookrightarrow & D & \hookrightarrow & Q \end{array}$$

be a virtual morphism over S and E an overconvergent isocrystal on $(X \subset Y/S)$. Let $Y \hookrightarrow P$ be a closed embedding into a formal S-scheme which is smooth in a neighborhood of X such that g extends to $u : P \to Q$.

Then, up to a canonical isomorphism, $Ru_{\mathrm{rig}}E_P$ only depends on our virtual morphism and not on P nor u. This is more generally true of $Ru_{\mathrm{rig},X'}E_P$ when X' is a closed subset of X. And this is also true for $Ru_{\mathrm{rig},c}E_P$ when $C = D$ and g is quasi-compact.

Proof Should be clear. □

We may therefore make the following definition:

Definition 7.4.4 *Let*

$$\begin{array}{ccccc} X & \hookrightarrow & Y & \hookrightarrow & P \\ \downarrow f & & \downarrow g & & \downarrow u \\ C & \hookrightarrow & D & \hookrightarrow & Q \end{array}$$

be a realization of a virtual morphism of S-frames and $E \in \mathrm{Isoc}^\dagger(X \subset Y/S)$.

The relative rigid cohomology *of E is the complex*

$$Rg_{\mathrm{rig}}E := Ru_{\mathrm{rig}}E_P$$

on $]D[_P$. When $Y = X$, we call it the convergent cohomology *and denote it by $Rg_{\mathrm{conv}}E$.*

If X' is a closed subvariety of X, the relative rigid cohomology of E with support in X' *is*

$$Rg_{\mathrm{rig},X'}E := Ru_{\mathrm{rig},X'}E_P.$$

Finally, if u is quasi-compact and $C = D$, the relative rigid cohomology with compact support *of E is*

$$Rg_{\mathrm{rig},c}E := Ru_{\mathrm{rig},c}E_P.$$

Note that, although the definition of the category is crystalline in nature, the definition of cohomology is not and this is not satisfactory. In particular, rigid cohomology is only defined for realizable virtual frames. In general, one must use simplicial techniques such as in [26] or ad hoc glueing methods.

If $(X \subset Y)$ is a virtual S-frame with structural map $g : Y \to S_k$ seen as a morphism to the frame $(S_k = S_k \subset S)$, we define for each $i \in \mathbf{N}$, the *absolute rigid cohomology* of E

$$\mathcal{H}^i_{\mathrm{rig}}(X \subset Y/S, E) := R^i g_{\mathrm{rig}} E.$$

More generally, there is the *absolute rigid cohomology with support in some closed subvariety X' of X.*

$$\mathcal{H}^i_{\mathrm{rig},X'}(X \subset Y/S, E) := R^i g_{\mathrm{rig},X'} E.$$

And we may also consider *absolute rigid cohomology with compact support*

$$\mathcal{H}^i_{\mathrm{rig},c}(X \subset Y/S, E) := R^i g_{\mathrm{rig},c} E.$$

In the case $S = \mathrm{Spf}\mathcal{V}$, we will consider the *absolute rigid cohomology* of E

$$R\Gamma_{\mathrm{rig}}(X \subset Y, E) := Rg_{\mathrm{rig}}E \quad \text{and} \quad H^i_{\mathrm{rig}}(X \subset Y, E) := R^i g_{\mathrm{rig}}E,$$

with support in a closed subset,

$$R\Gamma_{\mathrm{rig},X'}(X \subset Y, E) := Rg_{\mathrm{rig},X'}E$$

and

$$H^i_{\mathrm{rig},X'}(X \subset Y, E) := R^i g_{\mathrm{rig},X'}E$$

or with compact support

$$R\Gamma_{\mathrm{rig},c}(X \subset Y, E) := Rg_{\mathrm{rig},c}E \quad \text{and} \quad H^i_{\mathrm{rig},c}(X \subset Y, E) := R^i g_{\mathrm{rig},c}E.$$

Finally, if X is an S_k-variety with structural map $f : X \to S_k$ seen as the morphism from the virtual frame $(X = X)$ to the trivial frame $(S_k = S_k \subset S)$, we may also consider for each $i \in \mathbf{N}$, the *absolute convergent cohomology* of E

$$\mathcal{H}^i_{\mathrm{conv}}(X/S, E) := R^i g_{\mathrm{conv}} E$$

and in the case $S = \mathrm{Spf}\mathcal{V}$,

$$R\Gamma_{\mathrm{conv}}(X, E) := Rg_{\mathrm{conv}}E \quad \text{and} \quad H^i_{\mathrm{conv}}(X, E) := R^i g_{\mathrm{conv}}E.$$

Finally, when E is the trivial module with connection, we drop it from the H-notations.

Let us go back to the general case.

Proposition 7.4.5 *Let $(X \subset Y \subset P)$ and $(X \subset Y \subset P')$ be two realizations of the same virtual S-frame. If $E \in \mathrm{Isoc}^{\dagger}(X \subset Y/S)$, there exists an isomorphism*

$$\psi_{u,u'} : R\underline{\Gamma}_{\mathrm{rig}} E_P \simeq R\underline{\Gamma}_{\mathrm{rig}} E_{P'}$$

on Y. Moreover, if

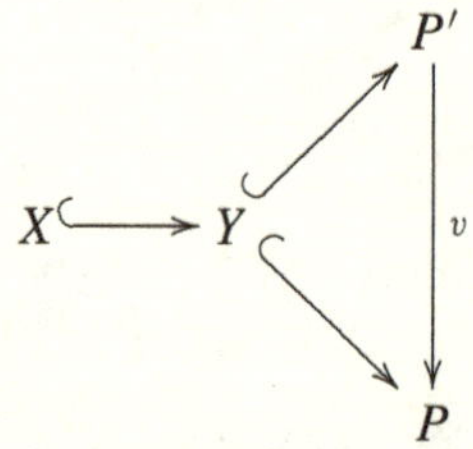

is any morphism of frames, the base change map is an isomorphism

$$v^* : R\underline{\Gamma}_{\mathrm{rig}} E_P \simeq R\underline{\Gamma}_{\mathrm{rig}} v^{\dagger} E_P$$

and $\psi_{u,u'} = \varphi_v \circ v^*$ where φ_v comes from the canonical map $v^{\dagger} E_P \simeq E_{P'}$.

The analogous results for $R\underline{\Gamma}_{\mathrm{rig},X'} E$ when X' is a subvariety of X and for $R\underline{\Gamma}_{\mathrm{rig},c} E$ are also true.

Proof Using Corollaries 6.5.12 and 6.5.18, the proof goes exactly as in Proposition 7.4.2. □

We may therefore make the following definition:

Definition 7.4.6 *Let $(X \subset Y \subset P/S)$ be a realization of a virtual S-frame and $E \in \mathrm{Isoc}^{\dagger}(X \subset Y/S)$. The* specialized de Rham complex *of E is*

$$R\underline{\Gamma}_{\mathrm{rig}} E := R\underline{\Gamma}_{\mathrm{rig}} E_P$$

on Y.

If X' is a closed subvariety of X, the specialized de Rham complex of E with support in X' *is*

$$R\underline{\Gamma}_{\mathrm{rig},X'} E := R\underline{\Gamma}_{\mathrm{rig},X'} E_P.$$

Finally, the specialized de Rham complex with compact support *of E is*

$$R\underline{\Gamma}_{\mathrm{rig},c} E := R\underline{\Gamma}_{\mathrm{rig},c} E_P.$$

Proposition 7.4.7 *Let $(X \subset Y)$ be a virtual S-frame and $E \in \mathrm{Isoc}^\dagger(X \subset Y/S)$. If*

$$\begin{array}{ccccc} X & \hookrightarrow & Y & & \\ \downarrow & & \downarrow g & & \\ C & \hookrightarrow & D & \hookrightarrow & Q \end{array}$$

is a virtual morphism, then

$$Rsp_* Rg_{\mathrm{rig}} E = Rg_* R\underline{\Gamma}_{\mathrm{rig}} E,$$
$$Rsp_* Rg_{\mathrm{rig},X'} E = Rg_* R\underline{\Gamma}_{\mathrm{rig},X'} E,$$

when X' is a closed subvariety of X and

$$Rsp_* Rg_{\mathrm{rig},c} E = Rg_* R\underline{\Gamma}_{\mathrm{rig},c} E$$

when g is quasi-compact and $C = D$.

Proof This is an immediate consequence of the corresponding result for frames proved in Propositions 6.2.4, 6.3.4 and 6.4.6. □

In particular, if $(X \subset Y)$ is a virtual frame and E an overconvergent isocrystal on $X \subset Y$, we have

$$R\Gamma_{\mathrm{rig}}(X \subset Y, E) = R\Gamma(Y, R\underline{\Gamma}_{\mathrm{rig}} E)$$
$$R\Gamma_{\mathrm{rig},X'}(X \subset Y, E) = R\Gamma(Y, R\underline{\Gamma}_{\mathrm{rig},X'} E)$$

when X' is a closed subvariety of X and

$$R\Gamma_{\mathrm{rig},c}(X \subset Y, E) = R\Gamma(Y, R\Gamma_{\mathrm{rig},c} E)$$

when Y is quasi-compact.

Proposition 7.4.8 *Let*

$$\begin{array}{ccccc} X & \hookrightarrow & Y & & \\ \downarrow & & \downarrow g & & \\ C & \hookrightarrow & D & \hookrightarrow & Q \end{array}$$

be a virtual morphism of S-frames and $E \in \mathrm{Isoc}^\dagger(X \subset Y/S)$.

(i) *If X' is a closed subvariety of X, there is a canonical morphism*

$$Rg_{\mathrm{rig},X'} E \to Rg_{\mathrm{rig}} E$$

which is an isomorphism when $X' = X$.

(ii) *If X' is a closed subvariety of X and U is an open neighborhood of X' in X, then*

$$Rg_{\mathrm{rig},X'}E = Rg_{\mathrm{rig},X'}E_{|U}.$$

(iii) *When $C = D$ and g is quasi-compact, there is a canonical morphism*

$$Rg_{\mathrm{rig},c}E \to Rg_{\mathrm{rig}}E$$

which is an isomorphism when $Y = X$.

Proof This is an immediate consequence of the corresponding result for frames in Propositions 6.3.2 and 6.4.12. □

Proposition 7.4.9 *Assume that we are given a commutative diagram of morphisms of (virtual) frames*

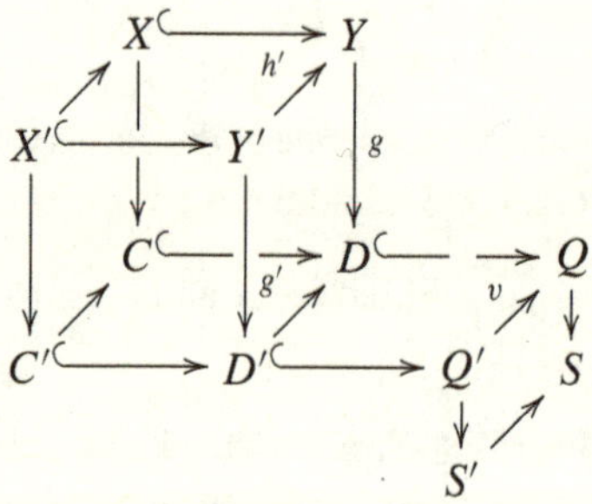

Then, if $E \in \mathrm{Iso}^\dagger(X \subset Y/S)$, there is a canonical base change map

$$h'*: Lv_K^* Rg_{\mathrm{rig}}E \to Rg'_{\mathrm{rig}}h'^*E$$

with $v_K :]D'[_{Q'} \to]D[_Q$. More generally, if X'' (resp. X''') is a closed subvariety of X (resp. X') with $f^{-1}(X'') \subset X'''$, there is a canonical base change map

$$h'*: Lv_K^* Rg_{\mathrm{rig},X''}E \to Rg'_{\mathrm{rig},X'''}h'^*E.$$

Finally, when $C = D$ and h' is cartesian and g is quasi-compact, we also have

$$h'*: Lv_K^* Rg_{\mathrm{rig},c}E \to Rg'_{\mathrm{rig},c}h'^*E.$$

Proof This is an immediate consequence of the corresponding result for frames proved in Propositions 6.2.6, 6.3.5 and 6.4.7. □

Proposition 7.4.10 *Let $(X \subset Y)$ be a virtual S-frame and E a finitely presented overconvergent isocrystal on $(X \subset Y/S)$. If $K \hookrightarrow K'$ is a finite field extension, there is a canonical isomorphism*

$$K' \otimes_K H^q_{\mathrm{rig}}(X \subset Y, E) \simeq H^q_{\mathrm{rig}}(X_k \subset Y_k, E_{K'}).$$

More generally, if X' is a closed subset of X, we have

$$K' \otimes_K H^q_{\mathrm{rig},X'}(X \subset Y, E) \simeq H^q_{\mathrm{rig},X'_k}(X_k \subset Y_k, E_{K'}).$$

Finally, if the Y is quasi-compact, we have

$$K' \otimes_K H^q_{\mathrm{rig},c}(X \subset Y, E) \simeq H^q_{\mathrm{rig},c}(X_k \subset Y_k, E_{K'}).$$

Proof This is an immediate consequence of the corresponding result for frames in Propositions 6.2.7, 6.3.10 and 6.4.8. □

Proposition 7.4.11 *If* $\mathrm{Char}(K) = 0$ *and* $(X \subset Y)$ *is a virtual frame over* $\mathcal{V}$, *then* $H^0_{\mathrm{rig}}(X \subset Y)$ *is a finite-dimensional space whose dimension is the number of geometrically connected components of* X.

Proof This follows from Proposition 6.2.11. □

Proposition 7.4.12 *Let*

$$\begin{array}{ccccc} X & \hookrightarrow & Y & & \\ \downarrow & & \downarrow g & & \\ C & \hookrightarrow & D & \hookrightarrow & Q \end{array}$$

be a virtual morphism of S-frames,

$$\begin{array}{ccccc} C' & \hookrightarrow & D' & \hookrightarrow & Q' \\ \downarrow & & \downarrow & & \downarrow \\ C & \hookrightarrow & D & \hookrightarrow & Q \end{array}$$

a cartesian open or mixed immersion of frames and

$$\begin{array}{ccccc} X' & \hookrightarrow & Y' & & \\ \downarrow & & \downarrow g' & & \\ C' & \hookrightarrow & D' & \hookrightarrow & Q' \end{array}$$

the pullback of the first morphism of frames along the cartesian open immersion.

If $E \in \mathrm{Isoc}^\dagger(X \subset Y/S)$, we have

$$(Rg_{\mathrm{rig}} E)_{|]D'[_{Q'}} \simeq Rg'_{\mathrm{rig}} E_{|Y'}.$$

More generally, we have

$$(Rg_{\mathrm{rig},X''} E)_{|]D'[_{Q'}} \simeq Rg'_{\mathrm{rig},X'''} E_{|Y'}.$$

when X'' is a closed subvariety of X and $X''' := X' \cap X''$, and

$$(Rg_{\mathrm{rig},c}E)_{|]D'[_{Q'}} \simeq Rg'_{\mathrm{rig},c}E_{|Y'}$$

when g is quasi-compact and $C = D$.

Proof This is an immediate consequence of the corresponding result for frames proved in Propositions 6.2.9, 6.3.6 and 6.4.9. □

Proposition 7.4.13 *Let*

$$\begin{array}{ccccc} X & \hookrightarrow & Y & & \\ \downarrow & & \downarrow g & & \\ C & \hookrightarrow & D & \hookrightarrow & Q \end{array}$$

be a virtual morphism of S-frames and $E \in \mathrm{Isoc}^\dagger(X \subset Y/S)$. Assume that we are given an open or finite closed covering

$$\begin{array}{ccc} X_i & \hookrightarrow & Y_i \\ \downarrow & & \downarrow \\ X & \hookrightarrow & Y \end{array}$$

Then, there is a canonical spectral sequence

$$E_1^{p,q} := \bigoplus_{|\underline{i}|=p} R^q g_{\underline{i}\mathrm{rig}} E_{|(X_{\underline{i}} \subset Y_{\underline{i}})} \Rightarrow Rg_{\mathrm{rig}}E.$$

More generally, if X' is a closed subvariety of X and $X'_{\underline{i}} := X' \cap X_{\underline{i}}$, there is a canonical spectral sequence

$$E_1^{p,q} := \bigoplus_{|\underline{i}|=p} R^q g_{\underline{i}\mathrm{rig},X_{\underline{i}}} E_{|(X_{\underline{i}} \subset Y_{\underline{i}})} \Rightarrow Rg_{\mathrm{rig},X'}E$$

and when g is quasi-compact, $C = D$ and the covering is cartesian, a canonical spectral sequence

$$E_1^{p,q} := \bigoplus_{|\underline{i}|=p} R^q g_{\underline{i}\mathrm{rig},c} E_{|Y_{\underline{i}}} \Rightarrow Rg_{\mathrm{rig},c}E.$$

Proof This is an immediate consequence of the corresponding result for frames proved in Propositions 6.2.10, 6.3.7 and 6.4.10. □

Proposition 7.4.14 *Let*

$$\begin{array}{ccccc} X & \hookrightarrow & Y & & \\ \downarrow & & \downarrow{\scriptstyle g} & & \\ C & \hookrightarrow & D & \hookrightarrow & Q \end{array}$$

be a virtual morphism of S-frames and $E \in \mathrm{Isoc}^{\dagger}(X \subset Y/S)$.

If $\sigma : K \hookrightarrow K'$ *is an isometric embedding, there is a canonical morphism*

$$(Rg_{\mathrm{rig}}E)^{\sigma} \to Rg^{\sigma}_{\mathrm{rig}}E^{\sigma}.$$

More generally, if X' *is a closed subvariety of* X, *we have*

$$(Rg_{\mathrm{rig},X'}E)^{\sigma} \to Rg^{\sigma}_{\mathrm{rig},X'^{\sigma}}E^{\sigma}$$

and when g is quasi-compact and $C = D$,

$$(Rg_{\mathrm{rig},c}E)^{\sigma} \to Rg^{\sigma}_{\mathrm{rig},c}E^{\sigma}.$$

Proof This is an immediate consequence of the corresponding result for frames proved in Propositions 6.2.8, 6.3.8 and 6.4.11.

□

Proposition 7.4.15 *Let*

$$\begin{array}{ccccc} X & \hookrightarrow & Y & & \\ \downarrow & & \downarrow{\scriptstyle g} & & \\ C & \hookrightarrow & D & \hookrightarrow & Q \end{array}$$

be a virtual morphism of S-frames. Let

$$X'' \hookrightarrow X' \hookrightarrow X$$

be a sequence of closed immersions, $U := X \setminus X''$ *and* $U' := X' \setminus X''$. *If* $E \in \mathrm{Isoc}^{\dagger}(X \subset Y/S)$, *there is an exact triangle*

$$Rg_{\mathrm{rig},X''}E \to Rg_{\mathrm{rig},X'}E \to Rg_{\mathrm{rig},U'}E_{|U} \to .$$

Proof This is an immediate consequence of the corresponding result for frames proved in Proposition 6.3.9. □

Proposition 7.4.16 *Let*

$$\begin{array}{ccccc} X & \hookrightarrow & Y & & \\ \downarrow & & \downarrow{\scriptstyle g} & & \\ D & = & D & \hookrightarrow & Q \end{array}$$

be a morphism of virtual S-frames with g quasi-compact and $E \in \mathrm{Isoc}^\dagger(X \subset Y/S)$. *Let* $X' \hookrightarrow X$ *be a closed immersion,* $U := X \setminus X'$, Y' *a closed subvariety of* Y *such that* $X' = Y \cap X$ *and* $g' : Y' \to D$ *the induced map.*

Then, there is an exact triangle

$$Rg_{\mathrm{rig},c}E_{|U} \to Rg_{\mathrm{rig},c}E \to Rg'_{\mathrm{rig},c}E_{|Y'} \to .$$

Proof This is an immediate consequence of the corresponding result for frames proved in Proposition 6.4.13. □

Definition 7.4.17 *A morphism of virtual frames*

$$\begin{array}{ccc} X' & \hookrightarrow & Y' \\ \downarrow{\scriptstyle f} & & \downarrow{\scriptstyle g} \\ X & \hookrightarrow & Y \end{array}$$

is proper (resp. projective, resp. finite) *if g is proper (resp. projective, resp. finite). A virtual frame* $(X \subset Y/S)$ *is* proper (resp. projective, resp. finite) *if the canonical morphism to the trivial virtual frame* $(S_k \subset S_k)$ *is proper (resp. projective, resp. finite).*

Of course, the latter just means that if Y is proper (resp. projective, resp. finite) over S_k.

Theorem 7.4.18 *Let*

$$\begin{array}{ccc} & & Y' \\ & \nearrow & \\ X & & \downarrow{\scriptstyle h} \\ & \searrow & \\ & & Y \end{array}$$

be a proper morphism of virtual S-frames. Then

(i) *The functor* $E \mapsto h^*E$ *is an equivalence of categories*

$$\mathrm{Isoc}^\dagger(X \subset Y/S) \simeq \mathrm{Isoc}^\dagger(X \subset Y'/S).$$

(ii) *Let* $E \in \mathrm{Isoc}^\dagger(X \subset Y/S)$ *and*

$$\begin{array}{ccccc} X & \hookrightarrow & Y & & \\ \downarrow & & \downarrow{\scriptstyle g} & & \\ C & \hookrightarrow & D & \hookrightarrow & Q \end{array}$$

a virtual morphism of S-frames.

If $g' = g \circ h$, *the base change map are isomorphisms*

$$Rg_{\mathrm{rig}} E \simeq Rg'_{\mathrm{rig}} h^* E.$$

More generally, if X' *is a closed subset of* X, *we get isomorphisms*

$$Rg_{X',\mathrm{rig}} E \simeq Rg'_{X',\mathrm{rig}} h^* E.$$

Finally, we also get isomorphisms

$$Rg_{\mathrm{rig},c} E \simeq Rg'_{\mathrm{rig},c} h^* E$$

when g *and* g' *are quasi-compact and* $C = D$.

(iii) *If* $E \in \mathrm{Isoc}^\dagger(X \subset Y/S)$, *the base change map are isomorphisms*

$$R\underline{\Gamma}_{\mathrm{rig}} E \simeq R\underline{\Gamma}_{\mathrm{rig}} h^* E,$$
$$R\underline{\Gamma}_{X',\mathrm{rig}} E \simeq R\underline{\Gamma}_{X',\mathrm{rig}} h^* E$$

when X' *is a closed subset of* X *and*

$$R\underline{\Gamma}_{\mathrm{rig},c} E \simeq R\underline{\Gamma}_{\mathrm{rig},c} h^* E$$

when g *and* g' *are quasi-compact and* $C = D$.

Proof All questions are local on X and on Y and only depend on a closed subset of Y containing X. In particular, we may assume X quasi-projective and by Chow's lemma (Corollary 5.7.14 of [75]), that h is projective. Also, thanks to Proposition 3.1.10, we may freely replace Y' by any closed subvariety that contains X. Now, Lemma 6.5.1 allows us to assume that h has an étale realization

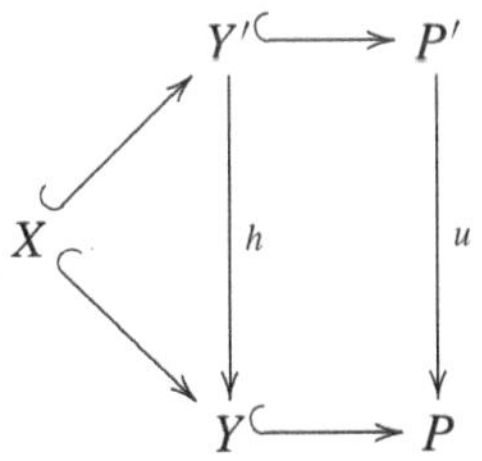

It induces therefore an isomorphism on strict neighborhoods and the theorem follows. □

8
Rigid cohomology

We fix some complete ultrametric field K of characteristic zero with $\mathcal{V}$, k and π as usual, and we denote by S a formal $\mathcal{V}$-scheme.

8.1 Overconvergent isocrystal on an algebraic variety

Definition 8.1.1 *Let X be an algebraic variety over S_k. An S-*frame over X *is an S-frame $(X' \subset Y' \subset P')$ endowed with a morphism of algebraic S_k-varieties $X' \to X$. A* morphism of S-frames over X *is a morphism of S-frames compatible with the morphisms to X:*

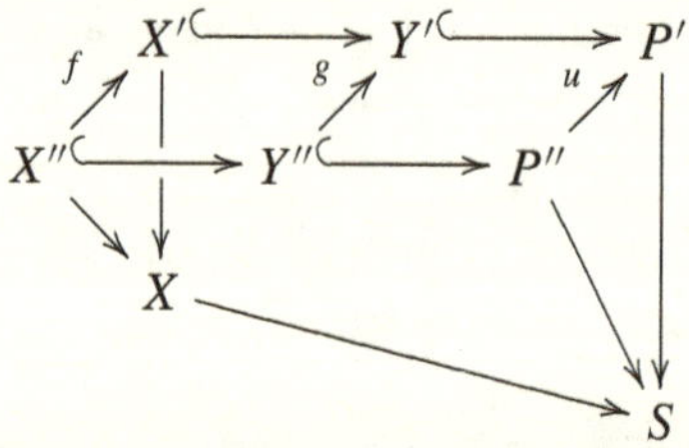

With these definitions, it is clear that S-frames over X form a category.

Proposition 8.1.2 *If X is an algebraic variety over S_k, then S-frames over X form a category with non empty finite inverse limits.*

Proof This is easily checked. □

Note however that, again, this category has no final object in general.

Definition 8.1.3 *A* (finitely presented) overconvergent isocrystal *on an algebraic S_k-variety X is a family of (coherent) $j^\dagger_{X'}\mathcal{O}_{]Y'[_{P'}}$-modules $E_{P'}$ for each*

S-frame $(X' \subset Y' \subset P')$ over X and, for each morphism of S-frames

$$\begin{array}{ccccc} X'' & \hookrightarrow & Y'' & \hookrightarrow & P'' \\ \downarrow{\scriptstyle f} & & \downarrow{\scriptstyle g} & & \downarrow{\scriptstyle u} \\ X' & \hookrightarrow & Y' & \hookrightarrow & P' \end{array}$$

over X, an isomorphism

$$\varphi_u : u^\dagger E_{P'} \simeq E_{P''}.$$

Moreover, these isomorphisms are subject to the cocycle condition

$$\varphi_{v \circ u} = \varphi_v \circ u'^* \varphi_u.$$

We call $E_{P'}$ the realization *of E on $(X' \subset Y' \subset P'/S')$.*

A morphism of overconvergent isocrystals *is a family of compatible morphisms of $j^\dagger_{X'} \mathcal{O}_{]Y'[_{P'}}$-modules.*

We will denote by $\mathrm{Isoc}^\dagger(X/S)$ the category of finitely presented overconvergent isocrystals on X/S.

Proposition 8.1.4 *Let $f : X' \to X$ be a morphism of algebraic varieties over some morphism of formal $\mathcal{V}$-schemes $S' \to S$. Then, any overconvergent isocrystal E on X/S defines by restriction an isocrystal*

$$f^* E := E_{|(X'/S')}$$

on X'/S'. And this is functorial in E. In particular, there is a pull-back functor

$$f^* : \mathrm{Isoc}^\dagger(X/S) \to \mathrm{Isoc}^\dagger(X'/S').$$

Proof Looking at the definition, this is again a triviality. □

It is important to note that the notion of finitely presented overconvergent isocrystal is of local nature on X:

Proposition 8.1.5 *Let X be an S_k-variety and $X = \cup X_i$ an open or a finite closed covering. Then, there is an equivalence of categories*

$$\mathrm{Isoc}^\dagger(X) \simeq \varprojlim \Big[\prod_i \mathrm{Isoc}^\dagger(X_i) \rightrightarrows \prod_{\{i,j\}} \mathrm{Isoc}^\dagger(X_{ij}) \\ \Rrightarrow \prod_{\{i,j,k\}} \mathrm{Isoc}^\dagger(X_{ijk}) \Big].$$

Proof As usual, this is an immediate consequence of Proposition 5.4.12. □

Actually, functoriality is far less rigid than it seems as the next result shows. In particular, we could have allowed a lot more morphisms in the category on which overconvergent isocrystals are defined. More precisely, we may have used *compatible pairs* as the next proposition shows.

Proposition 8.1.6 *Let $(X \subset Y \subset P)$ (resp. $(X' \subset Y' \subset P')$) be a frame and V (resp. V') a strict neighborhood of $]X[_P$ in $]Y[_P$ (resp. $]X'[_{P'}$ in $]Y'[_{P'}$). Let $f : X' \to X$ and $u : V' \to V$ be a pair of compatible morphisms. If E is an overconvergent isocrystal on X/S, there is a canonical isomorphism*

$$j^\dagger_{X'} u^* E_P \simeq E_{P'}.$$

Recall that the compatibility condition means that

$$\forall x \in V' \cap]X'[_{P'}, \quad f(\overline{x}) = \overline{u(x)}.$$

Proof Since the question is local (the isomorphisms will automatically glue), we may assume that the frames are quasi-compact.

Assume first that $f = \mathrm{Id}_X$ and that there exists a morphism of S-frames

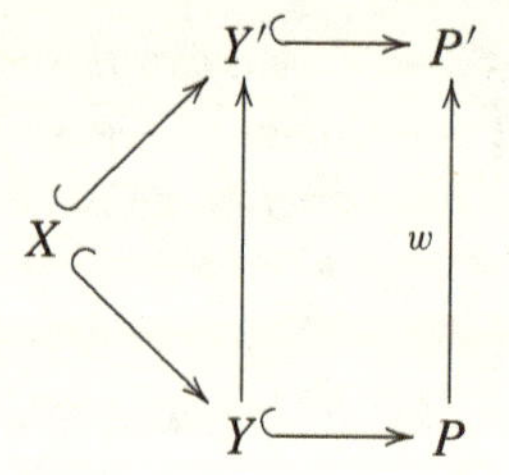

such that u is a section of w_K on V'. Then we have

$$u^*(w_K^* E_{P'})_{|V} = (E_{P'})_{|V'}$$

and applying $j^\dagger_X$, we get

$$j^\dagger_X u^*(w_K^* E_{P'})_{|V'} = E_{P'}.$$

On the other hand, it follows from Proposition 5.3.8 that

$$j^\dagger_X u^*(w_K^* E_{P'})_{|V'} = j^\dagger_X u^* j^\dagger_X w_K^* E_{P'} = j^\dagger_X u^* w^\dagger E_{P'} = j^\dagger_X u^* E_P.$$

We now come back to the general case and show that it reduces to the above particular case. In order to do so, we embed X' in $P' \times_S P$ by using the graph of f and consider the algebraic closure Y'' of X' in $Y \times_{S_k} Y' \subset P \times_S P'$. We

obtain two morphisms of frames

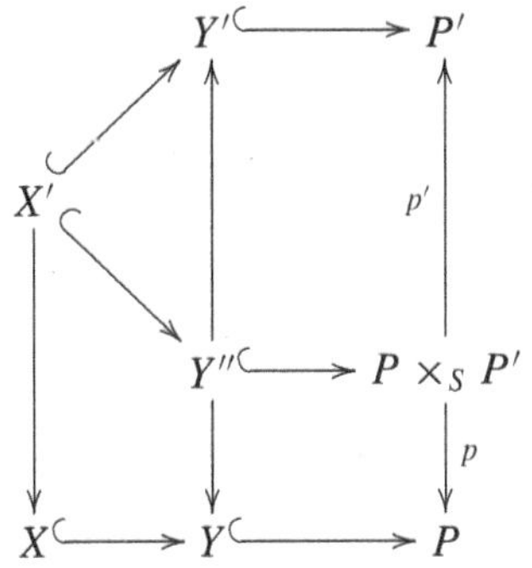

We may then consider the map $V' \to P_K \times_{S_K} P'_K$ induced by u and the inclusion map $V' \hookrightarrow P'_K$. Using Proposition 3.4.8, we may shrink V' and assume that this map induces a morphism $s : V' \to]Y''[_{P\times_S P'}$ which is section of p_K on V'. More precisely, Using the last assertion of Proposition 5.3.8 again, it results from the above particular case that

$$j^\dagger_{X'} u^* E_P = j^\dagger_{X'} s^* p^*_K E_P = j^\dagger_{X'} s^* j^\dagger_X p^*_K E_P$$
$$j^\dagger_{X'} s^* p^\dagger E_P = j^\dagger_{X'} s^* E_{P\times_S P'} \simeq E_{P'}.$$

□

Corollary 8.1.7 *In the situation of the proposition, we have the following*

(i) *If $E_P = j^\dagger_X \mathcal{E}$, then $E_{P'} = j^\dagger_{X'} u^* \mathcal{E}$.*
(ii) *If V is a strict neighborhood of $]X[_P$ in $]Y[_P$ and $\mathcal{E}$ is a coherent $\mathcal{O}_V$-module with an overconvergent integrable connection, then the connection of $u^*\mathcal{E}$ on $u^{-1}(V) \cap V'$ is also overconvergent.*

Proof The first assertion is again a consequence of Proposition 5.3.8 and the second one results from the definition. □

Proposition 8.1.8 *If $(X \subset Y)$ is a proper virtual S-frame, restriction induces an equivalence of categories*

$$\mathrm{Isoc}^\dagger(X/S) \simeq \mathrm{Isoc}^\dagger(X \subset Y/S).$$

Proof We are given some $E \in \mathrm{Isoc}^\dagger(X \subset Y/S)$ and an S-frame $(X' \subset Y' \subset P')$ over X and we need to give a meaning to $E_{P'}$. It is therefore sufficient to define some $E' \in \mathrm{Isoc}^\dagger(X' \subset Y'/S)$ and consider its realization on P'. But we may look at the restriction of E to $(X' \subset Y \times_{S_k} Y'/S)$ and we know from

Theorem 7.4.18 that

$$\mathrm{Isoc}^\dagger(X' \subset Y \times_{S_k} Y'/S) \simeq \mathrm{Isoc}^\dagger(X' \subset Y'/S).$$

□

Corollary 8.1.9 *Let X be an algebraic variety over S_k.*

(i) *If $X \hookrightarrow Y$ is an open embedding into a proper variety over S_k, then the category $\mathrm{Isoc}^\dagger(X \subset Y/S)$ only depends on X/S, up to natural equivalence.*
(ii) *If moreover $Y \hookrightarrow P$ is a closed embedding into a formal S-scheme which is smooth in a neighborhood of X, the category $\mathrm{Isoc}^\dagger(X \subset Y \subset P/S)$ only depends on X/S, up to natural equivalence.*

Proof The first assertion is a direct consequence of the proposition and the second one follows from Proposition 7.3.11. □

If $E \in \mathrm{Isoc}^\dagger(X/S)$, we will still denote by the same letter E its restriction to $(X \subset Y/S)$ or even $(X \subset Y \subset P/S)$.

Corollary 8.1.10 *In the definition of finitely presented overconvergent isocrystals over X/S, we get an equivalent category by considering only strictly local frames $(X' \subset Y' \subset P')$ with X' open in X.*

Proof Using Proposition 8.1.5, we may assume that X is affine and embed it into some projective compactification Y. Then, our assertion follows from Propositions 8.1.8 and 7.3.13. □

Corollary 8.1.11 *If X is an algebraic variety over S_k, then $\mathrm{Isoc}^\dagger(X/S)$ is an abelian category with tensor product and internal Hom. Moreover, if $X' \to X$ is any morphism of algebraic varieties over some morphism of formal schemes $S' \to S$, the pullback functor*

$$f^* : \mathrm{Isoc}^\dagger(X/S) \to \mathrm{Isoc}^\dagger(X'/S')$$

is exact.

Proof Using the same argument as in the previous proof, this follows from Propositions 7.3.14 and 7.1.4. □

Corollary 8.1.12 *Let X be an S_k-variety and $\sigma : K \hookrightarrow K'$ be an isometric embedding. Then there is a pullback functor*

$$\begin{array}{ccc} \mathrm{Isoc}^\dagger(X/S) & \longrightarrow & \mathrm{Isoc}^\dagger(X^\sigma/S^\sigma) \\ E & \longmapsto & E^\sigma \end{array}$$

such that for any frame $(X' \subset Y' \subset P')$ over X, we have $E^\sigma_{P'^\sigma} = (E_{P'})^\sigma$.

Proof Since the assertion is local on X, we may assume that X is realizable and use the proposition. □

Monsky–Washnitzer theory is a fundamental tool to study rigid cohomology. The first fundamental result is the following.

Proposition 8.1.13 *Let* $X = \operatorname{Spec} A$ *be a smooth affine* $\mathcal{V}$*-scheme and* Y *denotes the closure of* X *in some projective space for a given presentation of* A*. Then, there is an equivalence of categories*

$$\begin{array}{ccc} \mathrm{Isoc}^\dagger(X_k) & \xrightarrow{\simeq} & \mathrm{MIC}^\dagger(A_K^\dagger) \\ E & \longmapsto & M := \Gamma(\widehat{Y}_K^{\mathrm{rig}}, E_{\widehat{Y}}) \end{array}$$

between finitely presented overconvergent isocrystals on X *and coherent* $A_K^\dagger$*-modules with an overconvergent integrable connection.*

Proof By Proposition 8.1.8, there is an equivalence

$$\begin{array}{ccc} \mathrm{Isoc}^\dagger(X_k) & \xrightarrow{\simeq} & \mathrm{Isoc}^\dagger(X_k \subset Y_k \subset \widehat{Y}) \\ E & \longmapsto & E_{\widehat{Y}} \end{array}$$

and we saw in Proposition 7.2.13 that

$$\mathrm{Isoc}^\dagger(X_k \subset Y_k \subset \widehat{Y}) \simeq \mathrm{MIC}^\dagger(X_k \subset Y_k \subset \widehat{Y}).$$

Our assertion therefore follows from the equivalence of categories

$$\mathrm{MIC}^\dagger(X_k \subset Y_k \subset \widehat{Y}) \simeq \mathrm{MIC}^\dagger(A_K^\dagger)$$

of Definition 7.2.14. □

Now, we want to show that functoriality for overconvergent isocrystals is compatible with functoriality in Monsky–Washnitzer theory.

Proposition 8.1.14 *Let* $X = \operatorname{Spec} A$ *be a smooth affine* $\mathcal{V}$*-scheme,* Y *the closure of* X *in some projective space corresponding to a given presentation of* A *and* $V_\rho := X_K^{\mathrm{rig}} \cap B(0, \rho^+)$*. Define* A', X', Y', V'_ρ *in the same way. Let*

$$f : X'_k \to X_k$$

be a morphism of algebraic varieties.

Then any lifting $A^\dagger \to A'^\dagger$ *of* $f^* : A_k \to A'_k$ *induces a morphism* $u : V'_{\rho'} \to V_\rho$ *which is compatible to* f*.*

Moreover, if f is finite, we may assume that u is finite and then, the diagram

$$\begin{array}{ccc} V_{\rho'} & \xrightarrow{u} & V_\rho \\ \uparrow & & \uparrow \\ \widehat{X}'_K & \longrightarrow & \widehat{X}_K \end{array}$$

is cartesian.

Proof With our usual notations, it follows from Proposition 4.4.13 that the morphism of rings $A^\dagger \to A'^\dagger$ induces a morphism of rings $A_\rho \to A'_{\rho'}$ with $\rho, \rho' > 1$ that corresponds to a morphism of rigid analytic varieties $u : V'_{\rho'} \to V_\rho$. On the other hand the completion $\widehat{A} \to \widehat{A}'$ of our morphism $A^\dagger \to A'^\dagger$ induces a morphism of formal schemes $\widehat{X}' \to \widehat{X}$. By construction, this is a lifting of $f : X' \to X$ and the morphism induced on the generic fibers $\widehat{X}'_K \to \widehat{X}_K$ is simply the restriction of u.

When f is finite, so is its lifting $A^\dagger \to A'^\dagger$. One may therefore assume that the morphism $A_\rho \to A'_{\rho'}$ is also finite which means that u is. And since it is finite, we have

$$\widehat{A} \otimes_{A_\rho} A'_{\rho'} \simeq \widehat{A}',$$

which implies that the above diagram is cartesian. □

Note that actually, we should not actually write $V_{\rho'}$ in the cartesian diagram. It should be replaced by some affinoid open subset V with $V_{\rho'} \subset V \subset V_{\rho''}$. We do not want to bother with these details because, in the end, we only care about the direct system.

Proposition 8.1.15 *Let $X = \operatorname{Spec} A$ be a smooth affine $\mathcal{V}$-scheme and Y the closure of X in some projective space corresponding to a given presentation of A. Define A', X', Y' in the same way. Let*

$$f : X'_k \to X_k$$

be a morphism and $A^\dagger \to A'^\dagger$ a lifting of $f^* : A_k \to A'_k$. If $E \in \operatorname{Isoc}^\dagger(X_k/\mathcal{V})$ and if we set

$$M := \Gamma(Y_K^{\mathrm{rig}}, E_{\widehat{Y}}) \quad \text{and} \quad M' := \Gamma(Y'^{\mathrm{rig}}_K, (f^*E)_{\widehat{Y}'}),$$

there is a canonical isomorphism of modules with integrable connection

$$A'^\dagger_K \otimes_{A^\dagger_K} M \simeq M'.$$

Proof We saw in Lemma 8.1.14 that our lifting induces a morphism $u : V'_{\rho'} \to V_\rho$ which is compatible to f. Shrinking ρ and ρ' if necessary, we may assume that there exists an A_ρ-module of finite type with an integrable connection M_ρ such that $M = A_K^\dagger \otimes_{A_\rho} M_\rho$ and we will call $\mathcal{E}_\rho$ the corresponding coherent module on V_ρ. The corollary of Proposition 8.1.6 tells us that

$$j_{X'}^\dagger u^* \mathcal{E}_\rho = (f^* E)_{\widehat{Y'}}.$$

Taking global sections gives M' on the right-hand side and

$$\begin{aligned} \Gamma(Y'^{\mathrm{rig}}_K, j_{X'}^\dagger u^* \mathcal{E}_\rho) &= A'^\dagger_K \otimes_{A_{\rho'}} (A_{\rho'} \otimes_{A_\rho} M_\rho) \\ &= A'^\dagger_K \otimes_{A_K^\dagger} (A_K^\dagger \otimes_{A_\rho} M_\rho) = A'^\dagger_K \otimes_{A_K^\dagger} M \end{aligned}$$

on the left-hand side. □

8.2 Cohomology

Proposition 8.2.1 *Let*

$$\begin{array}{ccccc} X & \hookrightarrow & Y & & \\ \downarrow{\scriptstyle f} & & \downarrow{\scriptstyle g} & & \\ C & \hookrightarrow & D & \hookrightarrow & Q \end{array} \qquad \text{and} \qquad \begin{array}{ccccc} X & \hookrightarrow & Y' & & \\ \downarrow{\scriptstyle f} & & \downarrow{\scriptstyle g'} & & \\ C & \hookrightarrow & D & \hookrightarrow & Q \end{array}$$

be two virtual proper morphisms of S-frames.

If $E \in \mathrm{Isoc}^\dagger(X/S)$, there exists an isomorphism

$$\psi_{g,g'} : Rg_{\mathrm{rig}} E \simeq Rg'_{\mathrm{rig}} E.$$

Moreover, if

$$\begin{array}{ccc} & \nearrow & Y' \\ X & & \downarrow{\scriptstyle h} \\ & \searrow & Y \end{array}$$

is any morphism of virtual frames such that $g' = g \circ h$, then $\psi_{g,g'} = h^*$.

The analogous results for $Rg_{\mathrm{rig},X'} E$ when X' is a subvariety of X and for $Rg_{\mathrm{rig},c} E$ when g is quasi-compact and $C = D$ are also true.

Proof This works as usual. More precisely, if we consider the diagonal embedding

$$X \hookrightarrow Y \times_D Y',$$

the projections p_1 and p_2 are proper and induce an isomorphism on the cohomology of E by Theorem 7.4.18. We may then define

$$\psi_{g,g'} = (p_2^*)^{-1} \circ p_1^*.$$

If γ is the graph of h, we have $(p_2^*)^{-1} = \gamma^*$ and therefore

$$h^* = \gamma^* \circ p_1* = \psi_{g,g'}.$$

□

We can reformulate this proposition on the following form:

Corollary 8.2.2 *Let $(C \subset D \subset Q)$ be an S-frame, $f : X \to C$ a morphism of S_k-varieties and E be an overconvergent isocrystal on X/S. Let $X \hookrightarrow Y$ be an open embedding into a proper S_k-variety such that f extends to $g : Y \to D$.*

Then, up to a canonical isomorphism, $Rg_{\mathrm{rig}}E$ only depends on f and not on g nor u. This is more generally true of $Rg_{\mathrm{rig},X'}E$ when X' is a closed subset of X. And this is also true for $Rg_{\mathrm{rig},c}E$ when $C = D$.

Proof This follows from Propositions 8.2.1 and 7.4.2. □

We must also mention the following corollary to the above proposition.

Corollary 8.2.3 *Let $(C \subset D \subset Q)$ be an S-frame, $f : X \to C$ a morphism of S_k varieties and E be an overconvergent isocrystal on X/S. Let $X \hookrightarrow Y$ be an open embedding into a proper S_k-variety such that f extends to $g : Y \to D$. Let $Y \hookrightarrow P$ be a closed embedding into a formal S scheme which is smooth in a neighborhood of X such that g extends to $u : P \to Q$.*

Then, up to a canonical isomorphism, $Ru_{\mathrm{rig}}E_P$ only depend on f and not on g nor u. This is more generally true of $Ru_{\mathrm{rig},X'}E_P$ when X' is a closed subset of X. And this is also true for $Ru_{\mathrm{rig},c}E_P$ when $C = D$.

Proof This follows from Propositions 8.2.1 and 7.4.2. □

Definition 8.2.4 *A* realization of *an algebraic S_k-variety X is a proper smooth S-frame $(X \subset Y \subset P)$. In this case, we will say that X/S is* realizable.

Then, a realization *over $(C \subset D \subset Q)$ of a morphism of algebraic S_k-varieties $f : X \to C$ is a morphism of S-frames:*

$$\begin{array}{ccccc} X & \hookrightarrow & Y & \hookrightarrow & P \\ \downarrow{\scriptstyle f} & & \downarrow{\scriptstyle g} & & \downarrow{\scriptstyle u} \\ C & \hookrightarrow & D & \hookrightarrow & Q \end{array}$$

Note that a realizable variety over S_k is necessarily separated and of finite type over S_k. On the other hand, any quasi-projective variety over S_k is realizable.

Moreover, any morphism from a realizable variety to a proper smooth S-frame is realizable. More precisely, let us consider a proper smooth frame $(C \subset D \subset Q)$ and a morphism $f : X \to C$. If X is realizable, there exists a proper smooth S-frame $(X \subset Y \subset P)$ and we may consider the diagonal embedding $X \hookrightarrow P' := P \times_S Q$ and the algebraic closure Y' of X in $Y \times_{S_k} D$. We get a realization $(X \subset Y' \subset P')$ and a proper smooth morphism of S-frames

$$\begin{array}{ccccc} X & \hookrightarrow & Y' & \hookrightarrow & P' \\ \downarrow & & \downarrow & & \downarrow p_2 \\ C & \hookrightarrow & D & \hookrightarrow & Q. \end{array}$$

Unless otherwise specified, we will only consider realizable algebraic varieties and realizable morphisms.

Definition 8.2.5 *Let*

$$\begin{array}{ccccc} X & \hookrightarrow & Y & & \\ \downarrow f & & \downarrow g & & \\ C & \hookrightarrow & D & \hookrightarrow & Q \end{array}$$

be a virtual proper morphism of S-frames and $E \in \mathrm{Isoc}^\dagger(X/S)$. The relative rigid cohomology *of E is the complex*

$$Rf_{\mathrm{rig}}E := Rg_{\mathrm{rig}}E$$

on $]D[_P$.

If X' is a closed subvariety of X, the relative rigid cohomology of E with support in X' *is*

$$Rf_{\mathrm{rig},X'}E := Rg_{\mathrm{rig},X'}E.$$

Finally, if $C = D$, the relative rigid cohomology with compact support *of E is*

$$Rf_{\mathrm{rig},c}E := Rg_{\mathrm{rig},c}E.$$

Note again that, although the definition of the category is crystalline in nature, the definition of cohomology is not and this is not satisfactory. Moreover, rigid cohomology is only defined for realizable algebraic varieties over S_k. Recall that we will only consider realizable algebraic varieties and realizable morphisms between such. In general, one must use simplicial techniques such as in [26].

If X is a variety over S_k with structural map $f : X \to S_k$ seen as a morphism to the frame ($S_k = S_k \subset S$), we define for each $i \in \mathbf{N}$, the *absolute rigid cohomology* of E

$$\mathcal{H}^i_{\mathrm{rig}}(X/S, E) := R^i f_{\mathrm{rig}} E.$$

More generally, there is the *absolute rigid cohomology with support in some closed subvariety* X' of X.

$$\mathcal{H}^i_{\mathrm{rig},X'}(X/S, E) := R^i f_{\mathrm{rig},X'} E.$$

And we may also consider *absolute rigid geometry with compact support*

$$\mathcal{H}^i_{\mathrm{rig},c}(X/S, E) := R^i f_{\mathrm{rig},c} E.$$

In the case $S = \mathrm{Spf}\mathcal{V}$, we will consider the *absolute rigid cohomology* of E

$$R\Gamma_{\mathrm{rig}}(X, E) := Rf_{\mathrm{rig}} E \quad \text{and} \quad H^i_{\mathrm{rig}}(X, E) := R^i f_{\mathrm{rig}} E,$$

with support in a closed subset,

$$R\Gamma_{\mathrm{rig},X'}(X, E) := Rf_{\mathrm{rig},X'} E \quad \text{and} \quad H^i_{\mathrm{rig},X'}(X, E) := R^i f_{\mathrm{rig},X'} E$$

or with compact support

$$R\Gamma_{\mathrm{rig},c}(X, E) := Rf_{\mathrm{rig},c} E \quad \text{and} \quad H^i_{\mathrm{rig},c}(X, E) := R^i f_{\mathrm{rig},c} E.$$

Finally, when E is the trivial module with connection, we drop it from the H-notations.

We must mention now the easy but important following example, the proper smooth case. In general, we would need crystalline cohomology, but we will stick to a lifted situation.

Proposition 8.2.6

(i) *Let $u : \mathcal{X} \to \mathcal{C}$ be a morphism from a proper smooth formal scheme to any formal scheme over S and E an overconvergent isocrystal on $\mathcal{X}_k/S$. Then,*

$$Ru_{k,\mathrm{rig}} E = Ru_{K,\mathrm{dR}}(\mathcal{X}_K, E_{\mathcal{X}}).$$

(ii) *If X is a proper smooth scheme over $\mathcal{V}$, then,*

$$R\Gamma_{\mathrm{rig}}(X_k) = R\Gamma_{\mathrm{dR}}(X_K).$$

Proof For the first case, we may use the frame ($X_k \subset X_k \subset \mathcal{X}$) in order to compute rigid cohomology. The second case then is just a consequence of rigid analytic GAGA of [55]. □

It is not difficult to extend this result to the case of the complement of a relative normal crossing divisor with smooth components as shown in [25]. In particular, we can easily derive the rigid cohomology of a curve.

In the smooth affine case, we recover Monsky–Washnitzer cohomology:

Proposition 8.2.7 *Let $X = \operatorname{Spec} A$ be a smooth affine $\mathcal{V}$-scheme. If $E \in \operatorname{Isoc}^\dagger(X_k/\mathcal{V})$ and M denote the corresponding $A_K^\dagger$-module with connection, we have*

$$R\Gamma_{\text{rig}}(X_k, E) \simeq \Gamma_{\text{dR}}(M).$$

More generally, if X' is the hypersurface in X defined by some non zero $f \in A$, we have

$$R\Gamma_{\text{rig},X'_k}(X_k, E) \simeq \Gamma_{\text{dR}}(M^f)[-1].$$

Finally, we also have

$$R\Gamma_{\text{rig},c}(X_k, E) \simeq \Gamma_{\text{dR}}(M_c)[-d]$$

if X_k has pure dimension d.

Proof This follows from Propositions 6.2.12, 6.3.11 and 6.4.15. □

Recall also that, in general, when X is proper over C, an overconvergent isocrystal is the same thing as a convergent isocrystal and rigid cohomology is the same as convergent cohomology. In other words, when X is proper, there is no need to consider strict neighborhoods and one can work directly with tubes.

Proposition 8.2.8 *Let $(C \subset D \subset Q)$ be an S-frame, $X \to C$ be an S_k morphism of algebraic varieties and $E \in \operatorname{Isoc}^\dagger(X/S)$.*

(i) *If X' is a closed subvariety of X, there is a canonical morphism*

$$Rf_{\text{rig},X'}E \to Rf_{\text{rig}}E$$

which is an isomorphism when $X' = X$.

(ii) *If X' is a closed subvariety of X and U is an open neighborhood of X' in X, then*

$$Rf_{\text{rig},X'}E = Rf_{|U,\text{rig},X'}E_{|U}.$$

(iii) *When $C = D$, there is a canonical morphism*

$$Rg_{\text{rig},c}E \to Rg_{\text{rig}}E$$

which is an isomorphism when X is proper.

Proof This is an immediate consequence of the corresponding result for virtual frames shown in Proposition 7.4.8. □

Lemma 8.2.9 *Any proper (realizable) morphism $f : X' \to X$ of (realizable) S_k-varieties extends to a cartesian morphism of proper virtual S-frames*

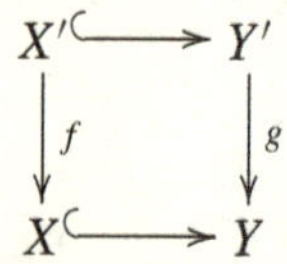

Proof Since we only consider realizable varieties and realizable morphisms, there exists such a morphism and we need to turn it into a cartesian one. We may assume that X' is dense in Y' and consider

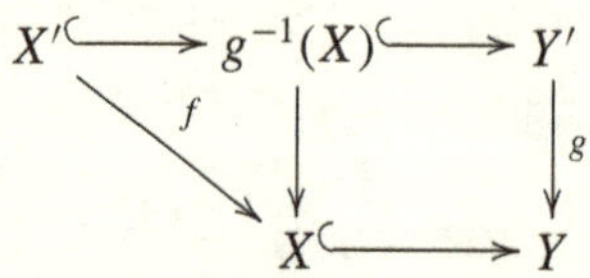

Since Y' and Y are proper, so is g. Since the square is cartesian, it follows that the middle arrow is proper too and since f is assumed to be proper, the open immersion $X' \hookrightarrow g^{-1}(X)$ is necessarily proper. Since X is dense in Y', we have $X' = g^{-1}(X)$ and we are done. □

Proposition 8.2.10 *Assume that we are given a commutative diagram*

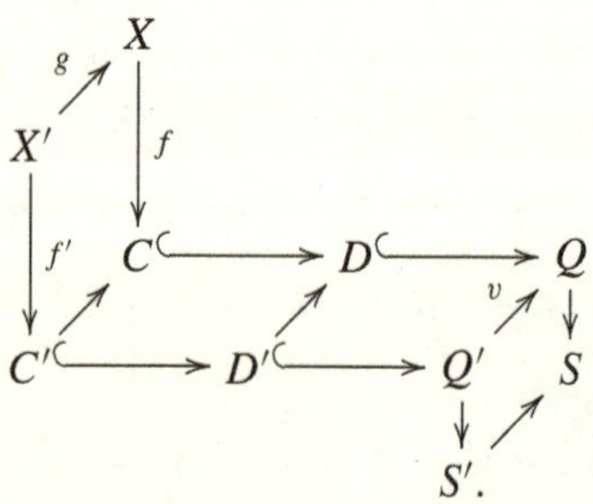

Then, if $E \in \mathrm{Isoc}^{\dagger}(X/S)$, there is a canonical base change map

$$g^* : Lv_K^* Rf_{\mathrm{rig}} E \to Rf'_{\mathrm{rig}} g^* E$$

with $v_K :]D'[_{Q'} \to]D[_Q$. More generally, if X'' (resp. X''') is a closed subvariety of X (resp. X') with $f^{-1}(X'') \subset X'''$, there is a canonical base change map

$$g'^* : Lv_K^* Rf_{\mathrm{rig},X''} E \to Rf'_{\mathrm{rig},X'''} g^* E.$$

Finally, when $C = D$ and g is proper, we also have

$$g^* : Lv_K^* Rf_{\mathrm{rig},c} E \to Rf'_{\mathrm{rig},c} g^* E.$$

Proof For the assertions without compact support, this is an immediate consequence of the corresponding result for frames proved in Proposition 7.4.9. And for the last assertion, we may use Lemma 8.2.9. □

For computations, it is necessary to give a more precise result.

Proposition 8.2.11 *Assume that we are given a commutative diagram with proper smooth realizations of X and X' over $(C \subset D \subset Q)$ and $(C' \subset D' \subset Q')$ respectively:*

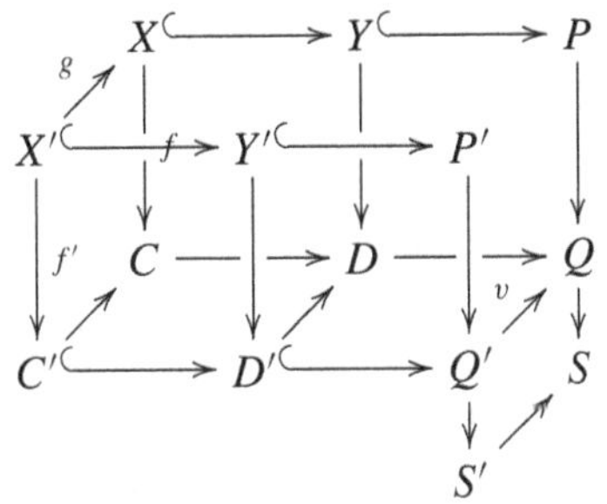

If $u : V' \to V$ is a morphism over Q_K between strict neighborhoods of $]X'[_{P'}$ in $]Y'[_{P'}$ and $]X[_P$ in $]Y[_P$ respectively which is compatible to f, then the base change morphism g^* in rigid cohomology is identical to the base change morphism in de Rham cohomology

$$u* : Lv_K^* Rp_{\mathrm{dR}} E_{P|V} \to Rp'_{\mathrm{dR}} E_{P'|V'}$$

where $p : V \to]D[_Q$ and $p' : V' \to]D'[_{Q'}$ denote the structural maps. More generally, if X'' (resp. X''') is a closed subvariety of X (resp. X') with $f^{-1}(X'') \subset X'''$, then the base change map for cohomology with support corresponds to

$$u* : Lv_K^* Rp_{\mathrm{dR}} \Gamma^\dagger_{X''} E_{P|V} \to Rp'_{\mathrm{dR}} \Gamma^\dagger_{X'''} E_{P'|V'}.$$

Finally, the analogous result holds for cohomology with compact support when $C = D$, $C' = D'$ and g is proper.

Proof Using the diagonal embedding again and the definition of the base change morphism, we may assume that

$$X' = X, \quad C' = C, \quad D' = D, \quad Q' = Q.$$

And we may also assume that there exists a morphism of frames

$$\begin{array}{ccccc} & & Y' & \hookrightarrow & P' \\ & \nearrow & \uparrow & & \uparrow u \\ X & & & & \\ & \searrow & \uparrow & & \uparrow \\ & & Y & \hookrightarrow & P \end{array}$$

over $(C \hookrightarrow D \hookrightarrow Q)$ such that u is a section of w_K over V'. In this situation, we have to show that the map

$$u* : Rp_{\mathrm{dR}} E_{P|V} \to Rp'_{\mathrm{dR}} E_{P'|V'}$$

is the canonical isomorphism in rigid cohomology. But, by definition, the canonical isomorphism $Rg'_{\mathrm{rig}} E \simeq Rg_{\mathrm{rig}} E$ is induced by w_K. Since u is a section of w_K on V, we see that the reverse isomorphism is induced by u. Both assertions with support are proved in the same way. □

Now, we want to show that functoriality for overconvergent isocrystals is compatible with functoriality in Monsky–Washnitzer theory.

Proposition 8.2.12 *Let $X = \mathrm{Spec} A$ and $X' = \mathrm{Spec} A'$ be two smooth affine $\mathcal{V}$-schemes. Let*

$$\varphi : X'_k \to X_k$$

be a morphism and $A^\dagger \to A'^\dagger$ a lifting of $\varphi^* : A_k \to A'_k$. If $E \in \mathrm{Isoc}^\dagger(X_k/\mathcal{V})$ and M denote the corresponding $A_K^\dagger$-module, then the diagram

$$\begin{array}{ccc} R\Gamma_{\mathrm{rig}}(X_k, E) & \xrightarrow{\varphi^*} & R\Gamma_{\mathrm{rig}}(X'_k, \varphi^* E) \\ \downarrow \simeq & & \downarrow \simeq \\ \Gamma_{\mathrm{dR}}(M) & \longrightarrow & \Gamma_{\mathrm{dR}}(A'^\dagger_K \otimes_{A_K^\dagger} M) \end{array}$$

is commutative.

More generally, if X'' is the hypersurface in X defined by some non zero $f \in A$ and X''' its inverse image in X', the diagram

$$\begin{array}{ccc} R\Gamma_{\mathrm{rig}, X''_k}(X_k, E) & \xrightarrow{\varphi^*} & R\Gamma_{\mathrm{rig}, X'''_k}(X'_k, \varphi^* E) \\ \downarrow \simeq & & \downarrow \simeq \\ \Gamma_{\mathrm{dR}}(M^f) & \longrightarrow & \Gamma_{\mathrm{dR}}((A'^\dagger_K \otimes_{A_K^\dagger} M)^{\varphi^*(f)}) \end{array}$$

is commutative.

Finally, when φ is finite, the diagram

$$\begin{array}{ccc} R\Gamma_{\mathrm{rig},c}(X_k, E) & \xrightarrow{\varphi^*} & R\Gamma_{\mathrm{rig},c}(X'_k, \varphi^* E) \\ \downarrow \simeq & & \downarrow \simeq \\ \Gamma_{\mathrm{dR}}(M_c) & \longrightarrow & \Gamma_{\mathrm{dR}}(A'^{\dagger}_K \otimes_{A^{\dagger}_K} M_c) \end{array}$$

is also commutative.

Proof Using Lemma 8.1.14, this immediately follows from the proposition. □

As an application of the proposition, we can prove the following result which is used in Berthelot's original proof to finiteness of rigid cohomology in [17].

Proposition 8.2.13 *If X is a smooth affine variety over k and*

$$\varphi : X' \to X$$

a finite and flat morphism, the canonical map

$$R\Gamma_{\mathrm{rig}}(X, E) \to R\Gamma_{\mathrm{rig}}(X', \varphi^* E),$$

has a retraction. More generally, if X'' is a hypersurface in X and $X''' = f^{-1}(X)$, the map

$$R\Gamma_{\mathrm{rig},X''}(X, E) \to R\Gamma_{\mathrm{rig},X'''}(X', \varphi^* E)$$

has a retraction. Finally, the map

$$R\Gamma_{\mathrm{rig},c}(X, E) \to R\Gamma_{\mathrm{rig},c}(X', \varphi^* E)$$

also has a retraction.

Proof We change notations and denote by X, X' and X'' smooth affine liftings of our varieties in order to use the notations of the previous proposition. We may assume X and X' are connected and denote by Q and Q' the fraction fields of $A^{\dagger}$ and $A'^{\dagger}$ respectively. We consider the semi-linear map

$$\Gamma_{\mathrm{dR}}(M) \to \Gamma_{\mathrm{dR}}(A'^{\dagger}_K \otimes_{A^{\dagger}_K} M)$$

and tensor with Q'. Since $A^{\dagger} \to A'^{\dagger}$ is generically étale (and finite), we obtain an isomorphism of complexes

$$Q' \otimes_{A^{\dagger}_K} \Gamma_{\mathrm{dR}}(M) \simeq Q' \otimes_{A'^{\dagger}_K} \Gamma_{\mathrm{dR}}(A'^{\dagger}_K \otimes_{A^{\dagger}_K} M).$$

Composing the inverse isomorphism with the trace map $\mathrm{tr}_{Q'/Q} \otimes \mathrm{Id}$ gives a morphism

$$\mathrm{tr} : Q' \otimes_{A'^{\dagger}_{K}} \Gamma_{\mathrm{dR}}(A'^{\dagger}_{K} \otimes_{A^{\dagger}_{K}} M) \to Q \otimes_{A^{\dagger}_{K}} \Gamma_{\mathrm{dR}}(M).$$

Then one shows that it induces a map

$$\mathrm{tr} : \Gamma_{\mathrm{dR}}(A'^{\dagger}_{K} \otimes_{A^{\dagger}_{K}} M) \to \Gamma_{\mathrm{dR}}(M)$$

which is a retraction for our original morphism. The point is to verify that $\Omega^{1}_{B^{\dagger}}$ is sent into $\Omega^{1}_{A^{\dagger}}$ (see [72] or [83] for the details). The same method works with M^{f} or M_{c}. □

Proposition 8.2.14 *Let X be an S_k-variety and E a finitely presented overconvergent isocrystal on $(X \subset Y/S)$. If $K \hookrightarrow K'$ is a finite field extension, there is a canonical isomorphism*

$$K' \otimes_{K} H^{q}_{\mathrm{rig}}(X, E) \simeq H^{q}_{\mathrm{rig}}(X_{k}, E_{K'}).$$

More generally, if X' a closed subset of X, we have

$$K' \otimes_{K} H^{q}_{\mathrm{rig},X'}(X, E) \simeq H^{q}_{\mathrm{rig},X'_{k}}(X_{k}, E_{K'}).$$

Finally, we have

$$K' \otimes_{K} H^{q}_{\mathrm{rig},c}(X, E) \simeq H^{q}_{\mathrm{rig},c}(X_{k}, E_{K'}).$$

Proof As usual, this is an immediate consequence of the corresponding result for virtual frames proved in Proposition 7.4.10. □

Proposition 8.2.15 *If* $\mathrm{Char} K = 0$ *and X is an algebraic variety over k, then $H^{0}_{\mathrm{rig}}(X)$ is a finite-dimensional space whose dimension is the number of geometrically connected components of X.*

Proof This follows from Proposition 7.4.11. □

Proposition 8.2.16 *Let $(C \subset D \subset Q)$ be an S-frame and $f : X \to C$ be an S_k-morphism of algebraic varieties. Let*

$$\begin{array}{ccccc} C' & \hookrightarrow & D' & \hookrightarrow & Q' \\ \downarrow & & \downarrow & & \downarrow \\ C & \hookrightarrow & D & \hookrightarrow & Q \end{array}$$

be a cartesian open or mixed immersion of frames and $f' : X' \to C'$ the pullback of f.

If $E \in \mathrm{Isoc}^\dagger(X/S)$, we have

$$(Rf_{\mathrm{rig}}E)_{|]D'[_{Q'}} \simeq Rf'_{\mathrm{rig}}E_{|X'}.$$

More generally, we have

$$(Rf_{\mathrm{rig},X''}E)_{|]D'[_{Q'}} \simeq Rf'_{\mathrm{rig},X'''}E_{|X'}$$

when X'' is a closed subvariety of X and $X''' := X' \cap X''$ and

$$(Rf_{\mathrm{rig},c}E)_{|]D'[_{Q'}} \simeq Rf'_{\mathrm{rig},c}E_{|X'}.$$

when $C = D$.

Proof This is an immediate consequence of the corresponding result for virtual frames proved in Proposition 7.4.12. □

Proposition 8.2.17 *Let $(C \subset D \subset Q)$ be an S-frame, $f : X \to C$ an S_k morphism of algebraic varieties and $E \in \mathrm{Isoc}^\dagger(X/S)$. If we are given an open or a finite closed covering $X = \cup X_i$, there is a canonical spectral sequence*

$$E_1^{p,q} := \bigoplus_{|\underline{i}|=p} R^q f_{\underline{i},\mathrm{rig}}E_{|X_{\underline{i}}} \Rightarrow Rf_{\mathrm{rig}}E.$$

More generally, if X' is a closed subvariety of X and $X'_{\underline{i}} := X' \cap X_{\underline{i}}$, there is a canonical spectral sequence

$$E_1^{p,q} := \bigoplus_{|\underline{i}|=p} R^q f_{\underline{i}\mathrm{rig},X_{\underline{i}}}E_{|X_{\underline{i}}} \Rightarrow Rf_{\mathrm{rig},X'}E.$$

Proof This is an immediate consequence of the corresponding result for virtual frames proved in Proposition 7.4.13. □

Note that this last proposition may be used in order to extend rigid cohomology to non realizable varieties.

Proposition 8.2.18 *Let $(C \subset D \subset Q)$ be an S-frame, $f : X \to C$ an S_k-morphism of algebraic varieties and $E \in \mathrm{Isoc}^\dagger(X/S)$ Then,*

(i) *If*

$$X'' \hookrightarrow X' \hookrightarrow X$$

is a sequence of closed immersions, $U := X \setminus X''$ and $U' := X' \setminus X''$, there is an exact triangle

$$Rf_{\mathrm{rig},X''}E \to Rf_{\mathrm{rig},X'}E \to Rf_{|U,\mathrm{rig},U'}E_{|U} \to .$$

(ii) *If $X' \hookrightarrow X$ is a closed immersion and $U := X \setminus X'$, there is an exact triangle*

$$Rf_{|U,\mathrm{rig},c}E_{|U} \to Rf_{\mathrm{rig},c}E \to Rf_{|X',\mathrm{rig},c}E_{|X'} \to .$$

Proof This is an immediate consequence of the corresponding result for virtual frames proved in Propositions 7.4.15 and 7.4.16. □

We also have the base change assertion.

Proposition 8.2.19 *Let $(C \subset D \subset Q)$ be an S-frame, $f : X \to C$ any morphism of algebraic varieties and $E \in \mathrm{Isoc}^\dagger(X \subset Y/S)$. If $\sigma : K \hookrightarrow K'$ is an isometric embedding, there are canonical morphisms*

$$(Rf_{\mathrm{rig}}E)^\sigma \to Rf^\sigma_{\mathrm{rig}}E^\sigma,$$
$$(Rf_{\mathrm{rig},c}E)^\sigma \to Rf^\sigma_{\mathrm{rig},c}E^\sigma$$

when $C = D$ and

$$(Ru_{\mathrm{rig},X}E)^\sigma \to Ru^\sigma_{\mathrm{rig},X'^\sigma}E^\sigma$$

when X' is a closed subvariety of X.

Proof This is a consequence of Proposition 7.4.14. □

It is not difficult to compute the rigid cohomology of a curve.

Proposition 8.2.20 *Let X be a smooth curve over k, d its geometric genus, c the geometric number of connected components and c' the geometric number of proper connected components. Finally, let v be the geometric number of (smooth) missing points.*

Then,

$$\dim_K H^0_{\mathrm{rig}}(X) = \dim_K H^2_{\mathrm{rig},c}(X) = c,$$
$$\dim_K H^1_{\mathrm{rig}}(X) = \dim_K H^1_{\mathrm{rig},c}(X) = 2d - v + c$$

and

$$\dim_K H^2_{\mathrm{rig}}(X) = \dim_K H^0_{\mathrm{rig},c}(X) = c'.$$

Proof We may extend the basis as we wish thanks to Proposition 8.2.14. Moreover, the formulas are additive. We may therefore assume that X is geometrically connected.

The case of $H^0_{\mathrm{rig}}(X)$ is already known from Proposition 8.2.15. Also, when X is proper, we may lift it and use the comparison Theorem 8.2.6. On the other hand, when X is affine, the cases of $H^0_{\mathrm{rig},c}(X)$ and $H^2_{\mathrm{rig}}(X)$ both follow from Monsky–Washnitzer theory. Moreover, we can embed X in a smooth projective curve Y, and there is a long exact sequence:

$$0 \to H^0_{\mathrm{rig},c}(Y) \to H^0_{\mathrm{rig},c}(Y \setminus X) \to H^1_{\mathrm{rig},c}(X) \to H^1_{\mathrm{rig},c}(Y) \to 0$$

and

$$H^2_{\mathrm{rig},c}(X) \simeq H^2_{\mathrm{rig},c}(Y).$$

The cohomology with compact support therefore follows from the proper smooth case.

Note that if $\mathcal{R}$ is the Robba ring on K, we have

$$H^0_{\mathrm{dR}}(\mathcal{R}) = K \quad \text{and} \quad H^1_{\mathrm{dR}}(\mathcal{R}) = K dt/t.$$

We can extend the basis and assume that all points outside X are rational and Proposition 6.4.16 reads

$$\begin{aligned} 0 \to K \to \oplus_{x \notin X} K \to H^1_{\mathrm{rig},c}(X) \to \\ H^1_{\mathrm{rig}}(X) \to \oplus_{x \notin X} K \to K \to 0 \end{aligned}$$

and it follows that

$$\dim_K H^1_{\mathrm{rig}}(X) = \dim_K H^1_{\mathrm{rig},c}(X) = 2d - v + 1.$$

□

There are also formulas for the rigid cohomology (with compact support) of singular curves but they involve the combinatorics of the components of the curve as shown in [60].

There exists a trace map in rigid cohomology as the next proposition shows.

Proposition 8.2.21 *Let X be an algebraic variety of dimension d over k. Then,*

(i) *We have $H^i_{\mathrm{rig},c}(X) = 0$ for $i > 2d$.*
(ii) *If U is a dense open subset of X, then the canonical map is an isomorphism $H^{2d}_{\mathrm{rig},c}(U) \simeq H^{2d}_{\mathrm{rig},c}(X)$ and the induced map*

$$\mathrm{tr}_X : H^{2d}_{\mathrm{rig},c}(X) \to K$$

only depends on X.

Proof We proceed by induction on the dimension of X. When X is empty the properties hold. Moreover, the first part of the second assertion follows from the long exact sequence of rigid cohomology with compact support since $X \setminus U$ has dimension strictly less than X. In order to prove the first assertion, we may therefore assume that X is smooth affine. After a finite extension of K, X becomes a disjoint union of smooth affine varieties of pure dimension at most d. We may therefore assume that X is smooth affine of pure dimension d in which case we may use Proposition 6.4.15. It remains to prove that the trace map is indeed independent of X. But this follows from Proposition 6.4.19. □

Definition 8.2.22 *With the notations of the proposition, the map*

$$\mathrm{tr}_X : H^{2d}_{\mathrm{rig},c}(X) \to K$$

is called the trace map.

Corollary 8.2.23 *Let X be an algebraic variety of dimension d over k and E be an overconvergent isocrystal on X with dual $E^\vee$. Then, there is a canonical pairing*

$$R\Gamma_{\mathrm{rig}}(X, E^\vee) \times R\Gamma_{\mathrm{rig},c}(X, E) \to K[-2d].$$

More generally, if X' is a closed subset of X, there is a canonical pairing

$$R\Gamma_{\mathrm{rig},X'}(X, E^\vee) \times R\Gamma_{\mathrm{rig},c}(X', E_{|X'}) \to K[-2d].$$

Proof This follows from Proposition 6.4.14. □

Lemma 8.2.24 *Let $X = \mathrm{Spec}A$ be a smooth affine $\mathcal{V}$-scheme and Y the closure of X in some projective space corresponding to a given presentation of A. Define A', X', Y' in the same way. Assume that X and X' have the same relative dimension d and let*

$$f : X'_k \to X_k$$

be a finite morphism. Then, any lifting $A^\dagger \to A'^\dagger$ of f induces a canonical map

$$f^* : \omega_{A,c} \to \omega_{A',c}.$$

Moreover, if f is flat and if $Z = \mathrm{Spec}C$ (resp. $Z' := \mathrm{Spec}C'$) is a smooth closed subvariety of X (resp. X') with $Z'_k := f^{-1}(Z_k)$, the diagram

$$\begin{array}{ccc} \omega_{C,c} & \longrightarrow & \omega_{C',c} \\ \downarrow & & \downarrow \\ \omega_{A,c} & \longrightarrow & \omega_{A',c} \end{array}$$

is commutative.

Proof We know from Proposition 8.1.14 that our lifting induces a finite map $u : V'_{\rho'} \to V_\rho$ between strict neighborhoods. We may consider the natural pull-back map $u^* : \Omega_{V_\rho} \to u_*\Omega_{V'_{\rho'}}$. Now, we use Proposition 5.2.13. Since the diagram

$$\begin{array}{ccc} V'_{\rho'} & \xrightarrow{u} & V_\rho \\ \uparrow & & \uparrow \\ \widehat{X}'_K & \longrightarrow & \widehat{X}_K \end{array}$$

is cartesian, u is finite and $\Omega^d_{V'_{\rho'}}$ is coherent, we have

$$H^d_{\widehat{X}'_K}(V'_\rho, \Omega^d_{V'_{\rho'}}) = H^d_{\widehat{X}_K}(V_\rho, u_*\Omega^d_{V'_{\rho'}})$$

and the above pullback map induces a morphism

$$H^d_{\widehat{X}_K}(V_\rho, \Omega^d_{V_\rho}) \to H^d_{\widehat{X}'_K}(V_\rho, \Omega^d_{V'_{\rho'}}).$$

We now turn to the second part. The third assertion of Proposition 5.4.19 tells us that the question is local. We may therefore assume that Z is defined by a regular sequence $f_1, \ldots, f_d$. Since f is flat, so is its lifting $A^\dagger \to A'^\dagger$ and it follows that $u : V'_{\rho'} \to V_\rho$ is flat. Thus, $u^*(f_1), \ldots, u^*(f_d)$ is a regular sequence on $V'_{\rho'}$ and by construction, it defines a closed subset $W'_{\rho'}$ which is a strict neighborhood of $\widehat{Z}'_K$. If we look at the definition of the Gysin map, we immediately see that there is a commutative diagram in the derived category

$$\begin{array}{ccc} \Omega^e_{W_\rho}[e] & \longrightarrow & u_*\Omega^e_{W'_{\rho'}}[e] \\ \downarrow & & \downarrow \\ \Omega^d_{V_\rho}[d] & \longrightarrow & u_*\Omega^d_{V'_{\rho'}}[d]. \end{array}$$

Our assertion follows. □

Proposition 8.2.25 *If $i : Z \hookrightarrow X$ is a closed immersion of smooth algebraic varieties of respective dimension d and e, there is a natural Gysin map*

$$H^{2d}_{\mathrm{rig},c}(Z) \to H^{2e}_{\mathrm{rig},c}(X).$$

More precisely, it is transitive and functorial with respect to finite flat morphisms.

Proof The question is clearly additive and we may therefore assume that X and Z are connected. Moreover, thanks to Proposition 8.2.21, we may replace X with any open subset that meets Z. In particular, we may assume X affine in which case our assertion results from the lemma because the Gysin map is compatible with differentials. □

8.3 Frobenius action

We still assume in this section that $\mathrm{Char} K = 0$ but we also make the additional hypothesis that $\mathrm{Char} k = p > 0$. We let $q = p^f$ with $f \in \mathbf{N} \setminus 0$.

Recall that we may always consider the (iterated) absolute Frobenius $F_X : X \to X$ on any algebraic k-variety X. It is the identity on the underlying space and we have

$$F_X^* : f \mapsto f^q$$

If we want to emphasize the role of q, we will write $F_{X/S}^f$.

If X is an algebraic S_k-variety, we will write $X^{(q/S)}$ for the pullback of X along F_{S_k} and

$$F_{X/S_k} : X \to X^{(q/S)}$$

for the relative Frobenius. Thus, we have a commutative diagram with cartesian square

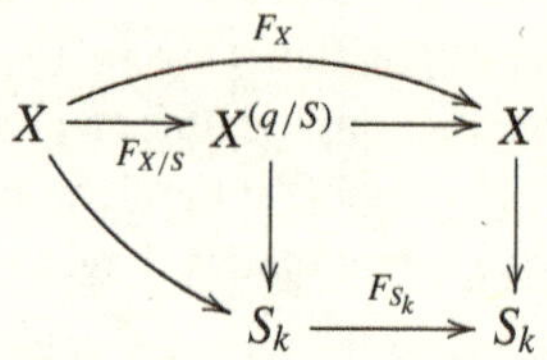

As before, if we want to emphasize the role of q, we can write $F_{X/S}^f$. In the case $S = \mathrm{Spf}\mathcal{V}$, we simply write $X^{(q/k)}$ and $F_{X/k}$.

A *Frobenius* on K is an isometry $\sigma : K \to K$ such that

$$\forall x \in K, |x| \leq 1 \Rightarrow |\sigma(x) - x^q| < 1.$$

A *relative Frobenius* on S is a lifting

$$F_{S/\mathcal{V}} : S \to S^\sigma$$

of the relative Frobenius $F_{S_k/k}$. The corresponding *absolute Frobenius* is the composite F_S of $F_{S/\mathcal{V}}$ with $S^\sigma \to S$. We will denote by

$$F_{S/K} : S_K \to S_K^\sigma$$

the generic fiber of $F_{S/\mathcal{V}}$ and if $M^\bullet$ is a complex of $\mathcal{O}_{S_K}$-modules on S_K, it will be convenient to write

$$LF_S^* M^\bullet = (LF_{S/K}^* M^\bullet)^\sigma.$$

But we should not forget that this depends both on σ and on the lifting of $F_{S/k}$.

Definition 8.3.1 *With the above notations, if $(X \subset Y/S)$ is a virtual frame, the pullback along the absolute Frobenius is the composite functor*

$$
\begin{aligned}
F^*_{(X \subset Y/S)} &: \mathrm{Isoc}^\dagger(X \subset Y/S) \\
&\to \mathrm{Isoc}^\dagger(X^{(q)} \subset Y^{(q)}/S^\sigma) \to \mathrm{Isoc}^\dagger(X \subset Y/S).
\end{aligned}
$$

Note that, by definition, if $(X \subset Y \subset P)$ is any S-frame and

$$
\begin{array}{ccccc}
X & \hookrightarrow & Y & \hookrightarrow & P' \\
\downarrow{\scriptstyle F_{X/k}} & & \downarrow{\scriptstyle F_{Y/k}} & & \downarrow{\scriptstyle F} \\
X^{(q)} & \hookrightarrow & Y^{(q)} & \hookrightarrow & P^\sigma
\end{array}
$$

is a morphism of frames that extend the relative Frobenius (necessarily cartesian), then

$$(F^*_{(X \subset Y/S)} E)_{P'} = F^*_K E^\sigma_P.$$

Definition 8.3.2 *With the above notations, an* overconvergent F-isocrystal *on a virtual frame $(X \subset Y/S)$ is a couple (E, Φ) where E is a finitely presented overconvergent isocrystal and*

$$\Phi_E : F^*_{X \subset Y/S} E \simeq E$$

an isomorphism.

Note that this definition depends on the choices of σ and φ. Note also that, in order to lighten the terminology, an overconvergent F-isocrystal is always assumed to be finitely presented.

Proposition 8.3.3 *With morphisms compatible with the data, overconvergent F-isocrystals form an abelian category $F{-}\mathrm{isoc}^\dagger(X \subset Y/S)$. Moreover, if E and E' are two overconvergent F-isocrystals, then $E \otimes E'$ and $\mathcal{H}om(E, E')$ have natural structures of overconvergent F-isocrystals. Finally, the forgetful functor*

$$F{-}\mathrm{isoc}^\dagger(X \subset Y/S) \to \mathrm{Isoc}^\dagger(X \subset Y/S)$$

is exact and faithful, and commutes with $\mathcal{H}om$ and $\otimes$.

Proof It is clear that the kernel (resp. cokernel) of a morphism of F-isocrystals inherits a Frobenius structure and that this structure makes it a kernel (resp. cokernel) in the category of F-isocrystal. One easily concludes that this is an abelian category and that the forgetful functor is exact. It is clearly faithful.

Finally, the Frobenius structure on $\mathcal{H}om(E, E')$ is given via the isomorphism

$$F^*\mathcal{H}om(E, E') \simeq \mathcal{H}om(F^*E, F^*E')$$

by

$$u \mapsto \Phi_{E'} \circ u \circ \Phi_E^{-1}.$$

And the Frobenius structure on $E \otimes E'$ is simply given via the isomorphism

$$F^*(E \otimes E') \simeq F^*E \otimes F^*E'$$

by $\Phi_E \otimes \Phi_{E'}$. □

In the particular case $Y = X$, we get the category $F-\mathrm{isoc}(X/S)$ of *convergent F-isocrystals* on X/S and in the case Y proper, we get the category $F-\mathrm{isoc}^\dagger(X/S)$ of *overconvergent F-isocrystals* on X/S.

In the Monsky–Washnitzer setting, the situation is very nice as we will see shortly. We first need some preliminary results. If A is a weakly complete $\mathcal{V}$-algebra, a *Frobenius endomorphism* of A is a lifting F^* of the q-th power of A_k over σ. If $\mathcal{M}$ is an A-module, we will write

$$\mathcal{M}^{(q)} := A \otimes_{F_K^*} \mathcal{M}.$$

Of course, this will apply in particular to an A_K-modules M, in which case we will have

$$M^{(q)} = A_K \otimes_{F_K^*} M.$$

Lemma 8.3.4 *Let A be a formally smooth weakly complete $\mathcal{V}$-algebra and M be a coherent A_K-module with an integrable connection. Let F^* be a Frobenius on A. Then, there exists an integer $r > 0$ and a coherent A-module $\mathcal{M}_r$ with an integrable connection ∇ such that*

$$M^{(q^r)} = K \otimes_{\mathcal{V}} \mathcal{M}_r.$$

Proof Since F^* induces the q-th power on A_k, it is clear that dF^* is zero mod $\mathfrak{m}$ and it follows that there exists $\alpha \in \mathfrak{m}$ with

$$dF^*(\Omega_A^1) \subset \alpha\Omega_A^1.$$

On the other hand, if we choose a finite set of generators s_j of M and let $\mathcal{M}$ denotes the sub A-module generated by the s_j's, there exists $r > 0$ such that

$$\nabla(\alpha^r\mathcal{M}) \subset \mathcal{M} \otimes \Omega_A^1.$$

We may then take $\mathcal{M}_r := \mathcal{M}^{(q^r)}$. □

Proposition 8.3.5 *Let A be a formally smooth weakly complete $\mathcal{V}$-algebra with étale coordinates $t_1, \ldots, t_m$ and M a coherent A_K-module with an integrable connection. Let F^* be a Frobenius on A and $\eta < |p|^{\frac{1}{p-1}}$. Then, there exists an integer $r > 0$ such that $M^{(q^r)}$ is η-convergent.*

Proof Using Lemma 8.3.4, this immediately follows from Proposition 4.4.15. □

Lemma 8.3.6 *Let A be a weakly complete $\mathcal{V}$-algebra with étale coordinates $t_1, \ldots, t_m$ and F^* a Frobenius on A. As usual, denote by $\tau_i = p_2^*(t_i) - p_1^*(t_i)$.*

Then, there exists $\delta < 1$ such that for each $\eta < 1$, there exists $\rho_0 > 1$ such that if $1 < \rho \leq \rho_0$, there exists $\rho' > 1$ such that F_K^* sends $A_\rho\{\underline{\tau}/\eta\}$ inside $A_{\rho'}\{\underline{\tau}/\eta'\}$ with $\eta' = \inf\{\eta^{1/q}, \eta/\delta\}$.

Proof There exists $\alpha \in \mathfrak{m}$ such that for each $i = 1, \ldots, N$,

$$F^*(t_i) = t_i^q + \alpha g_i$$

with $g_i \in A$ and we choose any $1 > \delta > |\alpha|$. Now, we choose some $\rho_0 < 1/|p|$. Using Corollary 7.2.16, we may shrink ρ_0 and assume that g_i is η'-convergent in A_{ρ_0}. In particular, if $\rho < \rho_0$, there exists M such that for $|\underline{k}| > M$, we have

$$\|\underline{\partial}^{[\underline{k}]}(g_i)\|_\rho \eta'^{|\underline{k}|} \leq \frac{\delta\eta'}{|\alpha|}.$$

On the other hand, if θ denotes the Taylor series, we have $\theta(g_i) \in A[[\tau]]$, and thus for all $\underline{k}$, $\|\underline{\partial}^{\underline{k}}(g_i)\|_1 \leq 1$. We may therefore take ρ' in such a way that whenever $|\underline{k}| \leq M$, we have

$$\|\underline{\partial}^{[\underline{k}]}(g_i)\|_{\rho'} \leq \frac{\delta}{|\alpha|}.$$

It follows that, for $\underline{k} \neq \underline{0}$, we have

$$\|\underline{\partial}^{[\underline{k}]}(g_i)\|_{\rho'} \eta'^{|\underline{k}|} \leq \frac{\delta\eta'}{|\alpha|}$$

which implies that

$$|\alpha| \|\theta(g_i) - g_i\|_{\rho',\eta'} \leq \delta\eta'.$$

Claim 8.3.7 *We can write*

$$F^*(\tau_i) = \tau_i^q + p\tau_i P(\tau_i) + \alpha(\theta(g_i) - g_i) \in A_\rho\{\underline{\tau}/\eta'\}$$

where P is a polynomial on $\mathcal{V}$.

This is easily checked: if, as usual, we take the first projection as structural morphism, then we have

$$\tau_i = p_2^*(t_i) - p_1^*(t_i) = \theta(t_i) - t_i.$$

Thus,

$$\begin{aligned} F^*(\tau_i) &= F^*(\theta(t_i) - t_i) = \theta(F^*(t_i)) - F^*(t_i) \\ &= \theta(t_i^q + \alpha g_i) - (t_i^q + \alpha g_i) = \theta(t_i)^q - t_i^q + \alpha(\theta(g_i) - g_i) \end{aligned}$$

and it is sufficient to recall that

$$\theta(t_i)^q = (t_i + \tau_i)^q = t_i^q + \tau_i^q + \sum_{k=1}^{q-1} \binom{q}{i} t_i^{q-k} \tau_i^k = t_i^q + \tau_i^q + p\tau_i P(\tau_i).$$

From the above formula, we see that

$$\begin{aligned} \|F^*(\tau_i)\|_{\rho',\eta'} &\leq \sup\{\|\tau_i^q\|_{\eta'}, |p|\|\tau_i\|\|P(\tau_i)\|_{\eta'}, |\alpha|\|\theta(g_i) - g_i\|_{\rho',\eta'}\} \\ &\leq \sup\{\eta'^q, |p|\eta', \delta\eta'\} < \eta. \end{aligned}$$

□

The next result is sometimes called *Dwork's trick* or *Dwork extension of radius of convergence along Frobenius.*

Proposition 8.3.8 *Let A be a weakly complete $\mathcal{V}$-algebra with étale coordinates and Frobenius. Then, there exists $\delta < 1$ such that, if M is an η-convergent A_K-module with an integrable connection, then $M^{(q)}$ is η'-convergent with $\eta' = \inf\{\eta^{1/q}, \eta/\delta\}$.*

Proof Thanks to the lemma, if we pullback the Taylor series in $M_\rho \otimes_{A_\rho} A_\rho\{\underline{\tau}/\eta\}$ by F^*, we fall inside $M_{\rho'}^{(q)} \otimes_{A_{\rho'}} A_{\rho'}\{\underline{\tau}/\eta'\}$. □

Corollary 8.3.9 *Let A be a weakly complete $\mathcal{V}$-algebra with étale coordinates and Frobenius. If M is an η-convergent A_K-module with an integrable connection and $\eta' < 1$, there exists $r > 0$ such that $M^{(q^r)}$ is η'-convergent.*

Proof We use the proposition and fix δ. Then, we define by induction $\eta_0 := \eta$ and $\eta_{r+1} = \inf\{\eta_r^{1/q}, \eta_r/\delta\}$ so that $M^{(q^r)}$ is η_r convergent. It is then sufficient to note that $\eta_r \to 1$. □

Let A be a formally smooth weakly complete $\mathcal{V}$-algebra with Frobenius F^*. If M is a topological A_K-module with an integrable connection, a *Frobenius* on M is a continuous horizontal morphism $\varphi : M^{(q)} \to M$. If it is an isomorphism, it is called a *strong Frobenius*. We will denote by $F-\mathrm{MIC}(A_K)$ the category of pairs (M, φ) where M is a coherent A_K-module with an integrable connection and φ is a strong Frobenius on M.

The next theorem says that the presence of Frobenius implies the overconvergence of the connection.

Theorem 8.3.10 *If $X = \operatorname{Spec} A$ is an affine $\mathcal{V}$-scheme, there is an equivalence of categories*

$$F\text{–}\mathrm{Isoc}^\dagger(X_k/\mathcal{V}) \simeq F\text{–}\mathrm{MIC}(A_K^\dagger).$$

Proof It immediately follows from Propositions 8.1.13 and 8.1.15 that there is a fully faithful functor

$$\begin{array}{ccc} F\text{–}\mathrm{Isoc}^\dagger(X_k/\mathcal{V}) & \longrightarrow & F\text{–}\mathrm{MIC}(A_K^\dagger) \\ E & \longmapsto & M \end{array}$$

and we only need to show that it is essentially surjective.

Also, the question is local on X by Proposition 4.3.15 and we may therefore assume that there are étale coordinates $t_1, \ldots, t_m$ on X. Thanks to Proposition 8.3.5, we know that M is η-convergent for some $\eta < 1$. It follows from Proposition 8.3.8 that M is η-convergent for any $\eta < 1$. We may now use Proposition 7.2.15. □

By functoriality, if E is an F-isocrystal on a virtual frame $(X \subset Y/S)$ and $g : Y \to S$ denotes the structural map, there are canonical morphisms

$$\phi : LF_S^* Rg_{\mathrm{rig}} E \to Rg_{\mathrm{rig}} E$$

and

$$\phi_c : LF_S^* Rg_{\mathrm{rig},c} E \to Rg_{\mathrm{rig},c} E.$$

Of course, if X' is a closed subset of X, we also have

$$\phi_{X'} : LF_S^* Rg_{X',\mathrm{rig}} E \to Rg_{X',\mathrm{rig}} E.$$

As a particular case, if X is an algebraic variety over S_k and $f : X \to S$ denotes the structural map, we get a morphism

$$\phi : LF_S^* Rf_{\mathrm{conv}} E \to Rf_{\mathrm{conv}} E$$

when E is a convergent isocrystal on X/S. More generally, we get when X' is a closed subset of X, a morphism

$$\phi_{X'} : LF_S^* Rf_{X',\mathrm{conv}} E \to Rf_{X',\mathrm{conv}} E.$$

Of course, we also have morphisms

$$\phi : LF_S^* Rf_{\mathrm{rig}} E \to Rf_{\mathrm{rig}} E,$$
$$\phi_{X'} : LF_S^* Rf_{X',\mathrm{rig}} E \to Rf_{X',\mathrm{rig}} E,$$

and

$$\phi_c : LF_S^* Rf_{\mathrm{rig},c} E \to Rf_{\mathrm{rig},c} E$$

when E is an overconvergent isocrystal on X/S.

Definition 8.3.11 *A* Frobenius *on a K-vector space E is a σ-linear endomorphism. An F*-isocrystal *on K is a vector space E endowed with a Frobenius.*

Thus, we see that if X is an algebraic variety over S_k and E is an overconvergent F-isocrystal on X, then $H^i_{\mathrm{rig}}(E)$ is an F-isocrystal on K. This is true more generally for $H^i_{X',\mathrm{rig}}(E)$ if X' is a closed subvariety of X as well as for $H^i_{\mathrm{rig},c}(E)$ in general.

Let A be a formally smooth weakly complete $\mathcal{V}$-algebra with a Frobenius F^*. Let M be a topological A_K-module with a continuous integrable connection and a Frobenius endomorphism φ. By functoriality, we obtain a Frobenius endomorphism on $\Gamma_{\mathrm{dR}}(M)$ given by

$$m \otimes \omega \mapsto \varphi(F^*(m)) \otimes dF^*(\omega).$$

Now, let $X := \mathrm{Spec} A$ be a smooth affine $\mathcal{V}$-scheme. If F^* is a Frobenius on $A^\dagger$, it induces a morphism $V_{\rho'} \to V_\rho$ from which we derive a continuous horizontal morphism $\varphi_c : A_c \to A_c$, the Frobenius of A_c. If M is a coherent $A_K^\dagger$-module with an integrable connection and a Frobenius φ, we may use the tensor product structure $M_c = M \otimes_{A_K^\dagger} A_c$ to endow M_c with a Frobenius. In the same way, if $X' = V(f)$ is a hypersurface in X, then there exists a Frobenius on $A_f^\dagger$ compatible with the Frobenius of $A^\dagger$ and it induces a Frobenius on the quotient $A^f \to A^f$. Again, we may use the tensor product structure $M^f = M \otimes_{A_K^\dagger} A_K^f$ to endow M^f with a Frobenius.

Proposition 8.3.12 *Let $X :=$* Spec*A be a smooth affine $\mathcal{V}$-scheme and E an overconvergent F-isocrystal on X_k. If M denotes the corresponding $A_K^\dagger$-module, the isomorphism*

$$R\Gamma_{\mathrm{rig}}(X_k, E) \simeq \Gamma_{\mathrm{dR}}(M)$$

is compatible with the Frobenius action. This is true more generally for

$$R\Gamma_{\mathrm{rig},X'_k}(X_k, E) \simeq \Gamma_{\mathrm{dR}}(M^f)$$

when X' is the hypersurface in X defined by some non zero $f \in A$, as well as for

$$R\Gamma_{\mathrm{rig},c}(X_k, E) \simeq \Gamma_{\mathrm{dR}}(M_c).$$

Proof This directly follows from Proposition 8.2.12. $\square$

We will denote $K(i)$ the F-isocrystal K endowed with the Frobenius $\alpha \mapsto q^{-i}\sigma(\alpha)$. More generally, if $s = i/j \in \mathbf{Q}$ with $j \in \mathbf{N}$ and i prime to j, then $K(s)$ is the F-isocrystal K^j with Frobenius matrix

$$\begin{bmatrix} 0 & \cdots & \cdots & 0 & q^{-i} \\ 1 & \ddots & & & 0 \\ 0 & \ddots & \ddots & & \vdots \\ \vdots & \ddots & \ddots & \ddots & \vdots \\ 0 & \cdots & 0 & 1 & 0 \end{bmatrix}.$$

Note that we may twist any F-isocrystal E on K by setting

$$E(i) := E \otimes_K K(i).$$

Proposition 8.3.13 *If X is an algebraic variety of pure dimension d over k, the trace map*

$$\mathrm{tr}_X : H^{2d}_{\mathrm{rig},c}(X) \to K(-d)$$

is a morphism of F-isocrystals.

Proof The question is additive and local and it is therefore sufficient to prove that if $X = \operatorname{Spec} A$ is a smooth affine connected $\mathcal{V}$-scheme of relative dimension d over K, the trace map

$$H^d_{\mathrm{dR}}(A_c) \to K(-d)$$

is a morphism of F-isocrystals. It means that the diagram

$$\begin{array}{ccc} H^d_{\mathrm{dR}}(A_c) & \xrightarrow{F^*} & H^d_{\mathrm{dR}}(A_c) \\ \downarrow{\scriptstyle tr} & & \downarrow{\scriptstyle tr} \\ K & \xrightarrow{q^d} & K \end{array}$$

is commutative.

But we know that, up to a constant, there exists a unique continuous linear map $H^d_{\mathrm{dR}}(A_c) \to K$. We may therefore choose a point $x \in X$, and compose with the Gysin map $K(x) \to H^d_{\mathrm{dR}}(A_c)$ before checking commutativity. Of course,

F^* lifts to a finite flat morphism $A^\dagger \to A^\dagger$ which induces a finite flat morphism $F : V_{\rho'} \to V_\rho$ and functoriality of Gysin with respect to finite flat maps gives a commutative diagram

$$\begin{array}{ccc} K(x) & \longrightarrow & \oplus_{F(\xi)=x_K} K(\xi) \\ \downarrow & & \downarrow \\ \omega_{A,c} & \xrightarrow{F^*} & \omega_{A,c} \end{array}$$

Since Char$K = 0$ and F has degree q^d, we see that

$$\sum_{F(\xi)=x_K} [K(\xi) : K] = q^d [K(x) : K]$$

and we are done. □

Note that we could not use Corollary 8.2.25 in order to prove this proposition because Frobenius is not separable in general. More precisely, if x is a point of an algebraic variety X over k, then $F^{-1}(x)$ is not reduced in general, and in particular, it is not smooth.

Corollary 8.3.14 *Let X be an algebraic variety of dimension d over k and E be an overconvergent F-isocrystal on X with dual $E^\vee$. Then, there is a canonical pairing*

$$R\Gamma_{\mathrm{rig}}(X, E^\vee) \times R\Gamma_{\mathrm{rig},c}(X, E) \to K[-2d](-d).$$

More generally, if X' is a closed subset of X, there is a canonical morphism

$$R\Gamma_{\mathrm{rig},X'}(X, E^\vee) \times R\Gamma_{\mathrm{rig},c}(X', E_{|X'}) \to K[-2d](-d).$$

Proof This follows from Proposition 8.2.23, because, by functoriality, the Poincaré pairing is compatible with Frobenius and we just proved that the trace map too is compatible. □

We also need to say a few words about Frobenius structures over Robba rings. A *Frobenius endomorphism* of $\mathcal{R}$ is a continuous lifting σ of the q-th power of $k((t))$ (extending our previous σ on K). If M is an $\mathcal{R}$-module, we will write

$$M^{(q)} := \mathcal{R} \otimes_\sigma M.$$

If M is has a connection, a *(strong) Frobenius* on M is a continuous horizontal isomorphism $\varphi : M^{(q)} \to M$. We will denote by $\sigma-\mathrm{MIC}(\mathcal{R})$ the category of pairs (M, φ) where M is a coherent $\mathcal{R}$-module with an integrable connection and φ is a Frobenius on M.

Proposition 8.3.15 *Let $\mathcal{Y}$ be a flat formal $\mathcal{V}$-scheme whose special fiber Y is a connected curve, X a non empty open subset of Y and x a smooth rational point in $Y \setminus X$. If (E, Φ) is an overconvergent F-isocrystal on $(X \subset Y)$, then E_x has a canonical Frobenius Φ_x and we get a functor*

$$\begin{array}{ccc} F\text{−MIC}(X \subset Y) & \longrightarrow & \sigma\text{−MIC}(\mathcal{R}) \\ (E, \Phi) & \longmapsto & (E_x, \Phi_x). \end{array}$$

Proof This immediately follows from the functoriality of the fiber functor $E \mapsto E_x$ of Proposition 6.1.19 since we know from Proposition 5.4.11 that $(F^*E)_x \simeq E_x^{(q)}$. □

We can now come back to our usual examples. We assume that the valuation is discrete.

We first consider a Dwork overconvergent isocrystal L_α with $|\alpha| = |p|^{\frac{1}{p-1}}$ on $\mathbf{A}^1_k$ (recall that the condition on α is necessary and sufficient for the connection to be overconvergent and non trivial). The realization of L_α on $\widehat{\mathbf{P}}^1_{\mathcal{V}}$ is the free rank one module whose connection is given by

$$\nabla(\theta) = -\alpha\theta \otimes dt.$$

We may choose the usual lifting of Frobenius on $\widehat{\mathbf{P}}^1_{\mathcal{V}}$ given by

$$(t_0, t_1) \mapsto (t_0^q, t_1^q)$$

inducing $t \mapsto t^q$ on $\mathbf{A}_K^{1,\mathrm{rig}}$ and $dt \mapsto qt^{q-1}dt$ on differentials. Then, the realization of $L_\alpha^{(q)}$ is the free rank one module whose connection is given by

$$\nabla(\theta^{(q)}) = -q\alpha t^{q-1}\theta^{(q)} \otimes dt.$$

A Frobenius isomorphism

$$\varphi : L_\alpha^{(q)} \simeq L_\alpha$$

corresponds to an invertible function h on some disk $\mathbf{D}(0, \rho^+)$ with $\rho > 1$ such that

$$-q\alpha t^{q-1}h\theta \otimes dt = h'\theta \otimes dt - \alpha h\theta \otimes dt.$$

It corresponds to the differential equation

$$\alpha(1 - qt^{q-1})h = h'$$

whose power series solution, up to multiplication by a non zero constant, is $\exp(\alpha(t - t^q))$. In general, this function only converges on $\mathbf{D}(0, 1^-)$. However, when π is a $(p-1)$-th root of $-p$, then Dwork showed that the series $\exp(\pi(t - t^q))$ is convergent on $\mathbf{D}(0, \rho^+)$ when $\rho < |p|^{\frac{1-p}{pq}}$. More precisely, we have

Claim 8.3.16 *The series* $\exp(\pi(t-t^q))$ *is (convergent and) bounded by* 1 *for* $|t| < |p|^{-\frac{p-1}{pq}}$.

First of all, we have

$$\exp(\pi(t-t^q)) =$$
$$\exp(\pi(t-t^p))\exp(\pi(t^p-(t^p)^p))\cdots\exp(\pi(t^{q/p}-(t^{q/p})^p)).$$

Thus, if we rewrite the condition as $|t^{q/p}| < |p|^{-\frac{p-1}{p^2}}$, we see that the general case will follow from the case $q = p$.

Now, we use the Artin–Hasse exponential

$$E(t) = \exp(\sum_{s=0}^{\infty} \frac{t^{p^s}}{p^s}) \in \mathbf{Z}_p[[t]]$$

and write

$$\exp(\pi(t-t^p)) = E(\pi t) \bullet \prod_{s=2}^{\infty} \exp - \frac{(\pi t)^{p^s}}{p^s}.$$

We assume that $|t| < |p|^{-\frac{p-1}{p^2}}$ and we show that all the terms in the infinite product are bounded by 1. For $E(\pi t)$, this is clear because

$$|\pi t| = |p|^{\frac{1}{p-1}}|t| < |p|^{\frac{1}{p-1}}|p|^{-\frac{p-1}{p^2}} = |p|^{\frac{2p-1}{(p-1)p^2}} < 1$$

and the Artin–Hasse series has coefficients in $\mathbf{Z}_p$. Now, for $s \geq 2$, we consider the n-th term in the series expansion of $\exp - \frac{(\pi t)^{p^s}}{p^s}$. We have

$$\left| \frac{(-\frac{(\pi t)^{p^s}}{p^s})^n}{n!} \right| = \frac{|\pi|^{np^s}|t|^{np^s}}{|p|^{ns}|n!|} < \frac{|p|^{\frac{np^s}{p-1}}(|p|^{-\frac{p-1}{p^2}})^{np^s}}{|p|^{ns}|p|^{\frac{n}{p-1}}}$$

and it only remains to check that

$$\frac{np^s}{p-1} - \frac{(p-1)np^s}{p^2} - ns - \frac{n}{p-1} > 0$$

when $s \geq 2$ which is left to the reader.

Thus, we find an overconvergent F-isocrystal, the *Dwork F-isocrystal* L_π given by

$$\nabla(\theta) = -\pi\theta \otimes dt, \quad \varphi : \theta^{(q)} \mapsto \exp \pi(t-t^q)\theta.$$

We can play the same game with a Kummer overconvergent isocrystal K_β with $\beta \in \mathbf{Z}_p \setminus \mathbf{Z}$ on $\mathbf{A}^1_k \setminus 0$ (recall that the condition on β is necessary and sufficient for the connection to be overconvergent and non trivial). The realization of K_α on $\widehat{\mathbf{P}}^1_{\mathcal{V}}$ is the free rank one module whose connection is given by

$$\nabla(\theta) = \beta\theta \otimes \frac{dt}{t}.$$

Then, the realization of $K_\beta^{(q)}$ is the free rank one module whose connection is given by

$$\nabla(\theta^{(q)}) = q\beta\theta^{(q)} \otimes \frac{dt}{t}.$$

In other words, $K_\beta^{(q)} = K_{q\beta}$.

A Frobenius isomorphism

$$\varphi : K_\beta^{(q)} \simeq K_\beta$$

corresponds to an invertible function h on some annulus $\mathbf{A}(0, (\frac{1}{\rho})^+, \rho^+)$ with $\rho > 1$ such that

$$q\beta h\theta \otimes \frac{dt}{t} = h'\theta \otimes dt + \beta h\theta \otimes \frac{dt}{t}.$$

It corresponds to the differential equation

$$\frac{(q-1)\beta}{t} h = h'$$

whose power series solution, up to multiplication by a non zero constant, is $t^{(q-1)\beta}$. In the case $\beta = \frac{i}{q-1}$ with $i \in \mathbf{Z}$, this is just t^i and we obtain an overconvergent F-isocrystal, the *Kummer F-isocrystal* $K_{i/q-1}$ given by

$$\nabla(\theta) = \frac{i}{(q-1)}\theta \otimes \frac{dt}{t}, \quad \varphi : \theta^{(q)} \mapsto t^i\theta.$$

Another example is given, when $p \neq 2$, by Kedlaya's crystal $K_{1/2,\bar{Q}}$ with Q a monic separable polynomial over $\mathcal{V}$ of degree $d > 0$. It is defined on the non zero locus U of $\bar{Q}$ in $\mathbf{A}^1_k$ and its realization on $\widehat{\mathbf{P}}^1_{\mathcal{V}}$ is the free module of rank one whose connection is given by

$$\nabla(\theta) = \frac{1}{2}\theta \otimes \frac{dQ}{Q}.$$

Of course, Q does not commute with the liftings of Frobenius: we have

$$Q^{(q)}(x) := Q(x^q) \neq Q(x)^q =: Q^q(x).$$

Thus, we have to be careful in computing the Frobenius structure. The realization of $K^{(q)}_{1/2,\bar{Q}}$ is the free rank one module with connection

$$\nabla(\theta^{(q)}) = \frac{1}{2}\theta^{(q)} \otimes \frac{dQ^{(q)}}{Q^{(q)}}$$

and Frobenius is given by

$$\varphi : \theta^{(q)} \mapsto Q^{\frac{q-1}{2}} (\frac{Q^{(q)}}{Q^q})^{1/2}\theta$$

as a – not so easy – computation shows. Of course, here, we let

$$(1+z)^{1/2} = \sum_{n\geq 0} \binom{1/2}{n} z^n$$

and use the fact that $|1 - \frac{Q^{(q)}}{Q^q}| = |\frac{Q^q - Q^{(q)}}{Q^q}| < 1$ in order to get overconvergence. Since all but the first cohomology group vanish and we have an explicit description

$$H^1_{\mathrm{rig}}(V, K_{\frac{1}{2},\bar{Q}}) = \{\frac{P(x)}{Q(x)}\theta \otimes dx, \deg P < d-1\},$$

one obtains an algorithm that computes the action of Frobenius on the rigid cohomology of $K_{\frac{1}{2},\bar{Q}}$. Now, since the plane analytic curve X defined by

$$y^2 = \bar{Q}(x) \quad \text{with} \quad y \neq 0.$$

satisfies

$$H^i_{\mathrm{rig}}(X) = H^i_{\mathrm{rig}}(U) \oplus H^i_{\mathrm{rig}}(U, K_{\frac{1}{2},Q})$$

we get an algorithm that computes the action of Frobenius on the rigid cohomology of X as in [52].

9
Conclusion

9.1 A brief history

As already mentioned, p-adic cohomology first showed up in an informal way in the work of Bernard Dwork ([39]) in the late 1950s. In the 1960s and 1970s, crystalline cohomology was developed by Pierre Berthelot (and others) and at the same time, Monsky–Washnitzer cohomology was developed by Paul Monsky (and others). The first one is suitable for proper smooth algebraic varieties and the second one for smooth affine varieties. Both theories were united by Pierre Berthelot in the early 1980s (see [11]), giving birth to rigid cohomology. The development of this theory can be split into three periods.

The first one is dedicated to foundations, and we gave it a rather complete treatment in this book. Actually, the rigid cohomology spaces of a given algebraic variety are not that hard to define. As we saw, we just need to compactify, embed into some smooth formal scheme and take the limit de Rham cohomology on strict neighborhoods. And in the quasi-projective case, this is even simpler. The main difficulty is to show that the results do not depend on the choices. The basic ingredient is the Fibration Theorem. Berthelot laid these foundations in the early 1980s.

The second period starts with Johan de Jong's alterations theorem ([35]). Roughly speaking, this theorem says that one can solve singularities in positive characteristic if one is willing to replace Zariski by étale topology. Most results concerning rigid cohomology with constant coefficients, such as finite dimensionality, can be derived from de Jong's theorem. This periods ends with Denis Pétrequin's thesis ([74]).

The third period starts with the conjecture of Richard Crew ([33]) which first appeared in the introduction of [32] in 1987. It claims that an F-isocrystal on a Robba ring is quasi-unipotent. As a consequence, we get finite dimensionality of rigid cohomology for overconvergent F-isocrystals on curves. This period is not over. The conjecture of Crew was first proved by Yves André ([2]) followed by Zogman Mebkhout ([65]). Both proofs rely heavily on Christol–Mebkhout theory of differential equations. At the same period, Kiran Kedlaya was able to prove the conjecture of Crew as a corollary of a Dieudonné–Manin theorem for F-isocrystals on a Robba ring. Actually, Kedlaya's methods allows him to prove finite dimensionality of rigid cohomology in general as well as many other amazing results. However, not much is known when there is no Frobenius structure. Also, we need a geometric monodromy theorem such as in Shiho's ([79]).

At the same time, Kedlaya was able to show in [52] that Rigid cohomology is suitable to compute the number of rational points of hyperelliptic curves. There are many other results in this directions due to Alan Lauder and Daqing Wan ([59]), Lauder alone ([58], [57]) or Jan Denef and Frederik Vercauteren ([37], [38]). There is also a thesis by Ralf Gerkmann on the subject ([46]).

Finally, let me mention arithmetic $\mathcal{D}$-modules. This theory was developed by Berthelot in the 1990s ([14], [18], [19]) in order to obtain a candidate for the constructible category of p-adic coefficients. Such a category should be stable under the standard operations and should also contain overconvergent F-isocrystals. Much remains to be done in this area and we can only hope that Kedlaya's techniques will be useful again.

9.2 Crystalline cohomology

Divided powers on an ideal I in a ring A were introduced in [77] as functions on I that behave like

$$x \mapsto x^{[n]} := x^n/n!.$$

More generally, if p is a fixed prime, A is a $\mathbf{Z}_{(p)}$-algebra and m an integer, a *divided power structure* of level m (see [14]) on I is an ideal J inside I with divided powers and such that $pI + I^{(p^m)} \subset J$ (one can then define partial divided powers on I that behave like

$$x \mapsto x^{\{n\}} := x^n/q!$$

where q is the integral part of n/m). The natural way to generalize this notion to the case $m = \infty$ is simply to require I to be nilpotent. These notions glue and one may consider the notion of divided power structure of level m, $(\mathfrak{a}, \mathfrak{b}, [\ ])$, on a scheme S over $\mathbf{Z}_{(p)}$ or a formal scheme S over $\mathbf{Z}_p$.

If X is a scheme over S (and $[\ ]$ extends to X), a *thickening* of level m of X is a embedding $X \hookrightarrow T$ whose ideal is endowed with a structure of level m.

The *crystalline site of level m* of X is the category of thickenings of level m of open subsets U of X (for $m = \infty$, we get the *infinitesimal site*).

A module on this site is given by a family of sheaves of $\mathcal{O}_T$-modules E_T with compatible transition maps $u^*E_T \to E_{T'}$ whenever $u : T' \to T$ is an S-morphism compatible with the structures. It is called a *crystal* if the transition maps are bijective. There is a standard notion of cohomology on a site and we may consider $H^*_{\text{cris},m}(X/S, E)$. The case $m = 0$ gives the usual crystalline cohomology introduced in [9] and [20]. The case $m < \infty$ is studied in the sequence of papers [61], [62] and [63].

For $m < \infty$, crystalline cohomology is computed by means of de Rham cohomology. More precisely, if E is a quasi-coherent crystal of level m on X/S and $i : X \hookrightarrow P$ is a closed embedding into a smooth formal S-scheme, then

$$E_P := \varprojlim (i_* E)_{P_n},$$

with $P_n := \mathbf{Z}/p^{n+1} \otimes P$, has a canonical integrable connection of level m and

$$R\Gamma_{\text{cris},m}(X/S, E) = R\Gamma_{\text{dR},m}(P/S, E_P)$$

(the notion of connection and de Rham cohomology both extend to higher level).

Assume that $p \in \mathfrak{a}$ and let S_0 be the corresponding scheme of characteristic p. If X is smooth over S_0 and $F : X \to X'$ denotes the relative Frobenius, then F^* induces an equivalence between m-crystals on X'/S and $m+1$-crystals on X/S. Moreover, we have an isomorphism on cohomology

$$R\Gamma_{\text{cris},m}(X'/S, E) \simeq R\Gamma_{\text{cris},m+1}(X/S, E).$$

This is called Frobenius descent for crystalline cohomology.

Now, we want to describe the link with rigid cohomology as in [12], [44] or [34], so we fix a complete ultrametric field of mixed characteristic p with valuation ring $\mathcal{V}$. For $m >> 0$, there are divided powers of level m on $\mathcal{V}$. Moreover, if X is an algebraic variety on k, any ∞-crystal E on $X/\mathcal{V}$ defines, for $m >> 0$, an m-crystal $E^{(m)}$ on $X/\mathcal{V}$. If X is a closed subvariety of a smooth formal $\mathcal{V}$-scheme P, and E is finitely presented, it defines a coherent module with a convergent integrable connection E_K characterized by $sp_* E_K = \varprojlim E^{(m)}_{P\mathbf{Q}}$.

Assume now that If X is a subvariety of a smooth formal $\mathcal{V}$-scheme P with proper Zariski closure $\overline{X}$, and that E extends to a finitely presented ∞-crystal $\overline{E}$ on $\overline{X}$. By restriction, $\overline{E}_K$ gives an overconvergent isocrystal E_K on $X/\mathcal{V}$. If we denote by $i : \overline{X} \setminus X \hookrightarrow \overline{X}$ the inclusion map, then we can make the *ad hoc* definition

$$R\Gamma_{\mathrm{cris},m,c}(X, E^{(m)}) := R\Gamma_{\mathrm{cris},m}(\overline{X}, \overline{E}^{(m)} \to i_*i^*\overline{E}^{(m)})$$

and we have an isomorphism

$$R\Gamma_{\mathrm{rig},c}(X, E_K) = R\varprojlim R\Gamma_{\mathrm{cris},m,c}(X, E^{(m)})_{\mathbf{Q}}.$$

It follows that, when X is proper and smooth, we have

$$R\Gamma_{\mathrm{rig}}(X, E_K) = R\Gamma_{\mathrm{cris},m}(X, E^{(m)})_{\mathbf{Q}}.$$

In general, the right-hand side might not be finite dimensional, and we only know ([34]) that for $m >> 0$ there is an isomorphism

$$R\Gamma_{\mathrm{rig},c}(X) = R\Gamma_{\mathrm{cris},m,c}(X)_{\mathbf{Q}}/F^{\infty}\text{−torsion}.$$

9.3 Alterations and applications

Let k be a perfect field. Then, de Jong's theorem ([35], [15]) states that if X is an integral algebraic variety over k and $Z \subset X$ a closed subset distinct from X, there exists a smooth projective variety X' over k, an open subset U' in X' and an alteration $\varphi : U' \to X$ such that $\varphi^{-1}(Z) \cup X' \setminus U'$ is (the support of) a strict normal crossing divisor. Recall that an *alteration* of integral k-varieties is a proper surjective morphism which is generically étale. Recall also that a *strict normal crossing divisor* is a finite union of integral subvarieties D_i such that any r-th intersection of the D_i's is regular of codimension r.

The first consequence of de Jong's theorem is finiteness of rigid cohomology as shown by Berthelot in [17] (see also [16]). We let K be a complete ultrametric field with valuation ring $\mathcal{V}$ and residue field k. We assume that the valuation is discrete but not anymore that k is perfect. Then, we have the following : If X is any algebraic variety over k, then

$$\forall i \geq 0, \quad \dim_K H^i_{\mathrm{rig},c}(X/K) < \infty.$$

Moreover, if X is smooth and $Z \subset X$ a closed subset, then

$$\forall i \geq 0, \quad \dim_K H^i_{\mathrm{rig},Z}(X/K) < \infty.$$

Finally, if X is of pure dimension d, Poincaré duality is an isomorphism

$$R\Gamma_{\mathrm{rig},Z}(X/K) \simeq R\Gamma_{\mathrm{rig},c}(Z/K)^{\vee}[-2d].$$

Note that the smoothness hypothesis can be removed as it was shown by Grosse-Kloenne ([48]) and Tsuzuki ([82]). It also seems that the discreteness hypothesis is not necessary either as it should appear in a yet unpublished article of Berkovich (using an idea of Gabber).

As a corollary of Poincaré duality, we obtain for Z a smooth closed subvariety of codimension c in X smooth, the existence of a canonical *Gysin isomorphism*

$$R\Gamma_{\mathrm{rig},Z}(X/K) \simeq R\Gamma_{\mathrm{rig}}(Z/K)[-2c].$$

Here is how Berthelot's proof of finiteness works. It relies on de Jong's theorem and finiteness of crystalline cohomology for proper smooth varieties. But, and this is the hard part, it is also necessary to prove the existence of the Gysin isomorphism in a liftable situation. The method uses $\mathcal{D}$-modules.

Actually, one shows that rigid cohomology satisfies Bloch–Ogus formalism ([74]).

9.4 The Crew conjecture

We assume here that $\mathcal{V}$ is a discrete valuation ring of mixed characteristic with fraction field K and residue field k.

A module with a connection on the Robba ring $\mathcal{R}$ on K is *unipotent* if it is a successive extension of the trivial one. An overconvergent isocrystal E on a curve X is *unipotent at infinity* if whenever $X \hookrightarrow P_k$ is a dense open immersion with P a flat formal $\mathcal{V}$-scheme and x a smooth point of $P_k \setminus X$, then E_x is unipotent on the Robba ring of x. It is said to be *quasi-unipotent at infinity* if there exists a finite étale map $\pi : X' \to X$ such that $\pi^* E$ is unipotent at infinity. The conjecture of Crew([33]) states that any overconvergent F-isocrystal E on a smooth algebraic curve is quasi-unipotent at infinity. This implies finite dimensionality of $H^i_{\mathrm{rig}}(X, E)$ and $H^i_{\mathrm{rig},c}(X, E)$ as well as Poincaré duality and a lot more.

The first proof due to André ([2]) is based on Christol and Mebkhout's study of differential equations. It should be mentioned at this point that Christol and Mebkhout were able to prove finite dimensionality of the cohomology of overconvergent F-isocrystals on smooth curves and recover Berthelot's finite dimensionality theorem several years before ([64]). Actually, Mebkhout also gave a direct proof of André's theorem in [65]. We will also consider later Kedlaya's approach which does not make use of Christol and Mebkhout's work and is more suitable to generalizations.

The notion of quasi-unipotence is local. More precisely, if $\mathcal{R}$ is the Robba ring on K, then the subring $\mathcal{E}^\dagger$ of formal series bounded on the outer side $(= \mathcal{R}^* \cup \{0\})$ is a Henselian field with residue field $k((t))$. Any finite separable extension of $k((t))$ has the form $k'((t'))$ and lifts to an unramified morphism $\mathcal{E}^\dagger \to \mathcal{E}'^\dagger$ that extends to a morphism of Robba rings $\mathcal{R} \to \mathcal{R}'$. A differential module M on $\mathcal{R}$ is said to be *quasi-unipotent* if there exists such an extension with $M' := \mathcal{R}' \otimes_{\mathcal{R}} M$ unipotent. Then, an overconvergent isocrystal E on a curve X is quasi-unipotent at infinity if whenever $X \hookrightarrow P_k$ is a dense open immersion with P a flat formal $\mathcal{V}$-scheme and x a smooth point of $P_k \setminus X$, then E_x is quasi-unipotent on the Robba ring of x. The local version of the conjecture of Crew states that any free module of finite rank over $\mathcal{R}$ with a connection and a strong Frobenius structure is quasi-unipotent.

If we have a linear action of $\mathrm{Gal}(k'((t'))/k((t))$ on some finite dimensional K-vector space V, then we get an action on the trivial differential module $M' := \mathcal{R}' \otimes_K V$ that descends to a differential module M on $\mathcal{R}$ with a strong Frobenius structure. On the other hand, a linear action of $\mathbf{G}_a$ on V will provide us with a nilpotent endomorphism N of V that will induce a connection $\nabla(1 \otimes v) = N(v)\frac{dt}{t}$ on $M := \mathcal{R} \otimes_K V$ turning it into a unipotent differential module that admits a strong Frobenius structure. We fix an algebraic closure $\overline{K}$ of K, set $\overline{\mathcal{R}} := \overline{K} \otimes_K \mathcal{R}$ and extend the Frobenius of $\mathcal{R}$ continuously on $\overline{\mathcal{R}}$. We consider the category $\mathrm{MCF}^\infty(\overline{\mathcal{R}})$ of free modules with connection that admit a strong Frobenius structure of some order. Then André's theorem states that we obtain a canonical equivalence of categories

$$\mathrm{Rep}_{\overline{K}}(I \times \mathbf{G}_a) \simeq \mathrm{MCF}^\infty(\overline{\mathcal{R}}).$$

According to the above considerations, this is clearly equivalent to the conjecture of Crew and the proof works as follows. The right-hand side is a Tannakian category corresponding to some affine group G and Christol–Mebkhout filtration on differential modules provides what is called a Hasse–Arf filtration on G. And such a filtration gives the above decomposition (and more).

9.5 Kedlaya's methods

We still assume that $\mathcal{V}$ is a discrete valuation ring of mixed characteristic with fraction field K and residue field k, and let π be a uniformizer. And we denote as usual by $\mathcal{R}$ the Robba ring of K endowed with a Frobenius σ. Then, Kedlaya's theorem ([53], or better, [54]) states that any σ-module M has a filtration by saturated σ-modules

$$0 = M_0 \subset \cdots \subset M_l = M$$

with quotients of pure slopes $s_1 < \ldots < s_l$ respectively. Of course, a *σ-module* is simply a free module of finite rank M with a strong Frobenius $\Phi : \sigma^* M \simeq M$

and we postpone for a while the definition of the slopes (recall also that a submodule M' is *saturated* if the quotient M/M' has no torsion). The first point is that, if M is endowed with a connection and the Frobenius is horizontal, then such a filtration is automatically preserved by the connection. Quasi-unipotence therefore reduces to the case of pure slope which easily results from the unit-root case (pure of slope zero) which is a theorem of Tsuzuki ([80]).

Kedlaya introduces an extended Robba ring, that we may denote $\overline{\mathcal{R}}$ that has also a Frobenius endomorphism σ and we may consider σ-modules on $\overline{\mathcal{R}}$. Again, we postpone for a while the definition of this ring but the first point is that a σ-module on $\mathcal{R}$ is said to be *pure of slope* s if $\overline{\mathcal{R}} \otimes_{\mathcal{R}} M$ is pure of slope s. Now, the definition of slope and purity becomes easy because there is a Diedonné–Manin decomposition theorem on $\overline{\mathcal{R}}$. More precisely, any σ-module on $\overline{\mathcal{R}}$ is a direct sum of σ-modules of the form

$$\overline{\mathcal{R}}(-s) := \overline{\mathcal{R}} \otimes_K K(-s).$$

It remains to recall that $K(-s)$ is the usual standard σ-module of slope $s \in \mathbf{Q}$ on K.

The construction of $\overline{\mathcal{R}}$ is as follows. The idea is to replace $k((t))$ by its complete algebraic closure C. First of all, we may express K as a finite extension of the fraction field of a Cohen ring W of k and we set $K(C) := K \otimes_W W(C)$ where $W(C)$ denotes the ring of Witt vectors. We find it convenient to consider, for $s \in k((t))$, the absolute value $|s| := |\pi|^{v(s)}$, and keep the same notation for its unique extension to C. Any element of $K(C)$ may be written uniquely $f = \sum_{i \in \mathbf{N}} \widetilde{s_i} \pi^i$ where $\widetilde{s}$ denotes the Teichmüller lifting of $s \in C$ and we set for $\eta < 1$,

$$\|f\|_\eta := \sup |s_i| \eta^{-i}.$$

Then, the subring

$$K_\eta(C) := \{f \in K(C), \quad \|f\|_\eta < \infty\}$$

is endowed with the family of semi-norms $\| - \|_{\eta'}$ for $1 > \eta' \geq \eta$ and we denote by $K_{\eta,\mathrm{an}}(C)$ its Fréchet completion. Finally, the extended Robba ring is simply $\overline{\mathcal{R}} := \cup K_{\eta,\mathrm{an}}(C)$. In order to embed $\mathcal{R}$ into $\overline{\mathcal{R}}$, one first remarks that

$$\{\sum_{i \in \mathbf{Z}} c_i t^i, \quad c_i \in W, \quad c_i \to 0 \text{ for } i \to -\infty\}$$

is a Cohen ring for $k((t))$. Therefore, the inclusion of $K((t))$ in C lifts to an injective ring homomorphism from this ring to $W(C)$ and we may extend it canonically to $\mathcal{R} \hookrightarrow \overline{\mathcal{R}}$.

Actually, using a relative variant of this local monodromy theorem, Kedlaya is able to prove finite dimensionality, Poincaré duality and Künneth formula with coefficients in an F-isocrystal.

9.6 Arithmetic $\mathcal{D}$-modules

This theory is developed by Berthelot in [14] and [18] (see also [19]) in order to extend the category of overconvergent F-isocrystals to a category of constructible coefficients. This category should be compatible to Grothendieck's six operations (direct image (naive and exceptional), inverse image (naive and exceptional), duality and external tensor product). The idea is to look for such a category inside the category of F-$\mathcal{D}^{\dagger}_{\mathbf{Q}}$-modules. We want to explain here what this category is and show its relation to rigid cohomology.

We fix a discrete valuation ring $\mathcal{V}$ of mixed characteristic and a (formal) scheme S on $\mathcal{V}$. Below, we only consider $m >> 0$ so that $\mathcal{V}$ has a level m structure. Any embedding $\mathcal{X} \hookrightarrow \mathcal{Y}$ of formal S-schemes factors canonically through a thickening $P^n_{\mathcal{Y},m}(\mathcal{X})$ of level m and order n called the *divided power envelope* (with structural sheaf $\mathcal{P}^n_{\mathcal{Y},m}(\mathcal{X})$). In particular, we may consider the divided power envelope $P^n_{\mathcal{X}/S,m}$ of level m and order n of the diagonal embedding of a formal S-scheme $\mathcal{X}$. When $\mathcal{X}$ is smooth over S, then

$$\mathcal{D}^{(m)}_{\mathcal{X}/S} := \cup \mathcal{H}om_{\mathcal{O}_\mathcal{X}}(\mathcal{P}^n_{\mathcal{X}/S,m}, \mathcal{O}_\mathcal{X})$$

has a natural ring structure. Then, we set

$$\mathcal{D}^{\dagger}_{\mathcal{X}/S} := \varinjlim \widehat{\mathcal{D}}^{(m)}_{\mathcal{X}/S}$$

and define the ring of arithmetic differential operators on X/S as

$$\mathcal{D}^{\dagger}_{\mathcal{X}/S,\mathbf{Q}} := \mathbf{Q} \otimes_{\mathbf{Z}} \mathcal{D}^{\dagger}_{\mathcal{X}/S}.$$

We consider now the case of a smooth formal S-scheme $\mathcal{X}$ with reduction X and fixed open subset U of X. If E is an overconvergent isocrystal on $U \subset X$, then $Rsp_* E_{\mathcal{X}_K}$ is in a natural way a complex of $\mathcal{D}^{\dagger}_{\mathcal{X}/S,\mathbf{Q}}$-modules and we have

$$Rp_{\mathrm{rig}} E = Rp_* R\mathcal{H}om_{\mathcal{D}^{\dagger}_{\mathcal{X}/S,\mathbf{Q}}}(\mathcal{O}_{\mathcal{X}_K}, Rsp_* E_{\mathcal{X}_K})$$

where $p : \mathcal{X} \to S$ denotes the structural map. More generally, if T is a closed subset of U, then $Rsp_* \underline{\Gamma}^{\dagger}_T E_{\mathcal{X}_K}$ is in a natural way a complex of $\mathcal{D}^{\dagger}_{\mathcal{X}/S,\mathbf{Q}}$-modules and we have

$$Rp_{\mathrm{rig},T} E = Rp_* R\mathcal{H}om_{\mathcal{D}^{\dagger}_{\mathcal{X}/S,\mathbf{Q}}}(\mathcal{O}_{\mathcal{X}_K}, Rsp_* \underline{\Gamma}^{\dagger}_T E_{\mathcal{X}_K}).$$

Actually, if $D := X \setminus U$ is a divisor and f a local defining equation, then

$$\widehat{\mathcal{B}}^{(m)}_{\mathcal{X}}(Z) := \mathcal{O}_{\mathcal{X}}\{T\}/(f^{p^{m+1}} - p)$$

has a natural structure of $\widehat{\mathcal{D}}^{(m)}_{\mathcal{X}/S}$-module and we have

$$Rsp_* j_U^\dagger \mathcal{O}_{\mathcal{X}_K} = \mathbf{Q} \otimes_{\mathbf{Z}} \varinjlim \widehat{\mathcal{B}}^{(m)}_{\mathcal{X}}(Z) \quad (=: \mathcal{O}_{\mathcal{X},\mathbf{Q}}({}^\dagger Z)).$$

It is shown in [12] that this is a coherent $\mathcal{D}^\dagger_{\mathcal{X}/S,\mathbf{Q}}$-module and this gives an alternative proof (also based on de Jong's theorem, though) of the finite dimensionality of the rigid cohomology of U.

If we set

$$\mathcal{D}^\dagger_{\mathcal{X}/S,\mathbf{Q}}({}^\dagger Z) = \varinjlim \widehat{\mathcal{B}}^{(m)}_{\mathcal{X}}(Z) \otimes_{\mathcal{O}_{\mathcal{X}}} \widehat{\mathcal{D}}^{(m)}_{\mathcal{X}/S,\mathbf{Q}},$$

then sp_* induces a fully faithful functor from the category of overconvergent isocrystals on $U \subset X$ into the category of coherent $\mathcal{D}^\dagger_{\mathcal{X}/S,\mathbf{Q}}({}^\dagger Z)$-modules. Moreover, it induces an equivalence between the category of overconvergent F-isocrystals on $U \subset X$ and the category of coherent F-$\mathcal{D}^\dagger_{\mathcal{X}/S,\mathbf{Q}}({}^\dagger Z)$-modules.

9.7 Log poles

Log schemes were introduced in order to extend to open varieties with a nice locus at infinity standard results about proper varieties ([51]). Together with resolution of singularities, this is a powerful way to approach many problems.

A *log (formal) scheme* is a (formal) scheme X endowed with a *log structure*, namely an additive monoid $\mathcal{M}_X$ on X and a homomorphism of monoids $\exp : \mathcal{M}_X \to \mathcal{O}_X$ inducing an isomorphism $\exp^{-1}(\mathcal{O}_X^*) \simeq \mathcal{O}_X^*$. Morphisms are defined as expected. The standard example is given by setting formally

$$\mathcal{M}_X := \{\log f,\ f \in \mathcal{O}_X \cap j_*\mathcal{O}_U\}$$

where U is a non empty open subset of X. When U is the complement of a normal crossing divisor given étale locally by $t_1 \cdots t_r = 0$, the log structure is "associated to" the free monoid M of rank r on formal generators $\log t_1, \ldots, \log t_r$.

In general, a log formal scheme is said to be *fine (saturated)* if, locally étale, the log structure is associated to a finitely generated integral (saturated) monoid M (the integral monoid M is said to be *saturated* if M^{gp}/M has no torsion). Also, a fine log formal scheme is said to be *of Zariski type* if the above condition holds locally for the Zariski topology. Finally, we will need the notion of *exact closed immersion* $i : X \hookrightarrow T$ of log schemes: it just means that $i^{-1}\mathcal{M}_T = \mathcal{M}_X$. We will call a morphism *formally log smooth* (or *log étale*) if it satisfies the usual local lifting condition with respect to exact closed immersions.

Following Ogus in [73], Shiho introduces, in [78] and [79], the *log convergent site*. We fix a discrete valuation ring $\mathcal{V}$ with residue field k and we endow the point with the trivial log structure. We let S be a fine log formal $\mathcal{V}$-scheme. If X is a fine log scheme over S_k, an *enlargement* of X is an exact closed immersion $X \hookrightarrow T$ into a flat fine log (formal) scheme over S such that $\mathbf{Z}/p \otimes_{\mathbf{Z}} T \subset X$. The *log convergent site* of X over S is the category of enlargements $Z \hookrightarrow T$ over S, where Z is a fine log scheme over X. We can choose any reasonable topology (Zariski, étale, flat) making sure that we only use "cartesian" coverings. We may consider the structural sheaf $\mathcal{O}_{X/S}$ and define *log convergent isocrystals* as crystals of $\mathbf{Q} \otimes_{\mathbf{Z}} \mathcal{O}_{X/S}$-modules.

Now, we focus on the case $S = \mathrm{Spf}\mathcal{V}$. Let U denote the locus of X where the log structure is trivial and $j : U \hookrightarrow X$ the inclusion map. Then any locally free log convergent isocrystal E of finite rank on $X/\mathcal{V}$ extends canonically to an overconvergent isocrystal $j^\dagger E$ on U/K. Assume moreover that X is saturated, of Zariski type, proper and log smooth. Assume also that E has a Frobenius structure. Then, using results from Baldassarri and Chiarellotto in [7] and [6], Shiho proves that

$$R\Gamma_{\mathrm{rig}}(U/K, j^\dagger E) \simeq R\Gamma_{\mathrm{log-conv}}(X/\mathcal{V}, E).$$

In order to derive Rigid cohomology from this theorem, it is necessary to find a way to compute log convergent cohomology. This is done through log crystalline cohomology. We choose some $m >> 0$ (it could be $m = \infty$) so that $\mathcal{V}$ has a level m structure and endow S with divided power structure of level m. If X is a fine log (formal) scheme over S, a log thickening of level m of X is an exact embedding $X \hookrightarrow T$ into a fine log (formal) scheme over S whose ideal is endowed with a divided power structure of level m. The *log crystalline site* of level m of X is the category of log thickenings of level m of étale U's over X into (formal) S-schemes. Log-crystals are defined as usual. And so does the cohomology.

We only consider the case $m = 0$ and $S = \mathrm{Spf}W$ where W is the ring of Witt vectors of k. If X is a quasi-compact fine log smooth log scheme over Spec$k$, then any log convergent isocrystal E on X/W gives rise to a log crystal up to isogeny E_0 on X/W. Moreover, we have

$$R\Gamma_{\mathrm{log-conv}}(X/W, E) \simeq \mathbf{Q} \otimes_{\mathbf{Z}} R\Gamma_{\mathrm{log-cris}}(X/W, E_0).$$

The same result should hold on $\mathcal{V}$ with $m >> 0$. Anyway, when X is proper, the right-hand side has finite-dimensional cohomology, and it follows that log convergent cohomology is also finite dimensional. As a corollary, we obtain finite dimensionality for rigid cohomology of overconvergent F-isocrystals that are "regular at infinity". This suggests another proof for Kedlaya's theorem: finiteness of rigid cohomology will be a consequence of a global monodromy theorem.

The conjecture of Shiho is as follows:

Let X be a smooth algebraic variety over k and E an overconvergent F-isocrystal on X. Then, after a finite extension of k, there exists a smooth projective variety X' over k, an open subset U' in X', an alteration $\varphi : U' \to X$ such that $X' \setminus U'$ is (the support of) a strict normal crossing divisor D and a log F-isocrystal E' on $(X, \mathcal{M}_D)$ such that $\varphi^* E \simeq j^\dagger E'$.

References

[1] *Théorie des topos et cohomologie étale des schémas. Tome 1: Théorie des topos.* Berlin, Springer-Verlag, 1972. Séminaire de Géométrie Algébrique du Bois-Marie 1963–1964 (SGA 4), Dirigé par M. Artin, A. Grothendieck, et J. L. Verdier. Avec la collaboration de N. Bourbaki, P. Deligne et B. Saint-Donat, Lecture Notes in Mathematics, Vol. 269.

[2] Y. André. Filtrations de type Hasse-Arf et monodromie p-adique. *Invent. Math.*, **148(2)**:285–317, 2002.

[3] Y. André. Représentations galoisiennes et opérateurs de Bessel p-adiques. *Ann. Inst. Fourier (Grenoble)*, **52(3)**:779–808, 2002.

[4] A. Arabia. Relèvements des algèbres lisses et de leurs morphismes. *Comment. Math. Helv.*, **76(4)**:607–39, 2001.

[5] F. Baldassarri. Cohomologie p-adique pour la fonction ${}_3F_2(\genfrac{}{}{0pt}{}{a,b_1,b_2}{c_1,c_2};\lambda)$. *Astérisque*, **(119–120)**:4, 51–110, 1984. p-adic cohomology.

[6] F. Baldassarri and B. Chiarellotto. Formal and p-adic theory of differential systems with logarithmic singularities depending upon parameters. *Duke Math. J.*, **72(1)**:241–300, 1993.

[7] F. Baldassarri and B. Chiarellotto. Algebraic versus rigid cohomology with logarithmic coefficients. In *Barsotti Symposium in Algebraic Geometry (Abano Terme, 1991)*, volume 15 of *Perspect. Math.*, pages 11–50. San Diego, CA, Academic Press, 1994.

[8] P. Berthelot. Cohomologie rigide et cohomologie à support propre, seconde partie. Incomplete.

[9] P. Berthelot. *Cohomologie cristalline des schémas de caractéristique $p > 0$.* Berlin, Springer-Verlag, 1974. Lecture Notes in Mathematics, Vol. 407.

[10] P. Berthelot. Cohomologie rigide et théorie de Dwork: le cas des sommes exponentielles. *Astérisque*, **(119–120)**:3, 17–49, 1984. p-adic cohomology.

[11] P. Berthelot. Géométrie rigide et cohomologie des variétés algébriques de caractéristique p. *Mém. Soc. Math. France (N.S.)*, **(23):3**, 7–32, 1986. Introductions aux cohomologies p-adiques (Luminy, 1984).

[12] P. Berthelot. Cohérence différentielle des algèbres de fonctions surconvergentes. *C. R. Acad. Sci. Paris Sér. I Math.*, **323(1)**:35–40, 1996.

[13] P. Berthelot. Cohomologie rigide et cohomologie à support propre, première partie. *Prépublication de l'IRMAR*, **96–03**:1–89, 1996.

[14] P. Berthelot. $\mathcal{D}$-modules arithmétiques. I. Opérateurs différentiels de niveau fini. *Ann. Sci. École Norm. Sup. (4)*, **29(2)**:185–272, 1996.

[15] P. Berthelot. Altérations de variétés algébriques (d'après A. J. de Jong). *Astérisque*, (**241**):Exp. No. 815, 5, 273–311, 1997. Séminaire Bourbaki, Vol. 1995/96.

[16] P. Berthelot. Dualité de Poincaré et formule de Künneth en cohomologie rigide. *C. R. Acad. Sci. Paris Sér. I Math.*, **325(5)**:493–8, 1997.

[17] P. Berthelot. Finitude et pureté cohomologique en cohomologie rigide. *Invent. Math.*, **128(2)**:329–77, 1997. With an appendix in English by Aise Johan de Jong.

[18] P. Berthelot. $\mathcal{D}$-modules arithmétiques. II. Descente par Frobenius. *Mém. Soc. Math. Fr. (N.S.)*, (**81**):vi+136, 2000.

[19] P. Berthelot. Introduction à la théorie arithmétique des $\mathcal{D}$-modules. *Astérisque*, (**279**):1–80, 2002. Cohomologies p-adiques et applications arithmétiques, II.

[20] P. Berthelot and A. Ogus. *Notes on Crystalline Cohomology*. Princeton, N.J., Princeton University Press, 1978.

[21] S. Bosch, U. Güntzer, and R. Remmert. *Non-Archimedean Analysis*, volume 261 of *Grundlehren der Mathematischen Wissenschaften [Fundamental Principles of Mathematical Sciences]*. Berlin, Springer-Verlag, 1984. A systematic approach to rigid analytic geometry.

[22] A. Chambert-Loir. Compter (rapidement) le nombre de solutions d'équations dans les corps finis. In *Séminaire Bourbaki*, page Exp. No. 968. Paris, 2006.

[23] B. Chiarellotto and B. Le Stum. Sur la pureté de la cohomologie cristalline. *C. R. Acad. Sci. Paris Sér. I Math.*, **326(8)**:961–3, 1998.

[24] B. Chiarellotto and B. Le Stum. Pentes en cohomologie rigide et F-isocristaux unipotents. *Manuscripta Math.*, **100(4)**:455–68, 1999.

[25] B. Chiarellotto and B. Le Stum. A comparison theorem for weights. *J. Reine Angew. Math.*, **546**:159–76, 2002.

[26] B. Chiarellotto and N. Tsuzuki. Cohomological descent of rigid cohomology for étale coverings. *Rend. Sem. Mat. Univ. Padova*, **109**:63–215, 2003.

[27] G. Christol and B. Dwork. Modules différentiels sur des couronnes. *Ann. Inst. Fourier (Grenoble)*, **44(3)**:663–701, 1994.

[28] G. Christol and Z. Mebkhout. Sur le théorème de l'indice des équations différentielles p-adiques. I. *Ann. Inst. Fourier (Grenoble)*, **43(5)**:1545–74, 1993.

[29] G. Christol and Z. Mebkhout. Sur le théorème de l'indice des équations différentielles p-adiques. II. *Ann. of Math. (2)*, **146(2)**:345–410, 1997.

[30] G. Christol and Z. Mebkhout. p-adic differential equations. In *Algebra and Number Theory (Fez)*, volume 208 of *Lecture Notes in Pure and Appl. Math.*, pages 105–16. New York, Dekker, 2000.

[31] G. Christol and Z. Mebkhout. Sur le théorème de l'indice des équations différentielles p-adiques. III. *Ann. of Math. (2)*, **151(2)**:385–457, 2000.

[32] R. Crew. F-isocrystals and p-adic representations. In *Algebraic Geometry, Bowdoin, 1985 (Brunswick, Maine, 1985)*, volume 46 of *Proc. Sympos. Pure Math.*, pages 111–38. Providence, RI, Amer. Math. Soc., 1987.

[33] R. Crew. Finiteness theorems for the cohomology of an overconvergent isocrystal on a curve. *Ann. Sci. École Norm. Sup. (4)*, **31(6)**:717–63, 1998.

[34] R. Crew. Crystalline cohomology of singular varieties. In *Geometric Aspects of Dwork Theory. Vol. I, II*, pages 451–62. Berlin, Walter de Gruyter GmbH & Co. KG, 2004.

[35] A. J. de Jong. Smoothness, semi-stability and alterations. *Inst. Hautes Études Sci. Publ. Math.*, (**83**):51–93, 1996.

[36] P. Deligne. La conjecture de Weil. I. *Inst. Hautes Études Sci. Publ. Math.*, (**43**):273–307, 1974.

[37] J. Denef and F. Vercauteren. Counting points on C_{ab} curves using Monsky-Washnitzer cohomology. *Finite Fields Appl.*, **12(1)**:78–102, 2006.

[38] J. Denef and F. Vercauteren. An extension of Kedlaya's algorithm to hyperelliptic curves in characteristic 2. *J. Cryptology*, **19(1)**:1–25, 2006.

[39] B. Dwork. On the rationality of the zeta function of an algebraic variety. *Amer. J. Math.*, **82**:631–48, 1960.

[40] B. Dwork. On p-adic differential equations. IV. Generalized hypergeometric functions as p-adic analytic functions in one variable. *Ann. Sci. École Norm. Sup. (4)*, **6**:295–315, 1973.

[41] B. Dwork. *Lectures on p-adic Differential Equations*, volume 253 of *Grundlehren der Mathematischen Wissenschaften [Fundamental Principles of Mathematical Science]*. New York, Springer-Verlag, 1982. With an appendix by Alan Adolphson.

[42] S. Edixhoven. Point counting after kedlaya. *Preprint*, 2003.

[43] J. Y. Étesse and B. Le Stum. Fonctions L associées aux F-isocristaux surconvergents. I. Interprétation cohomologique. *Math. Ann.*, **296(3)**:557–76, 1993.

[44] J. Y. Étesse and B. Le Stum. Fonctions L associées aux F-isocristaux surconvergents. II. Zéros et pôles unités. *Invent. Math.*, **127(1)**:1–31, 1997.

[45] J. Fresnel and M. van der Put. *Rigid Analytic Geometry and its Applications*, volume 218 of *Progress in Mathematics*. Boston, MA, Birkhäuser Boston Inc., 2004.

[46] R. Gerkmann. The p-adic Cohomology of Varieties over Finite Fields and Applications on the Computation of Zeta functions. PhD thesis, Essen, 2003.

[47] E. Grosse-Klönne. Rigid analytic spaces with overconvergent structure sheaf. *J. Reine Angew. Math.*, **519**:73–95, 2000.

[48] E. Grosse-Klönne. Finiteness of de Rham cohomology in rigid analysis. *Duke Math. J.*, **113(1)**:57–91, 2002.

[49] A. Grothendieck. Formule de Lefschetz et rationalité des fonctions L. In *Séminaire Bourbaki, Vol. 9*, pages Exp. No. 279, 41–55. Paris, Soc. Math. France, 1995.

[50] R. Hartshorne. *Residues and Duality*. Lecture notes of a seminar on the work of A. Grothendieck, given at Harvard 1963/64. With an appendix by P. Deligne. Lecture Notes in Mathematics, No. 20. Berlin, Springer-Verlag, 1966.

[51] K. Kato. Logarithmic structures of Fontaine-Illusie. In *Algebraic Analysis, Geometry, and Number Theory*, pages 191–224. Baltimore, MD, Johns Hopkins University Press, 1989.

[52] K. S. Kedlaya. Counting points on hyperelliptic curves using Monsky-Washnitzer cohomology. *J. Ramanujan Math. Soc.*, **16(4)**:323–38, 2001.

[53] K. S. Kedlaya. A p-adic local monodromy theorem. *Ann. of Math. (2)*, **160(1)**:93–184, 2004.

[54] K. S. Kedlaya. Slope filtrations revisited. *Doc. Math.*, **10**:447–525 (electronic), 2005.

[55] R. Kiehl. Der Endlichkeitssatz für eigentliche Abbildungen in der nichtarchimedischen Funktionentheorie. *Invent. Math.*, **2**:191–214, 1967.

[56] R. Kiehl. Die de Rham Kohomologie algebraischer Mannigfaltigkeiten über einem bewerteten Körper. *Inst. Hautes Études Sci. Publ. Math.*, **(33)**:5–20, 1967.

[57] A. G. B. Lauder. Rationality and meromorphy of zeta functions. *Finite Fields Appl.*, **11(3)**:491–510, 2005.

[58] A. G. B. Lauder. Rigid cohomology and p-adic point counting. *J. Théor. Nombres Bordeaux*, **17(1)**:169–80, 2005.

[59] A. G. B. Lauder and D. Wan. Computing zeta functions of Artin-Schreier curves over finite fields. *LMS J. Comput. Math.*, **5**:34–55 (electronic), 2002.

[60] B. Le Stum. Filtration par le poids sur la cohomologie de de Rham d'une courbe projective non singulière sur un corps ultramétrique complet. *Rend. Sem. Mat. Univ. Padova*, **93**:43–85, 1995.

[61] B. Le Stum and A. Quirós. Transversal crystals of finite level. *Ann. Inst. Fourier (Grenoble)*, **47(1)**:69–100, 1997.

[62] B. Le Stum and A. Quirós. The exact Poincaré lemma in crystalline cohomology of higher level. *J. Algebra*, **240(2)**:559–88, 2001.

[63] B. Le Stum and A. Quirós. The filtered poincaré lemma in higer level (with applications to algebraic groups). *Prépublication de l'IRMAR*, **04(44)**:25, 2004.

[64] Z. Mebkhout. Sur le théorème de finitude de la cohomologie p-adique d'une variété affine non singulière. *Amer. J. Math.*, **119(5)**:1027–81, 1997.

[65] Z. Mebkhout. Analogue p-adique du théorème de Turrittin et le théorème de la monodromie p-adique. *Invent. Math.*, **148(2)**:319–51, 2002.

[66] Z. Mebkhout. La théorie des équations différentielles p-adiques et le théorème de la monodromie p-adique. In *Proceedings of the International Conference on Algebraic Geometry and Singularities (Spanish) (Sevilla, 2001)*, volume 19, pages 623–65, 2003.

[67] D. Meredith. Weak formal schemes. *Nagoya Journal of Mathematics*, 1972.

[68] J. S. Milne. *Étale Cohomology*, volume 33 of *Princeton Mathematical Series*. Princeton, N.J., Princeton University Press, 1980.

[69] P. Monsky. Formal cohomology. II. The cohomology sequence of a pair. *Ann. of Math. (2)*, **88**:218–38, 1968.

[70] P. Monsky. *p-adic Analysis and Zeta Functions*, volume 4 of *Lectures in Mathematics, Department of Mathematics, Kyoto University*. Tokyo, Kinokuniya Book-Store Co. Ltd., 1970.

[71] P. Monsky. Formal cohomology. III. Fixed point theorems. *Ann. of Math. (2)*, **93**:315–43, 1971.

[72] P. Monsky and G. Washnitzer. Formal cohomology. I. *Ann. of Math. (2)*, **88**:181–217, 1968.

[73] A. Ogus. The convergent topos in characteristic p. In *The Grothendieck Festschrift, Vol. III*, volume 88 of *Progr. Math.*, pages 133–62. Boston, MA, Birkhäuser Boston, 1990.

[74] D. Petrequin. Classes de Chern et classes de cycles en cohomologie rigide. *Bull. Soc. Math. France*, **131(1)**:59–121, 2003.

[75] M. Raynaud and L. Gruson. Critères de platitude et de projectivité. Techniques de "platification" d'un module. *Invent. Math.*, **13**:1–89, 1971.

[76] P. Robba and G. Christol. *Équations différentielles p-adiques*. Actualités Mathématiques. [Current Mathematical Topics]. Hermann, Paris, 1994. Applications aux sommes exponentielles. [Applications to exponential sums].

[77] N. Roby. Les algèbres à puissances divisées. *Bull. Sci. Math. (2)*, **89**:75–91, 1965.

[78] A. Shiho. Crystalline fundamental groups. I. Isocrystals on log crystalline site and log convergent site. *J. Math. Sci. Univ. Tokyo*, **7(4)**:509–656, 2000.

[79] A. Shiho. Crystalline fundamental groups. II. Log convergent cohomology and rigid cohomology. *J. Math. Sci. Univ. Tokyo*, **9(1)**:1–163, 2002.

[80] N. Tsuzuki. Finite local monodromy of overconvergent unit-root F-isocrystals on a curve. *Amer. J. Math.*, **120(6)**:1165–90, 1998.

[81] N. Tsuzuki. Slope filtration of quasi-unipotent overconvergent F-isocrystals. *Ann. Inst. Fourier (Grenoble)*, **48(2)**:379–412, 1998.

[82] N. Tsuzuki. Cohomological descent of rigid cohomology for proper coverings. *Invent. Math.*, **151(1)**:101–33, 2003.

[83] M. van der Put. The cohomology of Monsky and Washnitzer. *Mém. Soc. Math. France (N.S.)*, **(23)**:4, 33–59, 1986. Introductions aux cohomologies p-adiques (Luminy, 1984).

[84] M. van der Put. Serre duality for rigid analytic spaces. *Indag. Math. (N.S.)*, **3(2)**:219–35, 1992.

[85] A. Weil. Numbers of solutions of equations in finite fields. *Bull. Amer. Math. Soc.*, **55**:497–508, 1949.

Index

For EU product safety concerns, contact us at Calle de José Abascal, 56–1°,
28003 Madrid, Spain or eugpsr@cambridge.org.

www.ingramcontent.com/pod-product-compliance
Lightning Source LLC
LaVergne TN
LVHW102008210926
848483LV00010B/158
* 9 7 8 0 5 2 1 8 7 5 2 4 0 *